刺参营养需求及配合饲料

王际英　李宝山　孙永智　宋志东　等　著

科学出版社
北京

内 容 简 介

近十几年来，刺参养殖规模迅速扩大，然而刺参营养学研究落后于产业发展，在一定程度上限制了刺参养殖业的良性发展。为了满足广大读者学习和掌握刺参饲料配制方法的需求，本书筛选总结了近年来的一些相对有应用价值的研究成果，内容涵盖刺参的营养组成、营养素的需求、刺参饲料原料的开发、功能性添加剂的应用，以及国际交流等。全书共分5章，作者在书中既对研究方法、条件进行了详细的说明，又深入浅出地对研究结论进行了论证，提出了针对实际问题的解决办法。本书兼顾了理论性和实用性，避免了传统教材内容的教条化，充分注重了文字的浅显易懂，使读者易学易用。

本书适合于广大刺参养殖从业者或企事业单位相关人员阅读，可作为养殖人员、配方设计师等专业人员培训用书，也可供行业中希望深入学习刺参营养学和饲料工艺学的研究人员作为参考书使用。

图书在版编目（CIP）数据

刺参营养需求及配合饲料 / 王际英等著. -- 北京：科学出版社，2024.10. -- ISBN 978-7-03-079545-8

Ⅰ．S968.9

中国国家版本馆 CIP 数据核字第 2024FQ9547 号

责任编辑：朱 瑾 习慧丽 / 责任校对：杨 赛
责任印制：肖 兴 / 封面设计：无极书装

科学出版社 出版
北京东黄城根北街 16 号
邮政编码：100717
http://www.sciencep.com

北京中石油彩色印刷有限责任公司印刷
科学出版社发行 各地新华书店经销

*

2024 年 10 月第 一 版 开本：720×1000 1/16
2024 年 10 第一次印刷 印张：17 3/4
字数：358 000

定价：238.00 元
（如有印装质量问题，我社负责调换）

《刺参营养需求及配合饲料》撰写人员名单

主要著者 王际英 李宝山 孙永智 宋志东

参与著者（按拼音排序）

曹体宏	陈　玮	盖芸芸	郝甜甜
胡　斌	黄炳山	贾戊辰	李　静
李　璐	李培玉	李志林	刘继臣
刘京熙	秦华伟	孙春晓	王　斌
王成强	王　丹	王福辰	王世信
王晓艳	王忠全	相智巍	邢红艳
张明亮	赵　岩	邹荣婕	左　震

前　　言

随着人们对健康生活及海洋美食的追求，海参作为一种含有丰富蛋白质、多种氨基酸以及 Ca、Zn、Cu 等矿物质的滋补食材，备受青睐，已经渐入寻常百姓家。虽然我国海参品种众多，如山东、河北和辽宁沿海的刺参（学名仿刺参），以及西沙群岛、海南岛和雷州半岛沿岸的梅花参、花刺参、绿刺参、白尼参、辐肛参等 20 多种可食用海参，但是只有 10 种具有较高的经济价值和养殖潜力。刺参的蛋白质含量高，还富含海参皂苷、多种不饱和脂肪酸、多种人体必需的矿物质等，因而被人们广泛认可，成为第五次海水养殖浪潮的代表性品种。本书以刺参为例进行研究，近年来，我国的刺参养殖产业迅猛发展，形成了以山东、辽宁沿海为主产区，以南北接力形式延伸到闽浙沿海的刺参增养殖产业群，刺参养殖产量逐年递增。根据《2023 中国渔业统计年鉴》，2022 年我国海参养殖总面积 250 356hm^2，总产量 248 508t，较 2021 年提高 11.59%。其中，山东养殖面积 82 217hm^2，养殖产量占全国总产量的 40.31%；其次是辽宁，养殖面积 158 308hm^2，养殖产量占全国总产量的 34.66%；福建近年来开拓"北参南养"接力养殖模式，虽然养殖面积仅 1680hm^2，但是养殖产量占全国总产量的 18.36%。

饲料是刺参养殖业健康发展的重要保障，也是养殖支出中占比最大的部分。营养是否均衡、能否在满足需求的情况下提高刺参的抗逆能力等因素决定了饲料在市场上被接受的程度。另外，在天然海域中，刺参喜食的大型海藻如鼠尾藻、马尾藻、海带等被大量采集，不少海域鼠尾藻、马尾藻资源已近枯竭。与此同时，由于刺参营养学数据不完善，市场上缺少刺参专用高效配合饲料，部分养殖企业采用自加工的低劣配合饲料进行投喂，导致刺参生长缓慢、污染严重。本书综合了作者所在科研团队多年来的科研成果和生产实践，并查阅近 20 年来的相关文献资料，详尽地介绍了刺参营养与饲料的理论进展和相关实验结果，旨在为满足刺参养殖业发展之需提供科学合理、营养均衡的配合饲料配方指导。

为了方便读者的阅读、理解和应用，第 1 章介绍了刺参的营养组成特点及差异，并探讨了不同生长阶段刺参体壁的营养组成变化规律，为配制刺参饲料提供思路和数据支撑。第 2 章从生长、生理生化等多方面探讨了刺参对宏量营养素的需求，如蛋白质、碳水化合物，以及对多种微量营养素的需求，为配制营养均衡的刺参饲料提供理论支持。第 3 章评价了一些海藻及新型蛋白原料在刺参养殖实验中的喂养效果，为新型饲料源开发提供理论支持和参考数据。第 4 章评价了几种功能性饲料添加剂在刺参饲料中的使用效果，包括酶制剂、棉子糖和半乳甘露寡糖等。第 5 章介绍了科研团队近年来对外国际交流的几项研究成果，包括刺参对蛋氨酸、维生素 E、磷的需求，以及酶解豆粕在刺参饲料中的应用。

与市面上的其他相关书籍相比，本书具有以下几个特点：①兼具科学性和实践性，全书围绕刺参饲料配制中的一些关键问题、热点问题进行论述，论述基础是科学实验所得到的大量数据，忠实于科学实验结果是本书的科学严谨性所在，而书中的实验是在模拟生产实践中或直接在生产实践中进行，其结果可以直接在实践中进行应用；②兼具理论性和创新性，本书通过多年科学实验、生产实践建立了较为系统的刺参营养学理论，很多理论在实践中获得了验证和发展，并且新的技术、新的思路催生了一些新的发现、新的理论的产生，在本书中均有体现；③兼具时效性和及时性，本书所论述的诸多实验及理论时间跨度虽然 20 多年，但是随着后期研究技术的不断更新、理论观点的不断发展，很多实验配方均实现了改动优化，力求符合现代渔业发展的需要，这也是保持本书科学严谨性的需要。

鉴于编者水平有限，不当之处在所难免，诚请各位读者批评指正。

目　录

第1章　刺参的营养组成 ·· 1
1.1　不同品种刺参的营养组成分析 ·· 1
1.2　不同阶段刺参的营养组成分析 ·· 8

第2章　刺参的营养需求 ·· 12
2.1　刺参对营养素的需求 ·· 12
2.2　刺参对氨基酸和脂肪酸的需求 ··· 47

第3章　刺参配合饲料原料的开发 ··· 87
3.1　海藻在刺参配合饲料中的应用 ··· 87
3.2　陆生植物蛋白原料在刺参配合饲料中的应用 ··· 111

第4章　刺参配合饲料中功能性饲料添加剂的筛选 ··· 127
4.1　刺参配合饲料中酶制剂的筛选 ··127
4.2　刺参配合饲料中糖制剂的筛选 ··142

第5章　国际交流 ·· 168
PART I: Optimal dietary methionine requirement for juvenile sea cucumber *Apostichopus japonicus* ·· 168

PART II: Requirement of vitamin E of growing sea cucumber *Apostichopus japonicus* ··· 181

PART III: Optimal dietary phosphorus requirement for juvenile sea cucumber *Apostichopus japonicus* ·· 192

PART IV: Application of hydrolyzed soybean meal in feed of sea cucumber *Apostichopus japonicus* ·· 209

主要参考文献 ··· 229

附录 ··· 274

第1章 刺参的营养组成

1.1 不同品种刺参的营养组成分析

白刺参俗称"白玉参",体内黑色素细胞缺乏,黑色素体不成熟,多种基因在黑色素合成信号通路上的表达受到抑制(马得友,2013)。目前为止,关于白刺参营养成分的研究尚未见报道。本节对相同养殖环境条件下的青刺参、白刺参体壁营养组成进行对比分析,从而探讨两种刺参营养品质与价值的差异(李忠清等,2016)。

1.1.1 材料与方法

1. 实验材料

实验所用青刺参、白刺参均为同一批次健康苗种,均放养在 3m×6m、水深 1m 左右的养殖池内,均设置 3 个平行,池底均放有波纹板支架及 60cm×60cm 的波纹板,每池均放养 90 头,密度为 5 头/m^2。实验在微流水环境中进行,采用充气增氧,保证溶解氧含量高于 6.5mg/L,控制水温在 18～21℃,pH 为 7.8～8.2,盐度为 28～32,亚硝酸氮、氨氮含量均低于 0.05mg/L,其他生长环境因子均相同,青刺参、白刺参在养殖期间均投喂相同的人工饵料。

2. 样品采集与处理

随机从每个养殖池内各捞取 20 头活体刺参,依次用消过毒的纱布吸干刺参表面的水分,再将其置于灭菌后的培养皿上称重,实验所取青刺参、白刺参的体质量分别为(41.21±4.96)g 和(44.30±5.02)g。依次解剖,去除内脏后测定其出皮率。去除石灰环并用过滤后的海水冲洗后称重。用剪刀将解剖后的青刺参、白刺参剪成小块,并用组织捣碎机捣碎。分别将捣碎后的青刺参、白刺参均分成两份,一份放在烘箱中(105℃)烘至恒重,存放于干燥器中备用,另一份直接存放于冰箱(-80℃)中保存备用。

3. 样品的测定与方法

营养成分测定方法见附录。从冰箱(-80℃)中取出备用样品,分别称约 1g 鲜样品放入微波消解罐中,加入 10ml 硝酸,经微波消解仪消解等处理后,用电感耦合等离子体质谱仪测定样品中 Ca、Fe、Mg、Cu、Cr、Zn、Mn 的含量,用原子荧光形态分析仪测定样品中 Se、Cd、Pb、As、Hg 的含量。

4. 数据统计分析

对青刺参、白刺参体壁中蛋白质营养价值的评定基于氨基酸标准模式和全鸡蛋蛋白模式（唐雪等，2011），计算公式如下：

蛋白质氨基酸评分（%）=待评蛋白质中氨基酸含量（mg/g）/参考蛋白质中同种氨基酸含量（mg/g）×100

化学评分（%）=待评蛋白质中氨基酸含量（mg/g）/鸡蛋蛋白质中同种氨基酸含量（mg/g）×100

采用 SPSS 19.0 软件对实验数据进行独立样本的 t 检验，确定其组间差异显著性（$P<0.05$），取置信水平 95%。统计数据以平均值±标准差表示。

1.1.2 结果分析

1. 蛋白质、粗脂肪、粗灰分、多糖的含量分析

从表 1-1 可知，白刺参出皮率显著高于青刺参（$P<0.05$），青刺参、白刺参体壁中的水分含量分别为 91.32%、90.37%，白刺参体壁中的粗灰分和多糖含量显著高于青刺参（$P<0.05$），青刺参、白刺参体壁中粗蛋白和粗脂肪的含量均无显著性差异（$P>0.05$）。

表 1-1　青刺参、白刺参的出皮率和体壁基本营养组成的比较（%）

	出皮率	水分	粗灰分（干基）	粗蛋白（干基）	粗脂肪（干基）	多糖（干基）
白刺参	66.08±2.42[a]	91.32±0.31[a]	32.06±0.11[a]	48.30±0.16	2.59±0.25	2.13±0.06[a]
青刺参	61.60±2.57[b]	90.37±0.13[b]	29.99±0.10[b]	47.72±0.40	2.67±0.18	1.85±0.07[b]

注：同列数据无字母或上标相同字母表示差异不显著（$P>0.05$），上标不同字母表示差异显著（$P<0.05$）。

2. 脂肪酸组成分析

由表 1-2 可知，青刺参体壁含 20 种脂肪酸，其中含饱和脂肪酸 7 种、单不饱和脂肪酸 5 种、多不饱和脂肪酸 8 种，总脂肪酸的含量为 67.52%，其中饱和脂肪酸、单不饱和脂肪酸和多不饱和脂肪酸的含量分别为 23.42%、21.65%和 22.67%；白刺参体壁含 20 种脂肪酸，其中含饱和脂肪酸 6 种、单不饱和脂肪酸 5 种、多不饱和脂肪酸 9 种，总脂肪酸的含量为 67.42%，其中饱和脂肪酸、单不饱和脂肪酸和多不饱和脂肪酸的含量分别为 21.33%、23.02%和 23.18%。

表 1-2 青刺参、白刺参体壁中脂肪酸的组成与含量（%）

脂肪酸	青刺参	白刺参
饱和脂肪酸		
C14:0	1.15±0.03	1.19±0.08
C15:0	0.30±0.03	0.27±0.02
C16:0	12.30±0.10[a]	11.61±0.14[b]
C17:0	1.29±0.04	1.34±0.05
C18:0	5.52±0.04[a]	5.21±0.06[b]
C20:0	1.38±0.08	1.30±0.06
C22:0	1.38±0.10	—
单不饱和脂肪酸		
C16:1n-7	8.20±0.27	8.25±0.14
C18:1n-7	5.73±0.14	5.49±0.14
C18:1n-9	5.67±0.04[a]	6.08±0.03[b]
C20:1n-7	2.33±0.05[a]	2.60±0.05[b]
C22:1n-9	0.41±0.03	0.43±0.03
多不饱和脂肪酸		
C18:2n-6	4.21±0.03	4.16±0.02
C18:3n-3	1.10±0.03	1.12±0.04
C18:3n-6	0.30±0.02	0.29±0.01
C20:2n-9	—	0.50±0.02
C20:3n-6	0.33±0.02	0.32±0.02
C20:4n-6（花生四烯酸）	6.73±0.10[a]	7.18±0.08[b]
C20:5n-3（二十碳五烯酸）	6.68±0.03[a]	6.12±0.11[b]
C22:5n-3（二十二碳五烯酸）	0.26±0.03	0.23±0.02
C22:6n-3（二十二碳六烯酸）	3.10±0.06[a]	3.60±0.16[b]
总量	67.52±0.61	67.42±0.17
∑饱和脂肪酸	23.42±0.18[a]	21.33±0.28[b]
∑单不饱和脂肪酸	21.65±0.50[a]	23.02±0.09[b]
∑多不饱和脂肪酸	22.67±0.61	23.18±0.57
∑n-3 多不饱和脂肪酸	11.35±0.14	11.18±0.20
∑n-6 多不饱和脂肪酸	11.50±0.16	11.63±0.23
n-3/n-6	0.99±0.01	0.96±0.02

注："—"表示未检出；同行数据无字母或上标相同字母表示差异不显著（$P>0.05$），上标不同字母表示差异显著（$P<0.05$）。

由统计学软件对青刺参、白刺参体壁中脂肪酸的含量进行差异显著性分析可知（表 1-2），青刺参、白刺参体壁中总脂肪酸及多不饱和脂肪酸的含量均无显著性差异（$P>0.05$），而饱和脂肪酸和单不饱和脂肪酸的含量均有显著性差异（$P<0.05$）。由具体脂肪酸组分对比分析可知，青刺参、白刺参体壁饱和脂肪酸中

C16:0 含量分别为 12.30%和 11.61%，C18:0 含量分别为 5.52%和 5.21%，青刺参体壁中 C16:0 和 C18:0 的含量均显著高于白刺参（$P<0.05$）；青刺参、白刺参体壁单不饱和脂肪酸中 C18:1n-9 的含量分别为 5.67%和 6.08%，C20:1n-7 的含量分别为 2.33%和 2.60%，青刺参体壁中 C18:1n-9 和 C20:1n-7 的含量均显著低于白刺参（$P<0.05$）；青刺参、白刺参体壁多不饱和脂肪酸中花生四烯酸的含量分别为 6.73%和 7.18%，二十碳五烯酸的含量分别为 6.68%和 6.12%，二十二碳六烯酸的含量分别为 3.10%和 3.60%，青刺参体壁中花生四烯酸和二十二碳六烯酸的含量均显著低于白刺参（$P<0.05$），但二十碳五烯酸的含量却显著高于白刺参（$P<0.05$）。青刺参、白刺参体壁中其他脂肪酸组分的含量均无显著性差异（$P>0.05$）。

3. 氨基酸组成分析

由表 1-3 可知，青刺参、白刺参的体壁中检测出 17 种氨基酸，其中含 7 种必需氨基酸、6 种鲜味氨基酸、6 种药效氨基酸。青刺参、白刺参体壁中所含氨基酸总量分别为 39.51%和 39.69%，无显著性差异（$P>0.05$）。其中，青刺参、白刺参体壁中必需氨基酸的含量分别为 12.46%和 12.54%，无显著性差异（$P>0.05$）；鲜味氨基酸的含量分别为 21.51%和 21.63%，无显著性差异（$P>0.05$）；药味氨基酸的含量也无显著性差异（$P>0.05$）。青刺参、白刺参体壁氨基酸组分中含量最高的均为谷氨酸，分别为 7.78%和 7.34%，而组氨酸含量分别为 0.61%和 0.54%，两者均无显著性差异（$P>0.05$）。青刺参体壁中天冬氨酸的含量显著低于白刺参（$P<0.05$），青刺参体壁中精氨酸的含量也显著低于白刺参（$P<0.05$），苯丙氨酸的含量显著高于白刺参（$P<0.05$），青刺参、白刺参体壁中其他氨基酸组分的含量均无显著性差异（$P>0.05$）。

表 1-3 青刺参、白刺参体壁中氨基酸的组成与含量（%）

氨基酸	青刺参	白刺参
天冬氨酸[②③]	4.01±0.09[a]	4.39±0.14[b]
苏氨酸[①]	2.33±0.04	2.53±0.10
丝氨酸[②]	2.04±0.04	2.18±0.08
谷氨酸[②③]	7.78±0.32	7.34±0.21
脯氨酸[②]	1.81±0.12	1.97±0.10
甘氨酸[②③]	3.77±0.52	4.02±0.28
丙氨酸[②]	2.03±0.16	2.18±0.22
半胱氨酸[③]	0.61±0.07	0.63±0.04
缬氨酸[①]	1.95±0.08	2.02±0.02
蛋氨酸[①]	0.86±0.06	0.84±0.16
异亮氨酸[①]	1.29±0.03	1.31±0.03
亮氨酸[①]	1.93±0.12	2.02±0.06
酪氨酸[③]	1.84±0.10	1.71±0.08

续表

氨基酸	青刺参	白刺参
苯丙氨酸①	1.75±0.06ᵃ	1.40±0.08ᵇ
赖氨酸①	1.93±0.16	1.83±0.04
组氨酸	0.61±0.09	0.54±0.05
精氨酸③	2.61±0.06ᵃ	3.04±0.08ᵇ
∑必需氨基酸	12.46±0.30	12.54±0.08
∑鲜味氨基酸	21.51±0.46	21.63±0.40
∑药效氨基酸	20.87±0.84	21.04±0.43
∑氨基酸	39.51±0.39	39.69±0.51
∑必需氨基酸/∑氨基酸	31.63	31.82
∑必需氨基酸/∑非必需氨基酸	46.06	46.19

注：①-必需氨基酸；②-鲜味氨基酸；③-药效氨基酸。同行数据无字母或上标相同字母表示差异不显著（$P>0.05$），上标不同字母表示差异显著（$P<0.05$）。

4. 氨基酸营养评价

为了更好地评价青刺参、白刺参体壁中氨基酸的营养价值，将氨基酸含量折算成每克蛋白质中氨基酸的量（mg）与 FAO/WHO 制定的氨基酸标准模式和全鸡蛋蛋白模式进行比较（唐雪等，2011），分别计算出青刺参、白刺参体壁中必需氨基酸的化学评分和氨基酸评分（表 1-4，表 1-5）。青刺参、白刺参体壁中苏氨酸的氨基酸评分分别为 123.74 分和 124.35 分，均超过 100 分，化学评分分别为 97.05 分和 97.53 分，也接近 100 分；苯丙氨酸+酪氨酸的氨基酸评分分别为 125.55 分和 107.49 分，也均超过 100 分，但是化学评分较低，分别为 75.33 分和 64.49 分。

表 1-4 青刺参体壁中化学评分和氨基酸评分

必需氨基酸	青刺参/(mg/g)	青刺参评分		全鸡蛋蛋白/(mg/g)	评分标准
		化学评分	氨基酸评分		
苏氨酸	49.52±0.63	97.05±1.24	123.74±1.58	51	40
缬氨酸	41.01±1.70	56.15±2.31	81.98±3.37	73	50
蛋氨酸+半胱氨酸	31.04±2.63	56.40±4.30	88.64±7.52	55	35
异亮氨酸	27.11±1.04	41.05±1.57*	67.74±2.60**	66	40
亮氨酸	40.53±2.61	46.04±2.95**	57.88±3.72*	88	70
苯丙氨酸+酪氨酸	75.33±2.95	75.33±2.95	125.55±4.92	100	60
赖氨酸	40.38±3.42	63.07±5.34	73.38±6.23	64	55

注："*"表示第一限制氨基酸；"**"表示第二限制氨基酸。

表 1-5 白刺参体壁中化学评分和氨基酸评分

必需氨基酸	白刺参/(mg/g)	白刺参评分		全鸡蛋蛋白/(mg/g)	评分标准
		化学评分	氨基酸评分		
苏氨酸	49.74±0.48	97.53±0.95	124.35±1.21	51	40

续表

必需氨基酸	白刺参/ （mg/g）	白刺参评分		全鸡蛋蛋白/ （mg/g）	评分标准
		化学评分	氨基酸评分		
缬氨酸	41.76±0.41	57.21±0.56	83.52±0.82	73	50
蛋氨酸+半胱氨酸	31.27±0.48	56.67±2.4	89.06±3.47	55	35
异亮氨酸	27.14±0.49	41.13±0.76*	67.86±1.24**	66	40
亮氨酸	41.78±1.25	47.48±1.43**	59.69±1.80*	88	70
苯丙氨酸+酪氨酸	64.49±3.89	64.49±3.90	107.49±6.49	100	60
赖氨酸	37.91±3.88	59.23±4.37	70.53±5.59	64	55

注："*"表示第一限制氨基酸；"**"表示第二限制氨基酸。

5. 矿物质及重金属元素含量的分析

分别对青刺参、白刺参体壁鲜样中 8 种矿物质和 4 种重金属元素的含量进行测定，分别见表 1-6 和表 1-7。青刺参、白刺参体壁鲜样中含有丰富的 Ca、Fe、Mg、Zn 等矿物质。青刺参、白刺参体壁中含量最高的矿物质均为 Mg 元素，含量分别为 1517.87mg/kg 和 1523.71mg/kg，无显著性差异（$P>0.05$）。青刺参、白刺参体壁鲜样中 Se 的含量分别为 0.06mg/kg 和 0.08mg/kg，青刺参体壁鲜样中含量显著低于白刺参（$P<0.05$）；青刺参、白刺参体壁鲜样中 Cr 的含量分别为 0.41mg/kg 和 0.37mg/kg，青刺参体壁鲜样中含量显著高于白刺参（$P<0.05$）；青刺参、白刺参体壁鲜样中 Mn 的含量分别为 0.56mg/kg 和 0.60mg/kg，青刺参体壁鲜样中含量显著低于白刺参（$P<0.05$）。青刺参、白刺参体壁鲜样中其他矿物质均无显著性差异（$P>0.05$）。青刺参、白刺参体壁鲜样中重金属元素 Cd、Pb、As、Hg 的含量均无显著性差异（$P>0.05$），且均符合国家食品卫生标准（GB 2762—2022）要求。

表 1-6　青刺参、白刺参体壁鲜样中矿物质含量的对比　　（单位：mg/kg）

矿物质	青刺参	白刺参
钙（Ca）	738.59±3.20	734.86±4.68
铁（Fe）	2.06±0.10	2.11±0.03
镁（Mg）	1517.87±57.72	1523.71±48.74
锌（Zn）	2.44±0.31	2.32±0.13
铜（Cu）	0.27±0.05	0.28±0.05
硒（Se）	0.06±0.01[a]	0.08±0.01[b]
铬（Cr）	0.41±0.01[a]	0.37±0.01[a]
锰（Mn）	0.56±0.02[a]	0.60±0.02[b]

注：同行数据无字母或上标相同字母表示差异不显著（$P>0.05$），上标不同字母表示差异显著（$P<0.05$）。

表 1-7　青刺参、白刺参体壁鲜样中重金属元素含量的对比　　（单位：mg/kg）

重金属元素	青刺参	白刺参
镉（Cd）	0.04±0.01	0.04±0.01

续表

重金属元素	青刺参	白刺参
铅（Pb）	0.14±0.01	0.15±0.01
砷（As）	0.02±0.00	0.02±0.00
汞（Hg）	<0.01	<0.01

注：砷（As）为非金属，但其化合物具有金属性，本书将其归入重金属一并统计。

1.1.3 讨论

出皮率是评价刺参品质是否优良的一个重要指标（Jiang et al.，2013；刘小芳等，2011；李丹彤等，2014），出皮率高说明刺参品质优良，本研究结果显示，白刺参出皮率比青刺参显著高 4.48%。海参体壁中的多糖类成分具有提高机体免疫力和抗癌等作用（马同江等，1982），白刺参体壁中多糖含量比青刺参高 0.28%，本研究测得多糖含量（表 1-1 中青刺参、白刺参体壁中多糖的含量分别为 1.85%、2.13%）介于展学孔等（2011）测得的多糖含量（1.25%~2.51%）。究其原因，一方面，不同种类、不同季节和不同养殖条件造成海参体壁中多糖的含量存在一定差异（刘小芳等，2011）；另一方面，目前测定多糖含量的方法不统一，因而得出的多糖含量不同。Jiang 等（2013）曾对相同养殖环境下青刺参、红刺参体壁的营养成分进行报道，其测得的红刺参的出皮率和粗脂肪含量与本研究白刺参的结果相近，而红刺参体壁中粗灰分含量低于白刺参，粗蛋白含量则高于白刺参。

由脂肪酸的测定结果比较分析可知，青刺参体壁中饱和脂肪酸的含量最高，白刺参体壁中多不饱和脂肪酸的含量最高。脂肪酸的营养价值主要体现在多不饱和脂肪酸上，多不饱和脂肪酸匮乏会导致机体心脏、大脑等器官严重发育不良（Samadi et al.，2006；Kitajka et al.，2002）。此外，青刺参体壁中的单不饱和脂肪酸显著低于白刺参，相关研究表明，单不饱和脂肪酸不仅有降低胆固醇的作用，还有促进人体类脂代谢的功能（Thomsen et al.，1999）。世界卫生组织（WHO）推荐，食物中 n-3/n-6 的值要大于 0.1 才对人体健康有益（Sánchez-Machado et al.，2004），青刺参、白刺参体壁中 n-3/n-6 值都接近 1，均符合健康食品的要求。脂肪酸组分中，青刺参体壁中花生四烯酸和二十二碳六烯酸的含量均显著低于白刺参。

李春艳和常亚青（2006）指出，青刺参、白刺参体壁中均检测出 17 种氨基酸，其中总氨基酸含量显著高于其他 20 余种海参，青刺参体壁中天冬氨酸和精氨酸含量均显著低于白刺参。根据 FAO/WHO 制定的必需氨基酸标准模式，青刺参、白刺参体壁的氨基酸评分分别为 57~125 分和 59~124 分，其中，青刺参、白刺参体壁中苏氨酸和苯丙氨酸+酪氨酸的氨基酸评分均超过 100。

青刺参、白刺参体壁中富含人体必需的矿物质，尤其是 Ca、Fe、Mg 的含量较高，这与刘小芳等（2011）、王哲平等（2012）、肖宝华等（2014）和董晓弟等（2013）对刺参矿物质的研究结果一致。此外，青刺参、白刺参体壁中 Zn 含量均高于海胆、鱼翅等海产品，而 As、Hg、Pb、Cd 重金属的含量低于其他海洋贝类（邓必阳和张展霞，1999；章超桦等，2000；杨宝灵等，2009）。白刺参体壁中 Mn 和 Se 的含量均显著高于青刺参。

《食品安全国家标准 食品中污染物限量》（GB 2762—2022）对水产品（鲜品）中重金属限量指标为：Pb≤1.0mg/kg、Cd≤0.1mg/kg、As≤0.5mg/kg、Hg≤0.5mg/kg。青刺参、白刺参体壁中重金属含量均明显低于食品安全国家标准。

1.1.4 结论

本节对相同养殖环境下青刺参、白刺参体壁营养成分进行了比较分析，结果显示，白刺参的出皮率、多糖含量、脂肪酸中不饱和脂肪酸含量、花生四烯酸含量、二十二碳六烯酸含量、氨基酸中天冬氨酸和精氨酸的含量以及矿物质 Mn 和 Se 含量均显著高于青刺参。

1.2 不同阶段刺参的营养组成分析

本研究对不同生长发育阶段刺参体壁营养成分及氨基酸组成比较分析（宋志东等，2009）。近年来，对于刺参体壁的营养成分已有很多学者研究过。例如，向怡卉等（2006）用氨基酸自动分析仪和气相色谱仪测定了刺参体壁及海参消化道中氨基酸、脂肪酸的组成和含量。但是，由于刺参生活的海洋环境、年龄、生长时期不同，其体内营养成分的化学组成及含量也不同。李丹彤等（2006）对 1 月和 5 月黄海獐子岛海域的野生刺参体壁中的营养组成进行了分析，发现刺参体壁中的水分含量存在显著性差异，粗蛋白、粗脂肪、粗灰分、糖分的含量等差异极显著。有关不同生长发育阶段刺参体内氨基酸组成的差异目前研究报道尚少，本节通过对稚参、幼参和成参体内的营养成分进行分析比较，探讨不同生长发育阶段刺参体内营养组成方面的差异。

1.2.1 材料与方法

1. 实验刺参

于山东烟台福山养殖场采集稚参（体长 0.5～1cm、1～2cm）、幼参（体长 2～7cm）、成参（体长 7cm 以上）各 4 头，分成 4 组，依次为 S4、S3、S2、S1，取回后分别称重、解剖、去除内脏团,用纱布吸干水分后再称重,然后用剪刀剪成小块,并用组织捣碎机捣碎,在烘箱中(105℃)烘至恒重后,分别对每个刺参进行水分、粗蛋白、粗脂肪、粗灰分、脂肪酸、氨基酸含量的测定。稚参因个体太小,难以取净内脏团,故在解剖后吸干水分直接匀浆测定。

2. 样品的测定

用 KOH 法提取糖,再用苯酚-硫酸法测定糖分,其他营养成分含量测定方法见附录。

3. 统计分析

实验数据用 SPSS 10.0 软件统计分析,首先对所有数据进行方差分析,如有显著性差异,则用邓肯多重范围检验(Duncan multiple-range test)进行多重比较分析。

1.2.2 结果分析与讨论

1. 营养成分分析

从表 1-8 中的数据来看,刺参的蛋白质含量比海水鱼的蛋白质含量(60%以上)低很多,脂肪含量也较低,但是水分和粗灰分含量较高。刺参含有大量的钙质内骨骼,推测这些钙质内骨骼是导致粗灰分含量较高的原因,而较高水分含量则是刺参这类棘皮动物的特点之一。刺参的营养成分指标受多种因素的影响,如年龄、个体大小及季节、水文因素等。本实验发现,个体较小的刺参,其水分、脂肪、粗灰分含量较高,蛋白质含量较低,但是随着个体的增长,蛋白质含量逐渐升高,水分、脂肪、粗灰分含量逐渐降低,这可能代表了刺参从幼体发育到成体的一种趋势。在从幼体发育到成体的过程中,刺参不断摄食藻类,导致蛋白质含量增加,而所摄食的食物中脂肪含量均很低,这就导致随着刺参个体长大脂肪含量反而降低。虽然内骨骼随着刺参的生长而变大,但是相比其他营养成分的积累速度要慢,这就导致粗灰分的含量反而降低。从饲料的营养角度来说,随着刺参的生长,应提高饲料中的蛋白质含量及钙磷含量,以满足其生长需求。

表 1-8 不同发育阶段刺参体壁的营养成分含量

项目	组别			
	S1	S2	S3	S4
水分	92.8	92.6	93.1	93.3
蛋白质(%DM)	49.58	43.92	38.26	39.10
脂肪(%DM)	2.97	3.32	4.61	3.93
粗灰分(%DM)	39.31	42.70	47.39	51.04

注:%DM 表示为该种物质所占干物质的比例。

2. 氨基酸组成分析

从表 1-9 可以看出，刺参的氨基酸组成有一定的规律性，如谷氨酸、天冬氨酸、甘氨酸、精氨酸等含量较高，半胱氨酸含量最低。4 个不同生长发育阶段（体长 0.5～1cm、1～2cm、2～7cm 和 7cm 以上）的刺参，其体壁的呈味氨基酸总量分别为 21.44%、16.85%、13.83%和 11.36%，药效氨基酸总量分别为 33.84%、29.50%、26.12%和 25.37%，能测到的 7 种人体必需氨基酸的总量分别为 11.44%、11.78%、11.38%和 11.24%。比较这 4 个不同生长发育阶段中刺参体壁的氨基酸组成，可以发现存在一定的规律性：随着刺参的生长，体壁大部分氨基酸的积累情况都发生显著的变化，如天冬氨酸、谷氨酸、丝氨酸、苏氨酸、甘氨酸、丙氨酸、脯氨酸及组氨酸，除了 S3 和 S4 组之间没有明显差异，其他组上述各种氨基酸的含量都随着生长发育明显升高。只有蛋氨酸和苯丙氨酸的含量较为稳定，各组间无明显差异。S2、S3、S4 组的精氨酸含量差异不大，而 S1 组的含量显著高于其他各组。异亮氨酸、亮氨酸、半胱氨酸、赖氨酸、酪氨酸等几种氨基酸的含量虽呈现一定的差异性，但因无明显规律性，故推测是个体原因造成。可以看出，刺参生长发育过程中氨基酸组成上的变化为：①天冬氨酸、谷氨酸、丝氨酸、苏氨酸、甘氨酸、丙氨酸、脯氨酸及组氨酸等几种氨基酸的含量在一定阶段内随着个体的生长呈现显著升高的趋势；②精氨酸在从稚参发育到幼参的过程中较为恒定，但发育到成参时含量显著提高；③刺参生长过程中，蛋氨酸和苯丙氨酸的含量较为稳定；④总氨基酸、呈味氨基酸、药效氨基酸等的含量都呈升高趋势，而人体必需氨基酸的含量则较为恒定。这说明随着刺参的生长，口味和药效价值增高，但是对人体的营养价值变化不大。

表 1-9　不同生长阶段刺参体壁的氨基酸组成（%）

氨基酸	组别			
	S1	S2	S3	S4
天冬氨酸	4.76±0.38^a	3.97±0.13^b	3.24±0.02^c	3.37±0.15^c
谷氨酸	7.56±0.77^a	6.46±0.06^b	5.41±0.05^c	5.10±0.22^c
丝氨酸	2.13±0.22^a	1.79±0.06^b	1.65±0.05^c	1.64±0.10^c
苏氨酸	2.31±0.20^a	1.96±0.09^b	1.72±0.05^c	1.69±0.10^c
精氨酸	3.40±0.44^a	2.62±0.116^b	2.25±0.07^c	2.12±0.15^b
甘氨酸	6.18±0.66^a	4.14±0.07^b	3.16±0.00^c	2.93±0.08^c
丙氨酸	2.94±0.31^a	2.28±0.04^b	2.02±0.03^c	1.96±0.08^c
脯氨酸	2.07±0.12^a	1.52±0.05^b	1.21±0.02^c	1.12±0.04^c
缬氨酸	1.90±0.11^a	1.85±0.05^a	1.69±0.07^b	1.70±0.08^b
蛋氨酸	0.72±0.00	0.76±0.04	0.70±0.03	0.76±0.12
异亮氨酸	1.55±0.11^{ac}	1.57±0.07^{ab}	1.45±0.05^{cd}	1.43±0.07^d
亮氨酸	2.15±0.13^{ac}	2.29±0.10^{ab}	2.17±0.09^c	2.22±0.07^c

续表

氨基酸	组别			
	S1	S2	S3	S4
苯丙氨酸	1.40±0.09	1.51±0.10	1.43±0.07	1.50±0.34
半胱氨酸	0.21±0.07^a	0.19±0.02^b	0.22±0.03^b	0.21±0.03^{ab}
赖氨酸	1.43±0.22^a	1.83±0.13^b	2.22±0.13^b	1.94±0.26^a
组氨酸	0.49±0.11^a	0.54±0.03^b	0.66±0.04^c	0.61±0.09^c
酪氨酸	1.00±0.53^{ab}	1.11±0.13^{bc}	1.59±0.04^c	1.47±0.13^a
∑氨基酸	42.20±4.49^a	36.40±1.28^b	32.79±0.82^c	31.78±2.11^c

注：同行数据无字母或上标相同字母表示差异不显著（$P>0.05$），上标不同字母表示差异显著（$P<0.05$）。

水产动物肌肉中氨基酸的组成受多种因素的影响，如年龄、地域、种类、季节及营养状况等。国内外关于这方面的研究仅有零星报道。例如，李行先等（2005）对丁鲹不同发育阶段鱼体常规养分和氨基酸组成进行比较分析后发现，成鱼肌肉中水分、粗蛋白和粗灰分的含量显著高于幼鱼和亲鱼，而肌肉中粗脂肪的含量显著低于幼鱼和亲鱼；幼鱼肌肉中组氨酸的含量显著高于成鱼；成鱼肌肉中脯氨酸的含量显著高于幼鱼，其他各种氨基酸的含量无显著性差异。李卫芬等（1998）对中华鳖不同发育阶段氨基酸组成的研究表明，幼鳖缬氨酸、亮氨酸、赖氨酸的含量显著低于稚鳖和成鳖；稚鳖的异亮氨酸含量高于幼鳖和成鳖。本文所测得的刺参营养成分的变化与上述动物具有很大的差异，这可能是动物种属性及营养的不同利用模式所致。

1.2.3 结论

通过对稚参、幼参、成参体壁的氨基酸组成进行研究，可以认为不同生长发育阶段的刺参对营养物质的需要是有差异的。随着刺参的生长发育，其体壁的营养组成发生一定程度的变化，如蛋白质含量、某些氨基酸的含量及氨基酸总量升高。对于饲料研究来说，应根据刺参的这种特性制定合理配方，以满足其发育过程中的营养需求。

第 2 章 刺参的营养需求

2.1 刺参对营养素的需求

2.1.1 刺参不同发育阶段对蛋白质需求量的研究

饲料中的蛋白质水平及氨基酸组成对养殖刺参不同阶段的生长极为重要。在一定的范围内,饲料中的蛋白质水平与刺参的体质量增长呈正相关关系,但是过高的蛋白质水平反而会造成蛋白质利用效率低下和饲料成本增高。除此之外,饲料的氨基酸组成也极为重要,平衡的氨基酸组成有利于刺参的消化和吸收,能够促进刺参的快速生长。关于刺参对蛋白质需求量的研究较少,朱伟等(2005)进行了刺参稚参对粗蛋白和粗脂肪需求量的研究后认为,刺参稚参对蛋白质的需求量为18.21%~24.18%,对粗脂肪的需求量为5%。王吉桥等(2007a)研究了饲料中不同蛋白源搭配下刺参的生长率,发现摄食玉米蛋白的刺参对粗蛋白的消化率最高(66.49%),得出了投喂以植物性蛋白为主的饲料时刺参的营养价值较高的结论。Okorie 等(2008)通过实验得出,刺参饲料中维生素 C 的最适含量为 100~105.3mg/kg。本实验根据刺参的不同生长阶段和食性特点设计配方,分别研究稚参、幼参及成参蛋白质的需求量,以及对氨基酸组成的需求特点(王际英等,2009)。

2.1.1.1 材料与方法

1. 饲料制备

综合考虑饲料的成本问题,参考刺参的食性及体壁肌肉的组成,以鱼粉、陆生植物、海藻粉作为主要蛋白源,这类蛋白品质品质好,易被刺参消化。刺参体壁中脂肪的含量很低,所以在饲料中对脂肪水平的设置较低,配制了粗蛋白含量分别为25.54%、28.79%、32.30%、35.49%和38.31%的 5 种实验饲料(表 2-1)。饲料成分经均匀混合后,超微粉碎至 500 目以上,然后经过微粒化处理,取 200 目筛下部分,装袋备用。

表 2-1 刺参实验饲料配方设计及营养成分实际测定值

成分		实验饲料组				
		D1	D2	D3	D4	D5
配方设计	优质脱脂鱼粉/%	5	5	3	2	2
	陆生植物/%	42	42	47	53	53

续表

成分		实验饲料组				
		D1	D2	D3	D4	D5
配方设计	海藻粉/%	8	8	8	5	4
	扇贝边粉/%	16	15	12	10	5
	谷朊粉/%	2	5	8	8	8
	小麦粉/%	20	15	12	10	10
	酵母/%	2	2	2	4	7
	豆粕/%	2	5	5	5	8
	复合维生素/%	2	2	2	2	2
	复合矿物质/%	1	1	1	1	1
实际测定值	粗蛋白/%	25.54	28.79	32.30	35.49	38.31
	粗脂肪/%	2.95	3.03	3.05	2.93	2.92
	水分/%	8.64	9.12	8.76	7.13	8.48
	能量/(kJ/g)	11.57	13.15	13.76	13.94	14.21

2. 饲养实验

饲养实验于2007年9月在山东烟台市福山区山东升索渔用饲料研究中心院内饲养实验室进行。实验用刺参取自蓬莱东方海洋育苗场。稚参体长约0.6cm，体质量约0.13g；幼参体长约4cm，体质量约0.98g；成参体长约7cm，体质量约2.95g。依照饲料配方将稚参、幼参、成参各分成5组，每组设3个平行，每个平行500头刺参。养殖容器为5m×5m×0.7m的水泥池，池内放置波纹板以供实验刺参栖息，用气石连续充气。每天9:00和17:00换水2/3，然后投喂饲料。饲料投喂稍过量，保持波纹板上有少量的剩饵，每周彻底清理实验池。养殖用水为经过沉淀、砂滤的深井海水，调节海水的盐度为31，水温控制在（18±1）℃。实验时间为60d。

3. 生长性能及营养成分测定

实验结束时，停喂3d，收集刺参，对每个实验组的刺参进行称量，然后将其放入冰箱（-80℃）中保存备用。分别统计特定生长率和存活率。

4. 氨基酸分析

对饲养效果较好的刺参饲料进行氨基酸分析，分析方法采用对甲氧基苯磺酰氯（DABS-Cl）柱前衍生法，选用贝克曼（BECKMAN）液相色谱，流动相为柠檬酸+乙腈，流速为1.4ml/min，检测波长为436nm，柱温为25℃。

5. 数据统计

实验数据用SPSS 10.0软件分析，先对所有数据进行方差分析，如有显著性差异，再用邓肯多重范围检验进行多重比较分析。

2.1.1.2 结果分析

对稚参、幼参、成参 60d 饲养实验的结果表明，处于不同生长阶段的刺参对饲料中蛋白质含量的要求也不相同。由表 2-2 可以看出，对于稚参，饲喂粗蛋白含量为 28.79%的 D2 组饲料，特定生长率（specific growth rate，SGR）最高，为（3.80±0.03）%/d，其次为 D3 和 D4 组饲料，D1 和 D5 组饲料的 SGR 显著低于其他 3 组（$P<0.05$）。对于幼参，饲喂粗蛋白含量为 32.30%的 D3 组饲料，日增重显著高于其他各组（$P<0.05$），其 SGR 为（1.44±0.05）%/d；D3 和 D4 组的 SGR 显著高于 D1 和 D2 组。对于成参，饲喂粗蛋白含量分别为 35.49%和 38.31%的 D4 和 D5 组饲料，其 SGR 分别为（0.84±0.04）%/d 和（0.79±0.06）%/d，两组之间的差异不显著（$P>0.05$），但这两组显著高于其他 3 组（$P<0.05$）。本实验的结果说明，刺参对粗蛋白的需求量随着生长发育而增大，即成参＞幼参＞稚参。粗蛋白含量最高的饲料并没有获得最大的增长率，说明刺参的生长在一定范围内是随着粗蛋白含量的升高而加快的，但是过高的粗蛋白含量可能通过其他途径反而对刺参的生长产生副作用。在本实验条件下，刺参的存活率因生长发育阶段不同而出现较大的变化，稚参为 35%~38%、幼参为 78%~83%、成参为 93%~96%，但在同一个生长阶段，各实验组刺参的存活率并无显著的差异（$P>0.05$）。在各生长阶段，刺参的脏壁比无显著性差异（$P>0.05$），但是有随着个体的增长而增大的趋势。

表 2-2　不同粗蛋白含量的饲料对实验刺参生长性能的影响

	组别	初始体质量/g	终末体质量/g	SGR/（%/d）	脏壁比/%	存活率/%
稚参	D1	0.13±0.04	0.85±0.25	3.13±0.06[ab]	0.12±0.02	38
	D2	0.13±0.03	1.27±0.17	3.80±0.03[c]	0.14±0.03	36
	D3	0.12±0.04	0.90±0.22	3.36±0.06[d]	0.12±0.01	35
	D4	0.14±0.06	1.02±0.16	3.31±0.07[d]	0.16±0.03	35
	D5	0.13±0.04	0.89±0.21	3.22±0.04[b]	0.12±0.02	37
幼参	D1	1.97±0.08	3.01±0.17	0.71±0.10[a]	0.19±0.03	83
	D2	1.92±0.06	3.14±0.25	0.82±0.09[a]	0.22±0.02	82
	D3	2.05±0.08	4.87±0.25	1.44±0.05[b]	0.22±0.01	79
	D4	1.93±0.12	4.39±0.16	1.37±0.03[c]	0.21±0.04	78
	D5	2.13±0.04	4.47±0.22	1.24±0.11[c]	0.20±0.03	80
成参	D1	6.45±0.13	8.97±0.28	0.55±0.09[a]	0.25±0.06	95
	D2	6.68±0.16	9.86±0.43	0.65±0.11[a]	0.27±0.03	93
	D3	6.39±0.21	9.15±0.44	0.60±0.06[a]	0.24±0.01	95
	D4	6.34±0.10	10.48±0.35	0.84±0.04[b]	0.24±0.05	96
	D5	6.42±0.10	10.33±0.49	0.79±0.06[b]	0.26±0.02	96

注：同列数据无字母或上标相同字母表示差异不显著（$P>0.05$），上标不同字母表示差异显著（$P<0.05$）。

对实验刺参机体成分的分析结果表明，饲料中粗蛋白含量的高低对各个发育阶段刺参的机体组成及营养成分的影响并不显著。实验刺参的粗蛋白含量为37.8%~41.8%，粗脂肪含量为6.52%~7.36%，粗灰分含量为34.6%~41.2%，水分含量为88.3%~91.5%（表2-3）。氨基酸的分析结果表明，在D2、D3和D4组3种饲养效果较好的饲料中，D3组饲料中氨基酸的总量及各种氨基酸的含量均较高（胱氨酸除外），其次是D4组，最低的是D2组；在氨基酸组成中，谷氨酸的含量最高，其次为天冬氨酸，胱氨酸的含量最低（表2-4）。

表2-3 实验刺参机体成分测定结果（%）

刺参	实验饲料组	粗蛋白（干物质）	粗脂肪（干物质）	粗灰分（干物质）	水分
稚参	D1	39.6±1.17	7.17±0.39	40.5±1.23	90.6±0.22
	D2	40.3±0.98	7.36±0.54	41.1±0.76	90.4±0.13
	D3	39.4±1.47	6.58±0.55	40.8±1.12	89.9±0.18
	D4	37.8±1.06	6.70±0.61	41.2±1.08	90.1±0.15
	D5	38.1±1.12	6.52±0.37	39.2±0.15	91.5±0.28
幼参	D1	40.6±1.28	7.13±0.58	34.6±0.57	90.6±0.34
	D2	41.4±1.54	6.95±0.44	37.8±0.86	90.4±0.26
	D3	41.8±0.87	6.93±0.38	37.1±1.04	89.9±0.37
	D4	40.9±1.75	7.02±0.49	37.0±1.17	90.1±0.31
	D5	38.4±1.92	7.21±0.61	35.1±0.45	91.1±0.19
成参	D1	39.2±1.19	7.23±0.36	36.5±0.43	88.3±0.25
	D2	40.5±0.95	7.07±0.52	38.3±0.85	91.1±0.42
	D3	41.0±0.94	6.97±0.63	38.2±0.96	90.2±0.30
	D4	41.5±1.25	6.82±0.87	40.0±1.23	91.2±0.16
	D5	40.6±1.02	7.16±0.41	39.4±0.73	91.3±0.37

表2-4 D2、D3、D4组饲料的氨基酸组成 （单位：mg/100g）

名称	实验饲料组			名称	实验饲料组		
	D2	D3	D4		D2	D3	D4
天冬氨酸	1.80	2.42	2.09	蛋氨酸	0.36	0.41	0.41
谷氨酸	2.77	4.31	3.84	异亮氨酸	0.96	1.27	1.18
丝氨酸	0.83	1.23	1.13	亮氨酸	1.46	2.05	1.96
苏氨酸	0.75	0.93	0.85	苯丙氨酸	0.82	1.28	1.20
精氨酸	1.02	1.87	1.68	胱氨酸	0.10	0.21	0.21
甘氨酸	0.98	1.35	1.25	赖氨酸	1.25	1.57	1.36
丙氨酸	1.32	1.39	1.30	组氨酸	0.31	0.53	0.51
脯氨酸	0.48	0.70	0.68	酪氨酸	0.40	0.70	0.56
缬氨酸	1.12	1.44	1.34				
∑氨基酸	16.73	23.69	21.55				

2.1.1.3 讨论

有关刺参蛋白质需求量研究的报道较少，仅有的研究主要集中于稚参和幼参的营养需求。一些实验表明，在脂肪含量为5%的条件下，刺参对饲料中蛋白质的需求量为18.21%~24.18%。而本实验中，在脂肪含量为3%的条件下，稚参、幼参、成参对粗蛋白的需求量分别达到28.79%、32.30%和35.49%，产生这种差异的原因可能与脂肪的含量有关。蛋白质、脂肪和碳水化合物是提供能量的三大能源物质，在脂肪含量不足的情况下，刺参对蛋白质的需求量提高。而刺参日增重的最大值并没有出现在粗蛋白含量最高的一组饲料中，其原因可能与王吉桥等（2007a）认为的一样，即当饲料中的蛋白质含量基本满足刺参各个阶段生长的需要时，过高的蛋白质含量反而对刺参的代谢造成负担。另外，Yang等（2005）指出，刺参的个体大小对消化食物及生长的最适温度有一定的影响。本次实验的温度为18℃，可能更接近稚参和幼参进行食物消化的最适温度，有助于生长。

脏壁比是衡量刺参机体可食部分生长速度的指标。朱伟等（2005）报道，蛋白质和脂肪配比最佳的饲料能够促进刺参体壁的生长，然而本实验并没有得到相同的结论，推测影响刺参体壁增长的因素不单是蛋白质和脂肪含量的高低及其配合的比例，可能与采用的蛋白源也有关。实验结束后，对刺参体壁成分组成的测定结果表明，其营养成分的组成相差不大，也说明蛋白质和脂肪的高低及配比并不能完全影响体壁增长及成分组成。但是，随着刺参个体的增长，脏壁比有增大的趋势，说明在刺参的生长发育过程中，内脏增大的速度要稍快于体壁。

动物对蛋白质的需求实质上就是对氨基酸的需求，因此饲料中氨基酸的组成对刺参的生长发育具有一定的影响。D3、D4、D2组饲料分别适合幼参、成参、稚参的营养需求，随着3种饲料中氨基酸总量的降低，生长速度却呈现成参＞幼参＞稚参的趋势，同饲料中蛋白质含量的升高趋势基本一致，这也充分说明除了氨基酸平衡对刺参的生长很重要，还有其他因素影响刺参的生长，导致成参日增重比稚参、幼参更大，各生长发育阶段刺参内脏的生长速度也出现了相似的趋向。对比向怡卉等（2006）报道的刺参体壁及消化道的氨基酸组成与本实验饲料中各种氨基酸组成的特点可以发现，对于几种含量比较高的氨基酸，如谷氨酸、天冬氨酸、亮氨酸、精氨酸、缬氨酸、赖氨酸等，饲料中的氨基酸组成与刺参消化道的氨基酸组成更为接近，这也是各个生长阶段内脏部分的比例保持较快增长的原因之一。除天冬氨酸、谷氨酸以外，本实验的饲料中苏氨酸、缬氨酸、亮氨酸、苯丙氨酸、赖氨酸、组氨酸、精氨酸等含量较高，而Sun等（2004）也发现，刺参摄食富含这几种氨基酸的饲料时体质量增长最大，因此可以断定这几种氨基酸在刺参的生长中起着极为重要的作用。研究人员对刺参配合饲料的研制工作起步较晚，在养殖生产中常用海藻粉、地瓜粉、杂鱼杂虾粉、麸皮等作为饵料投喂（常忠岳等，2003）。Liu 等（2009）研究了饲料中添加海泥和黄泥对刺参生长和能量收支的影响，认为添加 20%的海泥或黄泥能够获得较高的增长率。朱建新等（2007）用鼠尾藻干粉、鲜海带磨碎液、鲜石莼磨碎液和两种海参专用饲料养殖稚参、幼参，发现石莼磨碎液和市售"蛟龙牌"海参专用复合饲料效果要好于其他几组。徐宗法等（1999）也报道了用配合饲料饲养海参要好于纯海藻粉的实验结果。因此，营养搭配合理的配合饲料可以代替鲜饵，作为刺参的主要食物。

本实验中饲料选用海藻粉、扇贝边粉、谷朊粉、豆粕、酵母等作为主要的饲料原料，主要是考虑到海藻粉富含各种氨基酸，而有些氨基酸对水产动物具有很强的诱食作用，同时又具有黏合剂的作用；扇贝边粉不仅蛋白质含量高，还含有较多的不饱和脂肪酸（苏秀榕等，1997）；谷朊粉、豆粕、酵母等都具有较好的蛋白质组成，它们相互搭配后的氨基酸组成较为适合刺参的食性要求。

2.1.1.4 结论

以优质脱脂鱼粉、陆生植物、海藻粉、扇贝边粉、谷朊粉等蛋白源配制而成的刺参配合饲料，如果比例搭配得当，可以作为不同发育阶段刺参的专用配合饲料，在脂肪含量为 3%的条件下，粗蛋白含量分别为 28.79%、32.30%、35.49%的配合饲料，对稚参、幼参、成参的增重效果较好。

2.1.2 饲料中不同碳水化合物水平对刺参幼参生长和能量收支的影响

碳水化合物是水产动物饲料中廉价而重要的非蛋白质能量物质（Wilson，1994）。适宜的碳水化合物水平不仅可以提高饲料蛋白质利用效率，产生蛋白质节约效应（Azaza et al.，2013；Honorat et al.，2010；Vásquez-Torres and Arias-Castellanos，2013），还可以有效降低其他营养素的分解代谢，并为某些特殊化合物的合成提供代谢中间产物，促进机体的生长（Polakof et al.，2012；Vielma et al.，2003；Ye et al.，2009）。然而，过量的碳水化合物会造成脂肪沉积、血糖含量过高和免疫力下降等，从而抑制机体的生长（Chen et al.，2012b；Gao et al.，2010）。目前，有关水产动物摄食碳水化合物影响的报道主要集中在鱼类和甲壳类（Brauge et al.，1994；Mohanta et al.，2009；Oliveira et al.，2004；Radford et al.，2005）。研究表明，通过优化摄食碳水化合物水平可以增强机体糖代谢能力，提高能量利用效率（Hemre et al.，2002）。同时，对碳水化合物的利用能力因种类和摄食习性而有所差异（Li et al.，2014；Rawles and Gatlin，1998；Tran-Duy et al.，2008）。

有关不同碳水化合物水平对刺参生长的影响尚未见报道。刺参属于典型的杂食性动物，自然条件下主要靠楯形触手摄食栖息环境中的底栖微藻、小型底栖动物和沉降于水底的有机质碎屑等（Slater and Carton，2010；Gao et al.，2011），其天然饵料中含有较高水平的碳水化合物（Seo and Lee，2011；Seo et al.，2011）。本节拟通过设置饲料中碳水化合物不同梯度水平研究其对刺参生长、饲料利用、体组成及能量收支的影响，探求碳水化合物的最适需求量，并为开发高效刺参配合饲料奠定基础。

2.1.2.1 材料与方法

1. 饲料制备

以鱼粉、豆粕、海藻粉、鱿鱼油等为主要原料制备等蛋白质含量（19.93%）、等脂肪含量（2.63%）的5组配合饲料，调节糊化玉米淀粉含量使其碳水化合物水平分别为25.20%、34.92%、45.21%、55.41%和66.50%（表2-5）。海泥采集于潮间带，经干燥磨碎并高温煅烧去除有机质。饲料中含有 500mg/kg 三氧化二钇（Y_2O_3），作为测定表观消化率的惰性内标。原料混合均匀后经螺旋挤压机加工制成颗粒饲料，自然风干备用。

表 2-5 实验饲料组分及营养水平

项目		实验饲料组				
		D1	D2	D3	D4	D5
原料组分	鱼粉/%	8.0	8.0	8.0	8.0	8.0
	豆粕/%	27.0	27.0	27.0	27.0	27.0
	糊化玉米淀粉	0.0	10.0	20.0	30.0	40.0
	海藻粉/%	10.0	10.0	10.0	10.0	10.0
	小麦淀粉/%	6.0	6.0	6.0	6.0	6.0
	鱿鱼油/%	1.0	1.0	1.0	1.0	1.0
	卵磷脂/%	0.5	0.5	0.5	0.5	0.5
	矿物质预混料/%	1.0	1.0	1.0	1.0	1.0
	维生素预混料/%	1.0	1.0	1.0	1.0	1.0
	海泥/%	45.45	35.45	25.45	15.45	5.45
	三氧化二钇/%	0.05	0.05	0.05	0.05	0.05
营养水平（干物质）	粗蛋白/%	20.18	19.81	19.93	20.06	19.65
	粗脂肪/%	2.72	2.53	2.50	2.61	2.79
	粗灰分/%	51.90	42.74	32.26	21.92	11.06
	碳水化合物/%	25.20	34.92	45.21	55.41	66.50
	能量（kJ/g）	10.01	11.68	13.33	15.01	16.54
	蛋能比（mg/kJ）	19.79	17.28	14.95	13.09	11.98

注：①每千克矿物质预混料含 $MgSO_4 \cdot 7H_2O$ 90.0g、柠檬酸铁 18.0g、$ZnSO_4 \cdot 7H_2O$ 3.0g、$MnSO_4 \cdot H_2O$ 2.5g、$Ca(H_2PO_4)_2$ 160g、乳酸钙 25.8g、$CuCl_2$ 0.8g、$AlCl_3 \cdot 6H_2O$ 0.18g、NaCl 2.0g、KIO_3 0.04g、$CoCl_2 \cdot 6H_2O$ 0.07g；②每千克维生素预混料含抗坏血酸 150.0g、维生素E醋酸酯 10.0g、盐酸硫胺素 6.0g、维生素B_2 8.0g、盐酸吡哆醇 5.0g、烟酸 40.0g、肌醇 100.0g、生物素 0.3g、叶酸 1.5g、氨基苯甲酸 10.0g、维生素K 4.0g、维生素A 1.5g、维生素D_3 0.005g、维生素B_{12} 0.005g；③碳水化合物含量=100%−粗蛋白含量−粗脂肪含量−粗灰分含量（Seo and Lee，2011；Seo et al.，2011）。

2. 实验设计

实验用幼参购于山东东营市垦利区育苗场,初始体质量为(1.67±0.06)g,在暂养池中驯化3周,然后将其随机分成5组,每组设3个重复,每个重复40头刺参,于直径60cm、高80cm、水深50cm的圆柱形玻璃钢中进行饲养,实验周期为60d。实验采用循环水养殖系统,24h持续增氧。水温保持在(20±0.5)℃,盐度保持在28~30,pH保持在7.8~8.2。每天投喂一次,投喂量为幼参体质量的5%,投喂后20h采用虹吸法收集残饵和粪便,并于60℃条件下烘干保存。残饵量利用饲料在水中的浸出率进行校正(Pei et al., 2012)。

3. 样品采集与测定

实验开始前,随机取10头幼参称重作为初始体质量。实验结束前一天停止投喂,每桶取10头刺参称重,分别统计特定生长率、饲料系数、蛋白质效率、摄食率、排粪率、表观消化率。

饲料和粪便中三氧化二钇的含量采用电感耦合等离子体发射光谱仪测定。分别采用凯氏定氮法、索氏抽提法和马弗炉灼烧法测定饲料、刺参全参及其体组织中粗蛋白、粗脂肪和粗灰分的含量。样品中能量水平采用氧弹仪测定,氮含量采用元素分析仪测定。

刺参的能量收支符合 $C=G+F+U+R$(Pei et al., 2012),其中 C 为摄食能,G 为生长能,F 为粪能,U 为排泄能,R 为呼吸能。根据测定的饲料、粪便和幼参的能值,以及饲料摄食量、排粪量及刺参生长的干物质量,分别计算出摄食能、粪能和生长能。排泄能估算公式为:$U=(C_N-F_N-G_N)×24\,830$(Wang et al., 2003),其中,C_N 为摄食氮,F_N 为粪便氮,G_N 为生长氮,24 830 为刺参排泄每克氨氮的能值(J/g)。呼吸能由能量收支方程 $R=C-G-F-U$ 求出(安振华等,2008;袁秀堂等,2007)。

4. 统计分析

数据分析前均进行方差齐性检验。用 SPSS 16.0 软件进行单因素方差分析(one-way ANOVA)和 Turkey's 多重比较,以 $P<0.05$ 作为差异显著的标准(SPSS Inc, 2013)。基于特定生长率和饲料转化率的二次回归模型估算刺参碳水化合物的最适需求量(Robbins et al., 1979)。数值均采用平均值±标准差表示。

2.1.2.2 结果分析

1. 不同碳水化合物水平对刺参幼参生长和饲料利用的影响

不同碳水化合物水平的饲料投喂下刺参参生长和饲料利用情况见表2-6。实验周期内,不同摄食组间刺参幼参存活率无显著性差异($P>0.05$)。所有组刺参幼参均出现显著性增长,随着饲料中碳水化合物水平的升高,特定生长率呈现先上升后下降的趋势，D3 和 D4 组特定生长率显著高于其他各组（$P<0.05$）。基于特定生长率的二次回归模型估算刺参幼参饲料中碳水化合物的最适需求量为 50.15%（$y=-0.0008x^2+0.0767x-0.2190$，$r^2=0.96$）（图 2-1）。从 D1 组到 D3 组，饲料转化率显著增加，而 D3、D4 和 D5 组间饲料转化率无显著性差异（$P>0.05$）。基于饲料转化率和碳水化合物水平的二次回归关系估算刺参幼参饲料中碳水化合物的最适需求量为 53.95%（$y=-0.0017x^2+0.1829x-0.6000$，$r^2=0.99$）（图 2-2）。

表 2-6　不同碳水化合物水平的饲料投喂下刺参幼参生长和饲料利用情况（$n=30$）

指标	实验饲料组				
	D1	D2	D3	D4	D5
存活率/%	91.39±6.15	89.44±5.12	89.01±7.32	89.72±4.07	92.50±5.86
特定生长率/(%/d)	1.27±0.08[a]	1.47±0.12[b]	1.69±0.06[c]	1.72±0.06[c]	1.49±0.13[b]
饲料转化率/%	3.00±0.09[a]	3.63±0.21[b]	4.26±0.24[c]	4.32±0.26[c]	4.08±0.33[c]
蛋白质效率/%	0.15±0.01[a]	0.18±0.01[b]	0.21±0.02[c]	0.22±0.01[c]	0.21±0.02[bc]
摄食率/[g/(g·d)]	0.40±0.02[a]	0.38±0.01[ab]	0.37±0.01[bc]	0.37±0.02[bc]	0.34±0.02[c]
排粪率/[g/(g·d)]	0.33±0.01[a]	0.31±0.02[ab]	0.30±0.01[b]	0.29±0.02[b]	0.31±0.01[bc]
表观消化率/%	26.19±2.01[a]	27.88±1.07[ab]	30.97±0.86[c]	29.34±0.49[b]	27.01±1.23[a]

注：同行数据无字母或上标相同字母表示差异不显著（$P>0.05$），上标不同字母表示差异显著（$P<0.05$）。

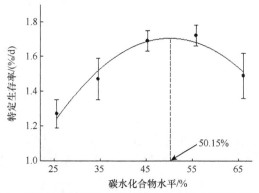

图 2-1　刺参幼参特定生长率与碳水化合物水平的回归分析

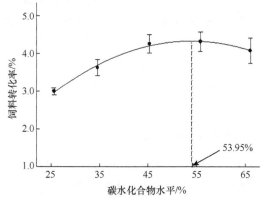

图 2-2　刺参幼参饲料转化率与碳水化合物水平的回归分析

与特定生长率相似，随着饲料中碳水化合物水平的升高，刺参幼参蛋白质效率和表观消化率也呈现先升高后下降的趋势，蛋白质效率在 D4 组最高（0.22），D3 组表观消化率显著高于其他各组（$P<0.05$）（表 2-6）。摄食率从 D1 组到 D5 组逐渐降低，不同摄食组间摄食率大多存在显著性差异（$P<0.05$）。此外，碳水化合物水平对刺参幼参排粪率的影响显著，D4 组排粪率最低，为 0.29g/（g·d）。

2. 不同碳水化合物水平对刺参幼参体组成的影响

不同碳水化合物水平的饲料投喂下刺参体组成见表 2-7。从 D3 组开始，全参粗脂肪含量显著升高，而 D3 组到 D5 组间无显著性差异。消化道粗脂肪含量在 D5 组显著升高（$P<0.05$）。碳水化合物水平对刺参体壁粗脂肪含量也存在显著影响，D4 和 D5 组显著高于 D1 和 D2 组（$P<0.05$）。随着饲料中碳水化合物水平的升高，仅体壁的粗蛋白含量在不同摄食组间出现显著性差异（$P<0.05$），且 D5 组最高（4.35%）。全参和体壁的粗灰分含量均呈现先下降后上升的趋势，而消化道的粗灰分含量在组间无显著性差异（$P>0.05$）。

表 2-7　不同碳水化合物水平的饲料投喂下刺参幼参体组成（%）（$n=30$）

刺参幼参部位	指标	实验饲料组				
		D1	D2	D3	D4	D5
全参	水分	89.59±2.01	91.14±1.77	90.04±1.05	91.45±1.55	90.38±1.71
	粗蛋白	4.11±0.20	4.31±0.25	4.29±0.17	4.17±0.10	4.23±0.16
	粗脂肪	0.15±0.01^a	0.18±0.02^{ab}	0.20±0.04^b	0.22±0.02^b	0.23±0.03^b
	粗灰分	2.51±0.06^{ab}	2.48±0.06^a	2.35±0.10^{abc}	2.21±0.15^c	2.33±0.08^{bc}
体壁	水分	92.14±1.98	91.88±1.32	90.67±2.38	91.02±1.45	92.37±2.00
	粗蛋白	3.98±0.15^a	4.10±0.11^{ab}	4.25±0.05^{bc}	4.31±0.15^{bc}	4.35±0.10^c
	粗脂肪	0.15±0.03^a	0.16±0.03^a	0.21±0.02^{ab}	0.21±0.01^b	0.22±0.02^b
	粗灰分	2.58±0.08^a	2.45±0.11^{ab}	2.40±0.08^b	2.34±0.06^b	2.59±0.10^a
消化道	水分	90.12±1.16	92.31±2.61	92.19±1.82	90.47±1.09	91.50±1.68
	粗蛋白	3.91±0.24	4.07±0.13	4.11±0.12	4.00±0.19	3.89±0.11
	粗脂肪	0.16±0.02^a	0.16±0.04^a	0.18±0.02^a	0.21±0.01^{ab}	0.22±0.01^b
	粗灰分	2.33±0.18	2.45±0.08	2.39±0.13	2.41±0.09	2.29±0.21

注：同行数据无字母或上标相同字母表示差异不显著（$P>0.05$），上标不同字母表示差异显著（$P<0.05$）。

3. 不同碳水化合物水平对刺参幼参能量收支的影响

不同碳水化合物水平的饲料投喂下刺参能量收支情况见表 2-8。随着饲料中碳水化合物水平的升高，刺参幼参摄食能显著增加，D4 和 D5 组摄食能无显著性差异。生长能及其占摄食能的比例均呈现先上升后下降的趋势，均在 D4 组获得最大值，分别为 0.47kJ/（g·d）和 8.46%。D4 和 D5 组粪能显著高于其他各组，而粪能占摄食能的比例则先下降后上升（$P<0.05$）。从 D1 组到 D5 组，排泄能逐渐增

加,而排泄能占摄食能的比例无显著性差异($P>0.05$)。从 D3 组到 D5 组,呼吸能显著高于 D1 和 D2 组,而组间无显著性差异,呼吸能占摄食能的比例先升高后降低,D3 组呼吸能最高,达 24.46%。

表 2-8　不同碳水化合物水平的饲料投喂下刺参能量收支情况

指标	实验饲料组				
	D1	D2	D3	D4	D5
摄食能/[kJ/(g·d)]	4.10±0.14^a	4.44±0.12^b	4.89±0.15^c	5.50±0.23^d	5.68±0.25^d
生长能占摄食能的比例/(%)	6.60±0.68^a	7.70±0.26^{bc}	8.32±0.61^{bc}	8.46±0.16^c	6.99±0.30^b
生长能/[kJ/(g·d)]	0.27±0.04^a	0.34±0.02^b	0.42±0.04^{cd}	0.47±0.01^d	0.40±0.03^{bc}
粪能占摄食能的比例/(%)	69.21±1.95^a	65.24±0.47^{ab}	61.08±2.86^b	63.43±3.05^{ab}	65.22±1.96^{ab}
粪能/[kJ/(g·d)]	2.80±0.18^a	2.90±0.08^a	2.98±0.07^a	3.49±0.25^b	3.71±0.28^b
排泄能占摄食能的比例/(%)	5.73±0.21	6.18±0.42	6.15±0.23	6.08±0.16	6.42±0.31
排泄能/[kJ/(g·d)]	0.23±0.01^a	0.27±0.02^b	0.30±0.01^{bc}	0.33±0.01^{cd}	0.36±0.03^d
呼吸能占摄食能的比例/(%)	18.46±2.47^a	20.88±0.28^{ab}	24.46±2.28^b	22.03±2.91^{ab}	21.37±2.56^{ab}
呼吸能/[kJ/(g·d)]	0.74±0.07^a	0.93±0.02^a	1.20±0.15^b	1.21±0.16^b	1.21±0.09^b

注:同行数据无字母或上标相同字母表示差异不显著($P>0.05$),上标不同字母表示差异显著($P<0.05$)。

2.1.2.3　讨论

目前,有关碳水化合物影响的研究主要集中于鱼类和甲壳类,而不同种类和摄食习性的水生动物对碳水化合物的利用能力存在差异(Brauge et al.,1994;Mohanta et al.,2009)。一般认为,杂食性和草食性水产动物可耐受较高水平的碳水化合物,并在一定范围内表现出明显的蛋白质节约效应(Mohanta et al.,2007;Peragón et al.,1999;Stone et al.,2003)。本研究中,在 25.61% 到 55.82% 的碳水化合物水平下,刺参幼参的特定生长率和饲料转化率均逐渐升高,但在 66.09% 的碳水化合物水平下出现下降趋势,说明摄食适宜的碳水化合物有利于促进刺参幼参生长、提高饲料利用效率,而摄食过量碳水化合物则会抑制刺参幼参机体生长(Li et al.,2013b;Tan et al.,2009;Wang et al.,2005)。基于特定生长率和饲料转化率的二次回归模型分析表明,刺参幼参饲料中碳水化合物的最适需求量为 50.15%~53.95%,这也说明杂食性的刺参具有较强的利用碳水化合物的能力(Seo et al.,2011)。

Kaushik 和 Médale(1994)报道,鲑鱼类可通过调节摄食率满足自身能量需求,且随着摄食碳水化合物水平的升高,其摄食率逐渐降低,本研究结果与其一致。刺参幼参摄食碳水化合物水平从 25.61% 增加到 45.31% 时,蛋白质效率逐渐升高,摄食能也显著增加,说明碳水化合物为刺参幼参的蛋白质节约效应提供了足够的能量,但随着饲料中碳水化合物水平的继续升高,蛋白质效率并无显著性差异。而 Tan 等(2009)研究发现,银鲫(*Carassius auratus gibelio*)摄食过量碳水化合物时,蛋白质效率出现明显下降趋势。本研究结果同样显示,摄食碳水化合物水平对刺参幼参的表观消化率具有显著影响。随着饲料中碳水化合物水平的升高，刺参幼参表观消化率先升高后下降，而排粪率先下降后上升，这说明摄食过量碳水化合物会降低刺参幼参机体对营养物质的利用效率，这与大多数杂食性鱼类和甲壳类的研究结果一致（Brauge et al., 1994; Mohapatra et al., 2003）。而对草食性水生动物而言，因其具有更高的碳水化合物耐受性，碳水化合物水平和消化率之间往往表现出非相关性或正相关性（Chen et al., 2012b; Gao et al., 2010）。Li 等（2014）研究发现，适宜的碳水化合物水平可诱导银鲫和草鱼（*Ctenopharyngodon idellus*）淀粉酶活性升高，增加机体表观消化率，但碳水化合物过量时又会抑制淀粉酶活性。

以往的研究表明，长期摄食较高水平的碳水化合物可能会造成水产动物机体内脂肪的沉积（Mohanta et al., 2009; Ren et al., 2011）。本研究中，刺参全参、体壁及消化道的粗脂肪含量随碳水化合物水平的升高均有不同程度的增加。Hemre 等（2002）发现，鱼类在高碳水化合物水平的饲料投喂下，戊糖磷酸代谢过程中的葡萄糖-6-磷酸脱氢酶活性显著升高。但是，通过此途径沉积的脂肪量相对较少，一种解释是高碳水化合物水平促进了脂肪的合成，同时降低了脂肪作为有氧代谢的能量供应。Shimeno 等（1993）的研究表明，尼罗罗非鱼（*Oreochromis niloticus*）可通过加快糖酵解、降低糖异生、促进糖原合成和脂肪沉积等过程来实现机体内糖代谢的稳态，维持正常生命活动。然而，过量摄食碳水化合物易造成糖代谢负担，诱发高血糖症，影响机体物质和能量代谢水平（Furuichi and Yone, 1981）。本研究中，在 66.09%碳水化合物水平的饲料投喂下，刺参幼参生长能占摄食能的比例显著降低，粪能、排泄能及呼吸能占摄食能的比例有所升高，表明过量摄食碳水化合物造成刺参幼参机体能量利用效率的下降。

2.1.2.4 结论

饲料中适宜的碳水化合物水平，可以促进刺参生长，提高饲料利用效率，产生蛋白质节约效应。幼参摄食碳水化合物的最适需求量为 50.15%～53.95%。

不同碳水化合物水平会显著影响刺参幼参的体组成，摄食过量的碳水化合物会抑制生长，并造成机体脂肪的沉积。

饲料中过量的碳水化合物会显著影响刺参能量代谢水平，降低机体能量利用效率。

2.1.3 刺参幼参对维生素 A 最适需求量的研究

维生素 A 是一种极其重要、动物体极易缺乏的维持机体正常代谢和机能所必需的脂溶性维生素。它是一元不饱和醇或具有醇类活性物质的总称，通常被称为视黄醇。维生素 A 能维持视觉（Moren et al., 2004），促进细胞分化及骨骼发育

（Hemre et al.，2004），也可以提高动物的生长速度、繁殖性能和疾病抵抗能力（Hernandez and Hardy，2020），同时也参与鱼类黏液分泌及脂肪代谢（蒋明，2007；Takeuchi et al.，1998）。研究发现，维生素 A 可以提高建鲤（杨奇慧和周小秋，2005）、大菱鲆（黄利娜等，2013）等水生动物的生长性能、免疫能力及繁殖性能，改善草鱼（张丽，2016）的肌肉品质。对褐牙鲆（王际英等，2010）、凡纳滨对虾（杨奇慧等，2007）等的研究证明，维生素 A 缺乏或不足时，会出现生长缓慢、皮肤出血、死亡率增加等症状。长期过量摄入维生素 A 也会导致鱼类骨骼畸形（Fernández et al.，2008），甚至产生毒性（Cuesta et al.，2002）。

已有的研究报道了刺参对维生素 B_6（李宝山等，2019）、维生素 C（Luo et al.，2014）、维生素 D（王丽丽等，2019）、维生素 E（Wang et al.，2015b）的需求，结果表明刺参对上述维生素的需求与鱼类差异较大。本实验通过在饲料中添加不同水平的维生素 A，探究其对刺参的生长、体成分、消化及非特异性免疫能力的影响，以期为刺参配合饲料中添加适量的维生素 A 提供参考（李宝山等，2023）。

2.1.3.1 材料与方法

1. 实验设计与饲料配制

以鱼粉和发酵豆粕为主要蛋白源，以鱼油和大豆卵磷脂为脂肪源，设计粗蛋白含量为22%（Liao et al.，2014）、粗脂肪含量为4%（Seo and Lee，2011）的基础饲料配方。在基础饲料中分别添加 0IU/kg、2000IU/kg、4000IU/kg、6000IU/kg、8000IU/kg、12 000IU/kg 的维生素 A 乙酸酯（浙江新维普添加剂有限公司，100 000IU/g），制成维生素 A 含量分别为 3250.00IU/kg、5187.00IU/kg、7054.00IU/kg、8970.00IU/kg、12 975.00IU/kg、16 400.00IU/kg 的 6 种等氮等能的实验饲料，分别命名为 D1、D2、D3、D4、D5、D6 组。饲料制作前将固体原料粉碎后过 200 目标准筛，按照配方比例进行称重，混匀后加入适量的蒸馏水和新鲜鱼油，然后用挤压机制成直径为 0.3cm、厚度为 0.05cm 左右的片状饲料，60℃烘干备用。实验饲料配方及营养组成见表 2-9。

表 2-9 实验饲料配方及营养组成（干物质）

项目		组别					
		D1	D2	D3	D4	D5	D6
原料组成	鱼粉/%	5.00	5.00	5.00	5.00	5.00	5.00
	发酵豆粕/%	10.00	10.00	10.00	10.00	10.00	10.00
	虾粉/%	15.00	15.00	15.00	15.00	15.00	15.00
	海藻粉/%	6.00	6.00	6.00	6.00	6.00	6.00
	小麦粉/%	23.00	23.00	23.00	23.00	23.00	23.00
	α-淀粉/%	2.00	2.00	2.00	2.00	2.00	2.00

续表

项目		组别					
		D1	D2	D3	D4	D5	D6
原料组成	鱼油/%	1.00	1.00	1.00	1.00	1.00	1.00
	大豆卵磷脂/%	1.00	1.00	1.00	1.00	1.00	1.00
	维生素预混料/%	0.50	0.50	0.50	0.50	0.50	0.50
	维生素A乙酸酯/(IU/kg)	0.00	2 000.00	4 000.00	6 000.00	8 000.00	12 000.00
	矿物质预混料/%	0.50	0.50	0.50	0.50	0.50	0.50
	海泥/%	36.00	36.00	36.00	36.00	36.00	36.00
营养组成	粗蛋白/%	24.00	24.00	23.72	24.28	23.75	23.71
	粗脂肪/%	4.56	4.65	4.61	4.54	4.72	4.70
	粗灰分/%	42.89	43.16	42.77	42.84	42.73	42.84
	能量/(kJ/g)	14.71	14.68	14.74	14.67	14.69	14.69
	维生素A/(IU/kg)	3 250.00	5 187.00	7 054.00	8 970.00	12 975.00	16 400.00

注：①维生素预混料、矿物质预混料配方参见文献（王际英等，2015）；②维生素预混料中不含维生素B_6。

2. 实验用刺参幼参及实验管理

实验在山东省海洋资源与环境研究院东营实验基地循环水养殖系统中进行，实验用刺参幼参购自山东安源种业科技有限公司。正式实验之前，将刺参幼参暂养于养殖系统中15d，暂养期投喂D1组饲料。待其完全适应养殖条件后，挑选大小均匀〔初始体质量为（15.48±0.01）g〕、参刺坚挺的刺参540头，随机放养于18个圆柱形玻璃钢养殖水槽（$h_{桶}$=80cm，$h_{水}$=60cm，$d_{桶}$=70cm）。每组饲料随机投喂3个水槽，养殖8周。每个水槽放置海参养殖筐2个，内嵌波纹板20张，每隔30d更换养殖筐1次。实验在微流水环境中进行，水温控制在17~19℃，水流速为2L/min，盐度为26~28，pH为7.6~8.2，采用充气增氧，保证溶氧量大于5.0mg/L，氨氮和亚硝酸盐的浓度低于0.05mg/L。每天8:00换水，换水量为养殖水体的1/3，每天16:00进行投喂，初始投喂量占体质量的2%，养殖期间依据刺参幼参的摄食情况调整投喂量，记录水温及刺参幼参死亡情况。每隔2d用虹吸法将残饵及粪便吸出，实验在弱光环境下进行。

3. 样品采集与处理

取样前将实验用刺参幼参控食48h，统计每桶刺参幼参数量。每桶随机取8头刺参幼参置于洁净的泡沫板上，用滤纸轻轻吸干体表水分，称量体质量。去除体腔液，分离肠道和体壁。去除肠道内容物，用滤纸擦净后称重，所有操作在冰盘上进行。样品在-20℃条件下保存，待测。

将肠道样品慢慢剪碎，称重，加入9倍体积预冷的生理盐水后匀浆，匀浆液离心10min（4℃，8000r/min）后分离上清液，分装于2ml的离心管中，-20℃保存，待测。

4. 测定指标与方法

统计刺参幼参增重率、特定生长率、存活率、肠体比。

饲料中的维生素 A 在通标标准技术服务（青岛）有限公司采用高效液相色谱法-皂化提取法（GB/T 17817-2010）测定；刺参幼参体壁羟脯氨酸采用南京建成生物工程研究所试剂盒测定；刺参幼参体壁维生素 A 含量采用上海酶联生物科技有限公司生产的酶联免疫吸附法（ELISA）试剂盒测定，测定方法参照试剂盒说明书。

采用紫外比色法测定刺参幼参肠道蛋白酶活性，采用比色法测定脂肪酶活性，采用碘-淀粉比色法测定淀粉酶活性，采用微量酶标法测定碱性磷酸酶活性，采用羟胺法测定超氧化物歧化酶活性，采用硫代巴比妥酸（TBA）法测定丙二醛含量，采用微板法测定谷丙转氨酶及谷草转氨酶的活性。所用试剂盒均购自南京建成生物工程研究所，碱性磷酸酶活性单位转化为国际单位，其他酶的活性定义及具体测定步骤参照试剂盒说明书。

5. 数据统计分析

实验所得数据采用 Microsoft Excel 2007 及 SPSS 11.0 统计软件进行单因素方差分析，差异显著（$P<0.05$）时用邓肯多重范围检验进行多重比较分析，结果以平均值±标准差表示。采用一元二次回归分析刺参幼参对维生素 A 的最适需求量。

2.1.3.2 结果分析

1. 维生素 A 对刺参幼参生长性能的影响

维生素 A 对刺参幼参生长性能的影响结果显示，刺参幼参存活率不受饲料中维生素 A 含量的影响（$P>0.05$）；随着饲料中维生素 A 含量的增加，增重率及特定生长率均呈现先升后降的趋势，D1 组显著低于其他组（$P<0.05$），D4 组与 D5 组显著高于其他组（$P<0.05$）；肠体比与增重率、特定生长率呈现相似趋势，D1 组与 D6 组无显著性差异（$P>0.05$），且 D1 组显著低于其他 4 组（$P<0.05$）（表 2-10）。

表 2-10 维生素 A 对刺参幼参生长性能的影响

生长指标	组别					
	D1	D2	D3	D4	D5	D6
初始体质量/g	15.47±0.12	15.48±0.12	15.47±0.11	15.48±0.76	15.50±0.09	15.49±0.12
终末体质量/g	19.22±0.31[a]	20.39±0.46[b]	19.59±0.98[b]	21.28±1.60[c]	22.07±0.64[c]	19.31±1.09[b]
增重率/%	24.34±0.31[a]	29.02±0.46[b]	30.18±0.98[b]	37.47±1.60[c]	35.94±0.64[c]	30.07±1.09[b]
特定生长率/（%/d）	0.37±0.01[a]	0.42±0.01[b]	0.46±0.02[b]	0.53±0.02[c]	0.51±0.01[c]	0.44±0.01[b]
肠体比/%	2.53±0.01[a]	3.36±0.16[b]	3.22±0.10[b]	3.16±0.84[b]	3.28±0.14[b]	3.04±0.12[ab]
存活率/%	86.67±1.33	88.00±2.31	91.33±1.76	85.56±1.11	84.89±2.47	90.00±1.15

注：同行数据无字母或上标相同字母表示差异不显著（$P>0.05$），上标不同字母表示差异显著（$P<0.05$）。

本实验条件下，以增重率为评价指标，经一元二次回归分析得出，初始体质量为 15.48g 的刺参幼参对饲料中维生素 A 的最适需求量为 10 000IU/kg（图 2-3）。

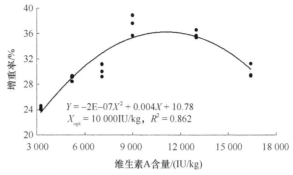

图 2-3 刺参幼参增重率与饲料中维生素 A 含量的回归分析

2. 维生素 A 对刺参幼参体壁基本成分的影响

维生素 A 对刺参幼参体壁基本成分的影响结果显示，水分及粗蛋白的含量受维生素 A 含量的影响不显著（$P>0.05$）。随着饲料中维生素 A 含量的增加，粗脂肪含量显著升高，D1 组显著低于其他组（$P<0.05$）；粗灰分含量先升高后平稳，D1 组显著低于其他组（$P<0.05$），D4 组达到最高值，与 D5、D6 组无显著性差异（$P>0.05$）；羟脯氨酸含量显著降低（$P<0.05$）；维生素 A 含量显著上升（$P<0.05$）（表 2-11）。

表 2-11 维生素 A 对刺参幼参体壁基本成分的影响

体成分	组别					
	D1	D2	D3	D4	D5	D6
水分/%	91.60±0.22	91.39±0.09	91.44±0.32	91.97±0.07	91.84±0.12	91.73±0.07
粗蛋白/%	45.01±0.08	45.16±0.07	45.21±0.15	45.21±0.08	45.47±0.19	45.44±0.27
粗脂肪/%	4.11±0.05^a	4.50±0.03^b	4.92±0.11^c	5.31±0.14^d	5.44±0.09^{de}	5.68±0.06^e
粗灰分/%	35.63±0.11^a	36.06±0.08^b	37.49±0.12^c	38.64±0.11^d	38.60±0.10^d	38.31±0.01^d
羟脯氨酸/（mg/kg）	723.85±3.47^e	616.45±6.93^d	449.12±0.60^c	414.11±7.49^b	402.96±6.86^b	340.92±3.65^a
维生素 A/（mg/kg）	269.37±4.53^a	266.90±7.85^a	276.47±4.31^{ab}	280.46±6.53^{ab}	290.13±3.22^{bc}	300.05±0.92^c

注：粗蛋白、粗脂肪、粗灰分为干基含量；同行数据无字母或上标相同字母表示差异不显著（$P>0.05$），上标不同字母表示差异显著（$P<0.05$）。

3. 维生素 A 对刺参幼参肠道消化酶活性的影响

刺参幼参肠道蛋白酶活性不受饲料中维生素 A 含量的影响（$P>0.05$）；脂肪酶活性随着维生素 A 含量的增加呈先上升后平稳趋势，D2~D6 组显著高于 D1 组（$P<0.05$）；淀粉酶活性呈现先升后降的趋势，D4 组达到最高值（表 2-12）。

表 2-12　维生素 A 对刺参幼参肠道消化酶活性的影响　　（单位：U/g prot）

消化酶	组别					
	D1	D2	D3	D4	D5	D6
蛋白酶	91.40±0.75	93.99±2.45	92.82±1.14	91.93±0.90	91.18±0.65	90.56±0.90
脂肪酶	24.87±0.76^a	31.36±0.10^b	32.63±0.75^c	37.00±0.48^d	35.55±0.53^e	36.33±0.66^d
淀粉酶	0.32±0.01^a	0.48±0.02^b	0.80±0.01^{de}	0.83±0.03^e	0.77±0.01^d	0.70±0.01^c

注：同行数据无字母或上标相同字母表示差异不显著（$P>0.05$），上标不同字母表示差异显著（$P<0.05$）。

4. 维生素 A 对刺参幼参肠道免疫酶活性的影响

随着饲料中维生素 A 含量的增加，刺参幼参肠道碱性磷酸酶活性呈现上升趋势，D1 组显著低于其他组（$P<0.05$）；谷草转氨酶活性先下降后升高，D4 组达到最低值；超氧化物歧化酶活性先升高后降低，D3 与 D4 组显著高于其他 4 组（$P<0.05$），但 2 组之间无显著性差异（$P>0.05$）；丙二醛含量呈现下降趋势，D5、D6 组显著低于其他 4 组（$P<0.05$）。维生素 A 对刺参幼参肠道谷丙转氨酶活性无显著影响（$P>0.05$）（表 2-13）。

表 2-13　维生素 A 对刺参幼参肠道免疫酶活性的影响

项目	组别					
	D1	D2	D3	D4	D5	D6
碱性磷酸酶活性/ （U/g prot）	0.68±0.03^a	1.00±0.04^b	1.68±0.03^c	1.59±0.02^c	2.10±0.06^d	2.48±0.03^e
谷草转氨酶活性/ （U/g prot）	5.38±0.38^d	3.53±0.11^c	3.53±0.02^c	2.53±0.02^a	2.79±0.05^b	2.95±0.06^b
谷丙转氨酶活性/ （U/g prot）	2.26±0.02	2.51±0.09	2.51±0.05	2.39±0.13	2.36±0.10	2.43±0.07
超氧化物歧化酶 活性/（U/mg prot）	91.03±1.57^a	121.42±3.24^c	175.89±7.85^d	170.38±4.92^d	110.10±5.21^{bc}	102.41±3.42^{ab}
丙二醛含量/ （nmol/mg prot）	1.43±0.02^b	1.41±0.04^b	1.43±0.03^b	1.37±0.06^b	1.15±0.05^a	1.02±0.03^a

注：同行数据无字母或上标相同字母表示差异不显著（$P>0.05$），上标不同字母表示差异显著（$P<0.05$）。

2.1.3.3　讨论

1. 维生素 A 对刺参幼参生长性能及体成分的影响

维生素 A 能够提高动物的生长速度并增强机体的免疫能力，饲料中维生素 A 含量过低和过高都会抑制动物的生长，甚至产生毒性。本研究表明，饲料中缺乏或有过量的维生素 A 均对刺参幼参存活率的影响不显著，然而，增重率及特定生长率均随着饲料中维生素 A 含量的增加呈现先升后降趋势。这与对建鲤（杨奇慧和周小秋，2005）及凡纳滨对虾（杨奇慧等，2007）等的研究结果一致。维生素 A 是维持成骨细胞及破骨细胞正常代谢的必需物质，当摄入维生素 A 过低或过高时，成骨细胞的溶解速度、旧骨细胞的活性都会下降，进而降低动物的生长速度。本实验中，刺参幼参的肠体比明显受饲料中维生素 A 含量的影响，且与生长趋势相同。这与维生素 A 可以保持消化道上皮细胞的完整性的功能有关，当维生素 A 缺乏或过量时，肠道上皮细胞就会受损，不利于肠道的正常发育（Warden et al.，1996）。

本实验条件下，刺参幼参对饲料中维生素 A 的最适需求量为 10 000IU/kg，与中华鳖（*Pelodiscus sinensis*）的需求量较为接近（Chen and Huang，2015），高于大多数已报道的鱼类的需求量（Hernandez and Hardy，2020），这既反映了动物种属的差异，又反映了动物组织分化的差异。刺参体内组织分化程度较低，不具备完整的消化系统，维生素 A 及其代谢产物能维持机体上皮的分化及黏液分泌，而上皮的完整性和黏液的分泌对维持刺参机体正常功能至关重要，因此其对维生素 A 的需求量可能会更高。此外，刺参的摄食较慢，且饲料呈细小颗粒状，在水中的散失也是影响实验数据的一个重要因素。

维生素 A 对不同水生动物体成分的影响存在较大差异。本实验中，维生素 A 对刺参幼参体壁水分及粗蛋白的含量无明显影响，这与 Hu 等（2006）对奥尼罗非鱼的研究结果不一致，可能是刺参与其他鱼类相比，饲料中蛋白质含量较低，维生素 A 对刺参幼参体内蛋白质的代谢影响不大（孙淑洁等，2012）。刺参幼参体壁中粗脂肪含量呈现上升趋势，粗灰分含量呈现先升高后平稳趋势，这与张璐等（2015）对日本花鲈的研究结果一致。维生素 A（乙酸酯）主要是以脂肪的形式存在，进入消化道后形成更多的乳糜微粒，延长了脂肪在肠道的停留时间，促进了机体对饲料中脂肪的吸收，进而引起体内脂肪积累量增加，导致体壁中粗脂肪含量升高。骨骼代谢是指骨骼形成及吸收的一个动态平衡过程，而水生动物的骨骼中 Ca、P 是主要元素，适宜水平的维生素 A 能够促进骨片中 Ca、P 等矿物质的积累，导致刺参幼参体壁中粗灰分含量上升，然而，高水平的维生素 A 会抑制矿物质的沉积，使体壁中粗灰分含量降低。刺参幼参体壁羟脯氨酸含量随维生素 A 含量的增加呈现下降趋势，这与张丽（2016）对草鱼的研究结果不一致，原因可能是维生素 A 在鱼体内转换成维甲酸后能抑制分解胶原的酶素，促进新的胶原产生，而刺参的代谢机制与鱼类存在较大差异。

2. 维生素 A 对刺参幼参肠道消化酶活性的影响

消化系统是动物对自身摄取的营养物质进行消化和吸收的主要场所，是为维持机体生命不断提供能量、蛋白质、脂肪等的重要器官。刺参为低等变温动物，消化器官发育不完善，但仍具有消化道，因此对营养物质的消化主要是在肠道中进行，而消化的实质主要是酶的消化，消化酶活性能够更直观地反映出营养物质被消化吸收的程度。本实验中，维生素 A 对刺参幼参肠道蛋白酶活性无显著影响，这与冯仁勇（2006）对幼建鲤的研究结果不一致，说明蛋白酶活性与蛋白质在机体内的沉积呈正相关关系。脂肪酶主要通过催化酰基甘油水解而发挥作用，在刺参的肠道中，脂肪酶与蛋白酶一样，属于同源性消化酶，稚参肠道中就可分泌这种酶。本实验中，刺参幼参肠道脂肪酶活性受维生素 A 的显著影响，说明维生素 A 通过提高肠道脂肪酶活性来增强消化能力，促进刺参幼参的生长，这与对老鼠的研究结果（肖露等，2017）一致。淀粉酶是消化酶中出现最早的一种酶，主要是水解淀粉和糖原。本实验仿刺参肠道淀粉酶活力呈先升后降趋势，说明饲料中适宜水平的维生素 A 对肠道中淀粉酶的分泌具有诱导作用。本实验中，刺参幼参肠道蛋白酶与脂肪酶活性的变化与体壁中粗蛋白及粗脂肪的沉积呈正相关关系，表明维生素 A 通过影响蛋白质与脂肪在刺参幼参肠道的消化吸收而影响其在体壁中的沉积。

3. 维生素 A 对刺参幼参肠道免疫酶活性的影响

免疫系统最重要的组成部分是免疫器官（杨先乐，1989），同时免疫器官也是维持机体正常免疫功能的组织学基础，免疫器官的发育程度受营养物质的影响显著（Erickson et al.，1980）。可将机体的抵抗力作为衡量营养物质对机体免疫作用的综合指标（Hernandez and Hardy，2020）。刺参为无脊椎动物，免疫机制主要是非特异性免疫，因此通过刺参肠道中非特异性免疫酶活性可直接判断机体的抗病能力。

碱性磷酸酶是骨骼、肠道等组织中广泛分布的一种酶，主要由成骨细胞产生，很多因素都会影响碱性磷酸酶的活性。已有研究表明，饲料中过低和过高水平的维生素 A 都能对碱性磷酸酶活性产生一定的影响。本实验中，刺参幼参肠道碱性磷酸酶活性总体呈现先升后降趋势，与索兰弟等（2002）对肉仔鸡血清中碱性磷酸酶活性的研究结果一致。这说明饲料中适宜含量的维生素 A 能够通过提高成骨细胞的活性，促进骨骼的正常发育，提高肠道碱性磷酸酶的活性（Hu et al.，2006）。

转氨酶是催化氨基酸与酮酸之间氨基转移的一类酶，普遍存在于动物组织中。在脊椎动物中，血清中转氨酶活性的高低通常被用来评价肝脏功能正常与否。刺参不具备肝脏这个重要的代谢器官，因此转氨酶活性的大小更可能反映了组织氨基酸代谢能力的高低。本实验中，刺参幼参肠道谷草转氨酶活性先降后升，谷丙转氨酶活性无显著变化。已有研究表明，谷丙转氨酶主要位于细胞质内，而谷草转氨酶主要位于线粒体内。因此，推测饲料中添加维生素 A 对蛋白质的代谢影响不大，但影响了能量代谢，这也与本实验中刺参幼参体壁基本成分、肠道消化酶活性等指标的变化相对应。

超氧化物歧化酶是一种二聚体酶，可以解毒、清除超氧阴离子自由基（Bannister et al., 1987），通过其活性可以判断机体清除氧自由基的能力（Anggraeni and Owens, 2000）。机体内丙二醛的含量可以反映出躯体的脂质过氧化程度，也能间接地反映机体细胞受氧自由基攻击的程度（Rudnicki et al., 2007），其含量过高时，对细胞的毒害作用就会产生，免疫力因此下降。本实验中，刺参幼参肠道超氧化物歧化酶活性随着维生素 A 含量的增加呈现先升后降趋势，丙二醛含量呈现下降趋势，表明适量的维生素 A 能够提高刺参幼参的抗氧化应激能力，这与对草鱼的研究结果（刘梦梅等，2017；Chen et al., 2016）一致。

2.1.3.4 结论

以增重率为评价指标，经一元二次回归分析可知，初始体质量为 15.48g 的刺参幼参对饲料中维生素 A 的最适需求量为 11 000IU/kg。饲料中添加适量的维生素 A 能够促进刺参幼参生长，提高机体脂肪代谢及抗氧化应激能力。

2.1.4 刺参幼参对维生素 B_6 最适需求量的研究

维生素 B_6 又称吡哆素，包括盐酸吡哆醇、吡哆醛及吡哆胺 3 种化学形式，在动物体内以磷酸酯的形式存在，是一种水溶性维生素。维生素 B_6 是 140 多种酶的辅酶，参与多种代谢反应，尤其是和氨基酸代谢有密切关系，是维持动物正常生理功能的一种必需营养素（Percudani and Peracchi, 2003）。此外，维生素 B_6 还参与糖异生、不饱和脂肪酸代谢，参与神经介质、核酸的合成（Vrolijk et al., 2017）。研究表明，饲料中添加适宜的维生素 B_6 会提高斑节对虾（Shiau and Wu, 2003）、尼罗罗非鱼（Teixeira et al., 2012）和印度鲶鱼（Mohamed, 2001）的生长性能，促进皱纹盘鲍（冯秀妮等，2006）、中华绒螯蟹（王玥等，2009）和凡纳滨对虾（李二超等，2010）对饲料蛋白质的利用，增加虹鳟（Maranesi et al., 2005）肌肉中二十二碳六烯酸的沉积。

研究表明，刺参对饲料中蛋白质（Liao et al., 2014）、脂肪（Seo and Lee, 2011）及碳水化合物（夏斌等，2015）的需求均较低，这与其消化道分化程度及摄食选择能力较低有关。维生素 B_6 是动物必需的一种营养素，对蛋白质、脂肪、糖类的吸收及利用影响巨大。目前未见刺参对饲料中维生素 B_6 需求量的报道。因此，本实验通过在基础饲料中添加不同量的维生素 B_6，饲喂刺参幼参，研究对其生长、体腔液代谢酶活性及消化生理的影响，从而得出刺参幼参对饲料中维生素 B_6 的最适需求量，以期为刺参代谢生理研究及饲料配制提供参考。

2.1.4.1 材料与方法

1. 实验饲料配制

以海藻粉、鱼粉及小麦粉为主要原料,设计粗蛋白含量为16%、粗脂肪含量为2%的基础饲料配方(Seo and Lee,2011)。在基础饲料中分别添加 0mg/kg、5mg/kg、10mg/kg、20mg/kg、40mg/kg、80mg/kg 的维生素 B_6(江西百盈生物技术有限公司,纯度≥99%),制作维生素 B_6 实测含量分别为 1.23mg/kg、5.29mg/kg、9.35mg/kg、17.47mg/kg、33.71mg/kg 和 66.17mg/kg 的 6 组等氮等能的实验饲料,分别命名为 D1、D2、D3、D4、D5、D6 组。实验饲料配方及营养组成见表 2-14。各原料粉碎后过 80 目标准筛,按配比称重,逐级混匀后加入蒸馏水,用小型饲料挤压机制作成条状饲料,自然风干后,用饲料破碎机破碎,筛取 20~40 目标准筛的颗粒,-20℃保存备用。

表 2-14 实验饲料配方及营养组成

	项目	D1	D2	D3	D4	D5	D6
原料组成	海泥/%	35.9	35.9	35.9	35.9	35.9	35.9
	维生素 B_6/(mg/kg)	0	5	10	20	40	80
	鱼粉/%	12	12	12	12	12	12
	海藻粉/%	36	36	36	36	36	36
	小麦粉/%	10	10	10	10	10	10
	多维(无维生素 B_6)/%	1	1	1	1	1	1
	多矿/%	1	1	1	1	1	1
	谷朊粉/%	3	3	3	3	3	3
	抗氧化剂/%	0.1	0.1	0.1	0.1	0.1	0.1
	大豆卵磷脂/%	1	1	1	1	1	1
营养组成	粗蛋白/%	16.21	15.83	15.94	15.84	15.61	16.06
	粗脂肪/%	2.08	2.13	2.04	1.97	2.01	2.09
	粗灰分/%	49.20	50.14	50.25	49.40	50.19	50.23
	能量/(kJ/g)	12.17	11.95	12.00	11.98	12.01	12.02
	维生素 B_6/(mg/kg)	1.23	5.29	9.35	17.47	33.71	66.17

注:多维、多矿配方参见文献(王际英等,2015),多维中不含维生素 B_6。

2. 实验设计与养殖管理

养殖实验在山东省海洋资源与环境研究院东营实验基地循环水养殖系统中进行。实验用刺参幼参购自山东安源种业科技有限公司。实验开始前,将 1000 头体质量约 12g 的刺参幼参放养于养殖系统中,饲喂 D1 组配合饲料,使其适应实验环境。21d 后,饥饿 24h,挑选体质量相近、参刺坚挺的刺参幼参 540 头,平均放置于 18 个养殖水槽中($h_桶$=80cm,$h_水$=60cm,$d_桶$=60cm),每个水槽放置刺参养殖筐一个。每种饲料随机投喂 3 个水槽，每日定时（16:00）投喂 1 次，初始投喂量约为刺参幼参体质量的 3%，投喂前在饲料中加入适量水润湿后，泼洒投喂。每天观察刺参幼参摄食情况，调整次日投喂量。实验采用流水养殖，控制水流流速约为 2L/min；每隔 2d 用虹吸法将残饵及粪便吸出，每隔 20d 更换养殖筐 1 次。养殖期间，控制水温为 14~17℃，pH 为 7.8~8.2，盐度为 30，溶氧量大于 6mg/L，氨氮、亚硝酸氮的浓度低于 0.05mg/L，养殖实验持续 84d。

3. 样品采集

养殖实验结束后，停饲 48h。将每个水槽中的刺参全部捞出计数，放置在洁净的白瓷盘上，待其恢复自然体长时称总重，然后随机选取 15 头，测量体长和体质量后解剖，收集体壁、肠道和体腔液，测量体壁重和肠道重，计算存活率、增重量、增重率、特定生长率、脏壁比及肠壁比。将体腔液在 4℃条件下以 500r/min 离心 10min 后收集上清液，采用超声波细胞破碎仪破碎 5min 后过 0.45μm 滤膜；肠道用 0.75%生理盐水冲洗干净。所有样品置于-20℃保存，待测。

在刺参幼参肠道前 1/3 处截取 0.3cm，用波恩试液固定，用于肠道组织切片的制作。

4. 样品测定

饲料中维生素 B_6 含量用高效液相色谱法测定，依据《添加剂预混合饲料中维生素 B_6 的测定 高效液相色谱法》（GB/T 14702—2018）。

体腔液中葡萄糖-6-磷酸脱氢酶、乳酸脱氢酶、异柠檬酸脱氢酶、乙酰辅酶 A 羧化酶、一氧化氮合酶、谷丙转氨酶、谷草转氨酶的活性采用 ELISA 试剂盒测定（上海纪宁实业有限公司），酶活性单位设定及测定方法见说明书。

肠道粗酶液采用吴永恒等（2012）的方法制备；蛋白酶、脂肪酶、淀粉酶及纤维素酶的含量采用南京建成生物工程研究所试剂盒测定，酶液中粗蛋白含量采用考马斯亮蓝法测定，酶活性单位设定及测定方法见说明书。

5. 肠道组织切片的制作与观察

固定后的样品经梯度酒精脱水、透明、浸蜡后包埋，在室温下切成 7μm 厚的切片，经苏木精-伊红染色后，用中性树脂封片。在×20 物镜下观测切片，选择典型视野，采用 Leica DM500 图像采集系统统计肠壁厚度及绒毛长度，每个样品观察 10 个非连续性的纵切片，其中肠壁厚度为肠外部至肌层与黏膜下层交接处的距离，绒毛长度为所有绒毛长度的平均值。

6. 数据统计

实验所得数据采用 Microsoft Excel 2007 及 SPSS 11.0 软件进行单因素方差分

析，结果以平均值±标准差表示，差异显著（$P<0.05$）时用邓肯多重范围检验进行多重比较分析。

2.1.4.2 结果分析

1. 维生素 B_6 对刺参幼参生长性能的影响

饲料中维生素 B_6 含量对刺参幼参存活率无显著影响（$P>0.05$）。随着饲料中维生素 B_6 含量的增加，刺参幼参的终末体质量、增重量、增重率和特定生长率均呈先上升后下降的趋势，D1 组显著低于其他组（$P<0.05$）；脏壁比和肠壁比均先升高后平稳，D3～D6 组无显著性差异（$P>0.05$）（表 2-15）。

表 2-15 维生素 B_6 对刺参幼参生长性能的影响（$n=3$）

指标	组别					
	D1	D2	D3	D4	D5	D6
初始体质量/g	12.25±0.23	12.26±0.12	12.24±0.07	12.22±0.08	12.21±0.12	12.18±0.07
终末体质量/g	16.91±0.77^a	19.28±0.33^b	20.24±0.17^c	21.25±0.63^d	22.33±0.31^e	21.59±0.54^{de}
增重量/g	4.66±0.54^a	7.03±0.21^b	7.99±0.12^c	9.03±0.58^d	10.12±0.23^e	9.41±0.51^{de}
增重率/%	38.01±3.79^a	57.29±1.18^b	65.32±0.76^c	73.91±4.45^d	82.89±2.61^e	77.23±4.11^d
特定生长率/(%/d)	0.58±0.05^a	0.81±0.01^b	0.91±0.01^c	0.99±0.04^d	1.08±0.03^e	1.02±0.04^{de}
脏壁比/%	16.30±1.10^a	18.66±0.96^b	19.62±0.63^{bc}	20.42±1.52^c	20.96±0.55^c	20.95±0.42^c
肠壁比/%	3.49±0.22^a	4.05±0.21^b	4.45±0.13^c	4.37±0.09^c	4.51±0.17^c	4.40±0.14^c
存活率/%	93.33±3.34	95.56±1.93	96.67±3.34	95.56±1.93	95.56±3.85	97.78±1.92

注：同行数据无字母或上标相同字母表示差异不显著（$P>0.05$），上标不同字母表示差异显著（$P<0.05$）。

以增重率为评价指标，经一元二次回归分析得出，初始体质量为 12.23g 的刺参幼参对饲料中维生素 B_6 的最适需求量为 45mg/kg（图 2-4）。

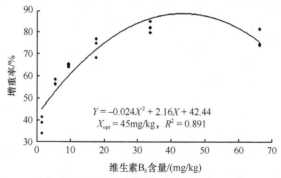

$$Y = -0.024X^2 + 2.16X + 42.44$$
$$X_{opt} = 45\text{mg/kg}, \ R^2 = 0.891$$

图 2-4 刺参幼参增重率与饲料中维生素 B_6 含量的回归分析

2. 维生素 B_6 对刺参幼参体壁基本成分的影响

饲料中维生素 B_6 含量对刺参幼参体壁中水分和粗灰分的含量无显著影响（$P>0.05$），粗蛋白含量先升高后降低；D1 组粗脂肪含量显著低于其他各组（$P<0.05$），

而其他各组之间无显著性差异（$P>0.05$）（表 2-16）。

表 2-16 维生素 B_6 对刺参幼参体壁基本成分的影响（$n=6$）（%）

指标	组别					
	D1	D2	D3	D4	D5	D6
水分	90.75±0.12	90.67±0.18	90.59±0.23	90.48±0.31	90.72±0.41	90.59±0.34
粗蛋白	4.14±0.10^{ab}	4.20±0.14^{ab}	4.41±0.07^c	4.37±0.08^{bc}	4.18±0.04^b	4.06±0.06^a
粗脂肪	0.13±0.01^a	0.16±0.01^b	0.18±0.01^b	0.17±0.00^b	0.17±0.01^b	0.15±0.00^b
粗灰分	3.10±0.07	3.30±0.28	3.08±0.06	3.29±0.21	3.18±0.25	3.41±0.23

注：同行数据无字母或上标相同字母表示差异不显著（$P>0.05$），上标不同字母表示差异显著（$P<0.05$）。

3. 维生素 B_6 对刺参幼参体腔液生理酶活性的影响

刺参幼参体腔液中葡萄糖-6-磷酸脱氢酶、异柠檬酸脱氢酶、乙酰辅酶 A 羧化酶及一氧化氮合酶的活性均随饲料中维生素 B_6 含量的增加呈现先升高后降低的趋势（$P<0.05$）；乳酸脱氢酶含量、谷丙转氨酶和谷草转氨酶的活性呈现上升趋势，D5、D6 组显著高于其他组（$P<0.05$），两组之间差异不显著（$P>0.05$）（表 2-17）。

表 2-17 维生素 B_6 对刺参幼参体腔液生理酶活性的影响（$n=6$）

指标	组别					
	D1	D2	D3	D4	D5	D6
葡萄糖-6-磷酸脱氢酶/（IU/L）	5.57±0.73^{bc}	6.08±0.42^c	5.47±0.39^{bc}	5.41±0.54^{ab}	5.01±0.32^{ab}	4.92±0.42^a
乳酸脱氢酶/（mg/L）	2.48±0.23^a	2.85±0.28^{ab}	3.02±0.40^b	3.14±0.52^b	3.74±0.15^c	4.14±0.42^c
异柠檬酸脱氢酶/（mIU/L）	730.14±33.945^b	753.33±57.63^{bc}	807.40±61.55^{bc}	829.34±10.18^c	582.08±14.79^a	599.06±55.92^a
乙酰辅酶 A 羧化酶/（ng/L）	90.10±5.17^c	113.36±7.81^d	105.87±11.01^d	107.69±8.13^d	79.57±5.38^b	65.17±7.43^a
一氧化氮合酶/（μmol/L）	10.07±0.35^a	11.13±0.58^b	13.08±0.71^c	12.24±0.69^b	10.12±0.52^a	9.53±0.25^a
谷丙转氨酶/（U/L）	2.05±0.31^a	2.26±0.11^{ab}	2.41±0.15^b	2.35±0.23^b	2.83±0.19^c	2.71±0.19^c
谷草转氨酶/（U/L）	4.54±0.49^a	4.91±0.48^a	5.09±0.23^a	5.13±0.32^a	6.10±0.39^b	5.77±0.49^b

注：同行数据无字母或上标相同字母表示差异不显著（$P>0.05$），上标不同字母表示差异显著（$P<0.05$）。

4. 维生素 B_6 对刺参幼参肠道消化酶活性的影响

刺参幼参肠道中蛋白酶、淀粉酶的活性均随着饲料中维生素 B_6 含量的增加而升高，D4~D6 组蛋白酶活性显著高于其他组（$P<0.05$），3 组之间无显著性差异（$P>0.05$），D5、D6 组淀粉酶活性显著高于其他组（$P<0.05$），两组之间差异不显著（$P>0.05$）；D1 组脂肪酶活性显著低于其他组（$P<0.05$），其他组之间差异不显著（$P>0.05$）；纤维素酶活性随着维生素 B_6 含量的增加而降低（$P<$

0.05)（表 2-18）。

表 2-18　维生素 B_6 对刺参幼参肠道消化酶活性的影响（$n=6$）（单位：U/mg prot）

指标 mg prot	组别					
	D1	D2	D3	D4	D5	D6
蛋白酶	44.85±3.58^a	58.74±3.25^b	63.29±4.71^c	74.52±4.96^d	73.94±5.33^d	75.39±5.50^d
脂肪酶	0.38±0.04^a	0.61±0.05^b	0.64±0.04^b	0.66±0.07^b	0.63±0.05^b	0.65±0.04^b
淀粉酶	5.58±0.57^a	6.69±0.71^b	8.34±0.78^c	10.25±0.46^d	11.91±0.83^e	12.08±0.95^e
纤维素酶	2.04±0.57^e	1.91±0.37^e	1.65±0.28^d	1.46±0.33^c	1.25±0.19^b	1.04±0.41^a

注：同行数据无字母或上标相同字母表示差异不显著（$P>0.05$），上标不同字母表示差异显著（$P<0.05$）。

5. 维生素 B_6 对刺参幼参肠道结构的影响

随着饲料中维生素 B_6 含量的增加，刺参幼参肠道的肠壁厚度及绒毛长度均呈现先升高后降低趋势，D5 组显著大于其他组（$P<0.05$）（表 2-19）。

表 2-19　维生素 B_6 对刺参幼参肠道结构的影响（$n=10$）（单位：μm）

指标	组别					
	D1	D2	D3	D4	D5	D6
肠壁厚度	3.25±0.46^a	3.88±0.52^a	5.17±0.34^b	7.41±0.69^c	8.45±0.72^d	7.37±0.59^c
绒毛长度	78.84±12.19^a	74.35±15.27^a	96.77±18.57^b	112.46±25.91^c	143.80±21.34^d	129.88±17.22^c

注：同行数据无字母或上标相同字母表示差异不显著（$P>0.05$），上标不同字母表示差异显著（$P<0.05$）。

饲料中缺乏维生素 B_6 时，刺参幼参肠道的绒毛长度变短、宽度变窄、数量变少，黏膜下层出现异常。D5 组刺参幼参肠道的绒毛致密、规则，肠壁厚度较大。D6 组刺参幼参肠道的黏膜下层、中央乳糜管均出现异常（图 2-5）。

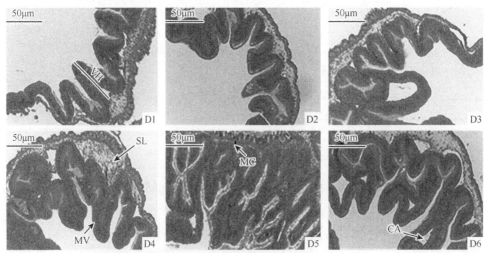

图 2-5　维生素 B_6 对刺参肠道组织形态的影响（×200）
VH-绒毛长度；MC-肠壁厚度；MV-微绒毛（肠上皮细胞）；CA-中央乳糜管；SL-黏膜下层

2.1.4.3 讨论

1. 刺参幼参对饲料中维生素 B_6 的需求量

本实验中,饲料中维生素 B_6 含量对刺参幼参的存活率无显著影响,但显著影响其增重率,表明维生素 B_6 缺乏会抑制刺参幼参的生长,但无致死性。对鱼类、甲壳动物及贝类的研究均表明,饲料中缺乏维生素 B_6 会影响动物的生长性能(Shiau and Wu,2003;Teixeira et al.,2012;Mohamed,2001;冯秀妮等,2006;王玥等,2009;李二超等,2010)。鱼类对维生素 B_6 的需求量为 1~15mg/kg(Teixeira et al.,2012;Mohamed,2001;刘凯等,2010),而斑节对虾对维生素 B_6 的需求量为 72~89mg/kg(Shiau and Wu,2003),中华绒螯蟹对维生素 B_6 的需求量为 81.675~85.74mg/kg(王玥等,2009),皱纹盘鲍对饲料中维生素 B_6 的最适需求量为 40mg/kg(朱伟,2001)。本实验得出,刺参对饲料中维生素 B_6 的最适需求量为 45mg/kg,均远高于鱼类的需求量(Mohamed,2001;刘凯等,2010;吴凡等,2018)。这体现了动物种属之间的差异,但更重要的是,这与动物的摄食习性有关,大部分鱼类摄食较为迅速,而虾、蟹、贝、参的摄食速度较慢,相当一部分维生素 B_6 在水中溶失,导致实验出现偏差。本实验中,随着饲料中维生素 B_6 含量的增加,刺参幼参的增重率升高后有下降趋势,这可能与高剂量的维生素 B_6 引起的细胞凋亡有关(Vrolijk et al.,2017)。此外,刺参体内不具备完善的吸收及转运系统,水管系统和血窦系统在营养物质的吸收及转运中起着重要作用(朱峰,2009),因此其对营养物质的吸收利用效率较低,由此导致机体对维生素 B_6 的生理需求量较高。

2. 维生素 B_6 对刺参幼参体成分的影响

维生素 B_6 是动物体内蛋白质、脂肪代谢酶的重要辅酶,广泛参与机体营养代谢。研究表明,饲料中适宜的维生素 B_6 含量显著提高组织粗蛋白、粗脂肪的含量(冯秀妮等,2006;王玥等,2009;Giri et al.,1997)。本实验中,刺参幼参体壁中粗蛋白含量先升后降,粗脂肪含量先升高后平稳,表明维生素 B_6 影响了刺参幼参体内蛋白质、脂肪的沉积。D6 组刺参幼参体壁中粗蛋白含量显著低于 D3、D4 和 D5 组,可能与维生素 B_6 的生理毒性有关,维生素 B_6 含量过高引起了细胞凋亡,因此降低了蛋白质代谢酶活性,减少了体壁蛋白质的沉积,而 D1 组粗蛋白含量与 D2 组无显著性差异,可能与实验开始前,刺参幼参体内储备维生素 B_6 有关。对中华绒螯蟹的研究表明,饲料中适宜含量的维生素 B_6 可以促进机体脂肪的沉积(王玥等,2009),这可能与维生素 B_6 可以促进机体蛋白质利用有关,当蛋白质超出其需要量时,部分氨基酸脱氨基代谢为 α-酮酸,继而转化为脂肪储存在体内(王镜岩等,2007)。但刺参与脊椎动物的营养需求及代谢机制存在较大差异,饲料及机体中粗脂肪的含量均较低，因此机体粗脂肪含量的升高可能是由于维生素B_6提高了刺参幼参对脂肪的利用能力，从而促进了机体粗脂肪的沉积。

3. 维生素B_6对刺参幼参体腔液生理酶活性的影响

葡萄糖-6-磷酸脱氢酶参与磷酸戊糖循环，通过调控还原型辅酶Ⅱ的含量维持细胞的能量；乳酸脱氢酶参与糖酵解和糖异生过程，催化乳酸生成丙酮酸和烟酰胺腺嘌呤二核苷酸（NADH）；异柠檬酸脱氢酶催化异柠檬酸生成α-酮戊二酸，是三羧酸循环（TCA）循环的限速酶；这3种酶均与糖及能量的代谢有关（王镜岩等，2007）。本实验中，随着饲料中维生素B_6含量的增加，葡萄糖-6-磷酸脱氢酶、异柠檬酸脱氢酶的活性呈现先升高后降低的趋势，而乳酸脱氢酶的含量则呈现上升趋势，表明饲料中维生素B_6含量的改变直接影响了刺参幼参体内磷酸戊糖循环、三羧酸循环、糖酵解及糖异生等生物过程的效率，从而影响了蛋白质、脂肪及糖三大类营养物质的代谢。当饲料中维生素B_6含量较低时，葡萄糖会被用来氧化供能；当维生素B_6含量超过其需求量时，部分蛋白质或脂肪可能被用来氧化供能，而糖类则沉积下来；只有饲料中维生素B_6含量适宜时，才能维持机体蛋白质、脂肪、糖三大类营养物质的正常代谢。乙酰辅酶A羧化酶催化乙酰辅酶A生成丙二酰辅酶A，是脂肪酸合成的关键酶（王镜岩等，2007）。随着饲料中维生素B_6含量的增加，刺参幼参脂肪酸合成能力先上升后下降。但从本实验测得的糖酵解或三羧酸循环酶活性推测，可能是由于维生素B_6提高了脂肪酸β-氧化供能的比例，从而降低了脂肪酸合成的能力。

谷丙转氨酶和谷草转氨酶是氨基酸代谢的2个重要酶类，它们的辅酶均为5′-磷酸吡哆醛（Percudani and Peracchi，2009）。本实验中，随着饲料中维生素B_6含量的增加，刺参幼参体腔液中谷丙转氨酶和谷草转氨酶的活性均显著上升，同样的现象在斑节对虾（Shiau and Wu，2003）、皱纹盘鲍（冯秀妮等，2006）、中华绒螯蟹（王玥等，2009）均有报道，表明维生素B_6显著提高了刺参体内氨基酸的代谢。但随着维生素B_6含量的增加，2种转氨酶活性出现一定程度的下降，这可能与维生素B_6的细胞毒性有关（Vrolijk et al.，2017）。

一氧化氮合酶催化精氨酸生产一氧化氮，一氧化氮是重要的生物信使分子，在心脑血管调节、神经调节、免疫调节等方面有十分重要的生物学作用（Hibbs et al.，1987；Jobgen et al.，2006）。本实验中，随着饲料中维生素B_6含量的增加，一氧化氮合酶活性有了一定程度的上升，这可能与底物精氨酸的浓度有关。维生素B_6含量的增加，增强了刺参体内氨基酸代谢，提供了更多的精氨酸（徐志昌等，1995）。但当维生素B_6含量超过一定数值后，一氧化氮合酶活性降低，这在其他动物尚未见报道，其机制也待更深入研究。

4. 维生素 B_6 对刺参幼参消化生理的影响

有研究报道，维生素 B_6 可以通过促进消化器官的生长发育，增强消化酶的分泌能力，从而提高消化吸收能力（何伟，2008）。刺参不具有专门的消化酶分泌器官，且主要消化酶位于肠道黏膜层的柱状细胞内。随着饲料中维生素 B_6 的添加，刺参肠道黏膜层环形褶皱逐渐致密，肠壁厚度逐渐增大，表明维生素 B_6 促进了刺参肠道的发育，维持了肠道结构完整性和功能的正常性。一方面，肠道发育良好可以促进消化酶的分泌，提高刺参的消化能力；另一方面，刺参消化能力的提高促进了机体对营养物质的利用，提高了机体对营养物质的需求，刺激了消化酶的合成与分泌。本实验中，从 D3 组开始，刺参幼参肠道蛋白酶和淀粉酶活性有了较大的提高，与肠道组织结构变化相吻合，表明刺参幼参对饲料中维生素 B_6 的需求量高于 9.35mg/kg，D4、D5 组刺参幼参肠道结构形态与吴永恒等（2012）报道的刺参正常肠道结构形态类似，表明刺参对饲料中维生素 B_6 的需求量为 17.47～33.71mg/kg，与本实验回归分析得出的结论较为符合。

2.1.4.4 结论

饲料中适宜含量的维生素 B_6 可以促进刺参的生长，促进肠道发育，提高机体营养物质代谢能力。以增重率为评价指标，刺参幼参饲料中适宜的维生素 B_6 含量为 45mg/kg。

2.1.5 维生素 D_3 对刺参幼参生长性能、体组成及抗氧化能力的影响

维生素 D_3 为脂溶性类固醇衍生物，是水生动物生长所必需的营养素（黎德兵等，2015）。水生动物自身不能合成或者合成量不能满足机体的需要，必须从食物中获取足量的维生素 D_3 来满足机体生理生化功能的需要（鹏翔和邵庆均，2010；Yanik et al.，2016）。已有研究表明，维生素 D_3 能够促进肠道黏膜上皮细胞 Ca 结合蛋白的形成，促进 Ca、P 的吸收，调节体内 Ca、P 平衡（李爱杰，1996）。此外，维生素 D_3 还参与协调提高机体的免疫能力（付京花等，2006）及抗氧化能力（Martinez-Álvarez et al.，2002）。目前，有关水生动物对维生素 D_3 最适需求量的研究主要集中在鲤鱼（张桐等，2011）、黄颡鱼（段鸣鸣等，2014）、鲈鱼（张璐等，2016）、草鱼（蒋明等，2009）等少数鱼类，有关刺参幼参对维生素 D_3 的最适需求量未见报道。

目前，维生素在刺参饲料中的应用研究较少，仅见维生素 C（Luo et al.，2014）和维生素 E（汪将，2015）对刺参生长和免疫特性影响的研究。因此，本实验通过在饲料中添加不同水平的维生素 D_3，研究其对刺参幼参生长性能、消化酶活性

及免疫能力的影响,以期为刺参对饲料中维生素 D_3 的需求量研究提供参考。

2.1.5.1 材料与方法

1. 实验设计与饲料配制

实验分为 6 个处理组,每个处理组 3 个重复,每个重复 30 头刺参,大小为(15.43±0.14)g。实验所用维生素 D_3 含量为 500 000IU/g(浙江新维普添加剂有限公司)。以鱼粉和发酵豆粕为蛋白源,以鱼油和大豆卵磷脂为脂肪源,配制粗蛋白含量为 22%、粗脂肪含量为 4% 的基础饲料,在基础饲料中分别添加 0IU/kg、250IU/kg、500IU/kg、1000IU/kg、2000IU/kg、3000IU/kg 的维生素 D_3,配制成 6 种实验饲料,分别命名为 D1、D2、D3、D4、D5、D6 组。固体原料经超微粉碎后过 200 目标准筛,按配比称重,加入新鲜鱼油及适量的蒸馏水混匀,用小型颗粒饲料挤压机制成直径为 0.3cm、厚度为 0.05cm 的片状饲料,60℃烘干备用。实验饲料配方及营养组成见表 2-20。

表 2-20 实验饲料配方及营养组成(干物质)

项目		组别					
		D1	D2	D3	D4	D5	D6
原料组成	鱼粉/%	5.00	5.00	5.00	5.00	5.00	5.00
	发酵豆粕/%	9.00	9.00	9.00	9.00	9.00	9.00
	虾粉/%	13.00	13.00	13.00	13.00	13.00	13.00
	藻粉/%	12.00	12.00	12.00	12.00	12.00	12.00
	小麦粉/%	20.00	20.00	20.00	20.00	20.00	20.00
	α-淀粉/%	2.00	2.00	2.00	2.00	2.00	2.00
	鱼油/%	1.00	1.00	1.00	1.00	1.00	1.00
	大豆卵磷脂/%	1.00	1.00	1.00	1.00	1.00	1.00
	维生素预混料/%	0.50	0.50	0.50	0.50	0.50	0.50
	维生素 D_3/(IU/kg)	0	250.00	500.00	1000.00	2000.00	3000.00
	矿物质预混料/%	0.50	0.50	0.50	0.50	0.50	0.50
	海泥/%	36.00	36.00	36.00	36.00	36.00	36.00
营养组成	粗蛋白/%	22.69	22.49	22.42	22.82	22.52	22.17
	粗脂肪/%	4.06	4.46	4.72	4.05	4.04	4.28
	粗灰分/%	44.16	44.30	44.30	44.14	44.07	44.1
	能量/(kJ/g)	14.15	14.21	14.18	14.35	14.33	14.46
	钙/%	2.43	2.32	2.53	2.51	2.48	2.51
	磷/%	0.85	0.83	0.85	0.82	0.83	0.81
	维生素 D_3/(IU/kg)	95.00	334.00	570.00	1076.00	2063.00	3081.00

注:①每千克维生素预混料含维生素 A 38.0mg、α-生育酚 210.0mg、维生素 B_1 115.0mg、维生素 B_2 380.0mg、盐酸吡哆醇 88.0mg、泛酸 368.0mg、烟酸 1030.0 mg、生物素 10.0mg、叶酸 20.0mg、维生素 B_{12} 1.3mg、肌醇 4000.0mg、抗坏血酸 500.0mg;②矿物质预混料配方组成见文献(王际英等,2014)。

2. 实验用刺参及实验管理

养殖实验在山东省海洋资源与环境研究院东营实验基地循环水养殖系统中进行，实验用刺参幼参购自蓬莱安源水产有限公司。挑选大小均匀、健康无病的刺参幼参于养殖系统中暂养 15d，其间投喂基础饲料。待其完全适应饲养条件后随机分为 6 个处理组，每个处理组 3 个重复，每个重复 30 头刺参，平均体质量为（15.43±0.14）g，随机放养于 18 个圆柱形养殖桶（$h_{桶}$=80cm，$h_{水}$=65cm，$d_{桶}$=60cm），实验周期为 8 周。每桶放置海参养殖筐 2 个，内嵌波纹板 20 张，每隔 30d 更换养殖筐一次。养殖期间，控制水温为 17～19℃，水流流速为 2L/min，盐度为 26～28，pH 为 7.6～8.2，溶氧量大于 5.0mg/L，氨氮和亚硝酸盐浓度低于 0.05mg/L。每天 8:00 进行换水，换水量为 1/3，每天 16:00 投饵一次，初始投喂量占体质量的 2%，每天观察刺参幼参的摄食情况及时调整投喂量，并记录水温及死亡情况。每 3d 吸底 1 次，用虹吸法将残饵及粪便吸出，养殖实验在弱光环境下进行。

3. 样品采集与处理

养殖实验结束后，控食 48h，统计每桶刺参幼参的数量并称重，用于存活率、增重率、特定生长率及肠体比的计算。每桶随机选取 8 头刺参幼参置于洁净的泡沫板上，用滤纸轻轻将刺参幼参体表的水分吸干后分别称重。去除体腔液后，置于冰盘上分离肠道和体壁。去除肠道内容物，用定性滤纸擦净后分别称重，将肠道样品和体壁放在–20℃条件下冷冻保存，待测。

4. 测定指标与方法

1）营养成分的测定

饲料中的维生素 D_3 在通标标准技术服务（青岛）有限公司采用高效液相色谱法-皂化提取法，依据《饲料中维生素 D_3 的测定 高效液相色谱法》（GB/T 17818—2010）测定。测定步骤为：称取样品进行皂化，将皂化液转移到分液漏斗中进行提取，然后将提取液蒸干浓缩，加入甲醇溶解，过滤，得到待测液，用高效液相色谱仪检测；体壁 Mg、Fe、Mn 含量采用浓硝酸微波消解后用电感耦合等离子体质谱仪测定；体壁羟脯氨酸活性采用南京建成生物工程研究所试剂盒测定；体壁维生素 D_3 含量采用上海酶联生物科技有限公司生产的 ELISA 试剂盒测定，具体测定步骤参照试剂盒说明书。

2）抗氧化酶活性的测定

取肠道样品，剪碎，加入 9 倍体积冰冷的生理盐水（0.9%），制成 10%的匀浆，在 4℃条件下，8000r/min 离心 10min，取出上清液，分装于 2ml 的离心管中，保存于冰箱（−80℃）中，待测。

碱性磷酸酶、总超氧化物歧化酶、过氧化氢酶的活性、丙二醛含量以及总抗氧化能力均采用南京建成生物工程研究所生产的试剂盒测定，酶的活性定义及具体测定步骤参照试剂盒说明书。

5. 数据统计分析

采用 SPSS 17.0 软件进行单因素方差分析，当处理组之间差异显著（$P<0.05$）时，用邓肯多重范围检验进行多重比较分析，统计数据以平均值±标准误表示。

2.1.5.2 结果分析

1. 维生素 D_3 对刺参幼参生长性能和形体指标的影响

饲料中维生素 D_3 对刺参幼参的存活率无显著影响（$P>0.05$）；随着饲料中维生素 D_3 含量的增加，增重率及特定生长率呈现先上升后下降的趋势，均在 D4 组达到最高值，D1 组显著低于其他组（$P<0.05$）。D3、D4、D5 组肠体比显著大于其他 3 组（$P<0.05$），但 3 组之间差异不显著（$P>0.05$）（表 2-21）。

表 2-21 维生素 D_3 对刺参幼参生长性能及形体指标的影响

生长性能	组别					
	D1	D2	D3	D4	D5	D6
初始体质量/g	15.41±0.10	15.44±0.10	15.48±0.02	15.40±0.08	15.45±0.13	15.41±0.11
终末体质量/g	18.97±0.34^a	21.54±0.28^b	23.10±0.22^c	25.41±0.31^d	25.49±0.16^d	20.85±0.03^b
存活率/%	91.33±2.69	86.44±2.79	88.67±2.40	86.67±1.93	86.22±3.47	91.78±0.97
增重率/%	20.55±0.78^a	34.85±1.24^b	49.41±1.38^c	65.43±1.23^d	63.70±1.54^d	34.6±0.51^b
特定生长率/(%/d)	0.32±0.01^a	0.49±0.00^b	0.67±0.02^c	0.84±0.01^d	0.82±0.02^d	0.50±0.01^b
肠体比/%	2.70±0.08^a	2.73±0.01^a	3.38±0.06^b	3.37±0.06^b	3.70±0.13^b	2.75±0.16^a

注：同行数据无字母或上标相同字母表示差异不显著（$P>0.05$），上标不同字母表示差异显著（$P<0.05$）。

以增重率为评价指标，经一元二次回归分析得出，初始体质量为 15.43g 的刺参幼参对饲料中维生素 D_3 的最适需求量为 1587.5IU/kg（图 2-6）。

2. 维生素 D_3 对刺参幼参体组成的影响

饲料中维生素 D_3 显著影响刺参幼参体壁粗灰分、Ca、P、Mg、Fe 及羟脯氨酸的含量（$P<0.05$），但对水分、粗蛋白、粗脂肪及 Mn 的含量影响不显著（$P>0.05$）（表 2-47，表 2-48）。随着饲料中维生素 D_3 含量的增加，粗灰分含量呈现先上升后下降的趋势，D3、D4、D5 组无显著性差异（$P>0.05$），D1、D2、D6 组显著低于 D4、D5 组（$P<0.05$），但与 D3 组无显著性差异（$P>0.05$）；羟脯氨酸

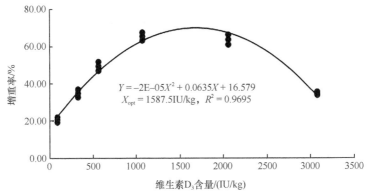

图 2-6 刺参幼参增重率与饲料中维生素 D_3 含量的回归分析

含量呈现先下降后上升的趋势，D4 组达到最低值，D1、D6 组无显著性差异，且显著高于其他 4 组（$P<0.05$）；刺参幼参体壁维生素 D_3 沉积量与羟脯氨酸含量呈现相反趋势，D4 组显著高于其他组（$P<0.05$），D4、D5 组显著高于 D1、D2 组（$P<0.05$），D5、D6 组与 D3 组间无显著性差异（$P>0.05$）（表 2-22）。

表 2-22 维生素 D_3 对刺参幼参体壁基本成分的影响

体成分	组别					
	D1	D2	D3	D4	D5	D6
水分/%	91.56±0.22	91.60±0.27	91.22±0.18	91.52±0.33	91.29±0.47	91.60±0.28
粗蛋白/%	45.88±0.02	45.68±0.00	45.73±0.13	46.01±0.24	45.59±0.12	45.57±0.12
粗脂肪/%	3.67±0.14	3.77±0.17	3.47±0.12	3.46±0.08	3.42±0.07	3.51±0.12
粗灰分/%	35.82±0.06^a	36.23±0.58^a	36.48±0.05^{ab}	37.89±0.97^b	38.06±0.20^b	36.07±0.38^a
羟脯氨酸/（μg/g）	1000.23±7.96^e	923.15±3.18^d	659.22±13.78^b	488.18±4.96^a	816.70±11.90^c	984.81±2.23^e
维生素 D_3/（μg/kg）	224.95±4.73^a	223.59±10.12^a	232.89±1.96^{ab}	273.90±3.95^c	245.35±4.32^b	240.03±7.33^{ab}

注：粗蛋白、粗脂肪、粗灰分为干基含量；同行数据无字母或上标相同字母表示差异不显著（$P>0.05$），上标不同字母表示差异显著（$P<0.05$）。

饲料中维生素 D_3 显著提高了刺参幼参体壁 Ca 含量，各组显著高于 D1 组（$P<0.05$）；随着饲料中维生素 D_3 含量的增加，P 含量呈现先上升后平稳的趋势，D1、D2 组无显著性差异（$P>0.05$），D4、D5、D6 组显著高于 D1、D2 组（$P<0.05$），且 3 组间无显著性差异（$P>0.05$），D3 组显著高于 D1 组（$P<0.05$），与其他 4 组间无显著性差异（$P>0.05$）；Mg 含量呈现上升趋势，D1 组显著低于其他组（$P<0.05$）；Fe 含量呈现先升后降趋势，D5、D6 组显著低于其他组（$P<0.05$），D5、D6 组间无显著性差异（$P>0.05$）；Mn 含量呈现下降趋势，但各组间差异不显著（$P>0.05$）（表 2-23）。

表 2-23 维生素 D_3 对刺参幼参体壁矿物元素含量的影响

体成分	组别					
	D1	D2	D3	D4	D5	D5
Ca/（mg/g）	12.39±0.05^a	13.29±0.01^{bc}	13.50±0.17^{cd}	13.71±0.12^{cd}	13.89±0.06^d	13.20±0.29^b
P/（mg/g）	9.07±0.01^a	9.12±0.01^{ab}	9.25±0.09^{bc}	9.28±0.02^c	9.36±0.01^c	9.34±0.08^c
Mg/（mg/g）	8.74±0.17^a	9.95±0.20^b	10.46±0.44^{bc}	10.86±0.35^{cd}	11.36±0.14^d	12.17±0.11^e
Fe/（mg/kg）	41.28±0.18^b	48.63±1.67^c	50.40±2.42^c	48.12±2.77^c	29.45±0.08^a	30.06±1.11^a
Mn/（mg/kg）	5.42±0.37	5.46±0.01	5.28±0.28	5.05±0.23	5.12±0.15	5.16±0.25

注：钙、磷、镁、铁、锰为干基含量；同行数据无字母或上标相同字母表示差异不显著（$P>0.05$），上标不同字母表示差异显著（$P<0.05$）。

3. 维生素 D_3 对刺参幼参肠道抗氧化酶活性的影响

随着饲料中维生素 D_3 含量的增加，刺参幼参肠道碱性磷酸酶及超氧化物歧化酶的活性均呈现先上升后下降的趋势，碱性磷酸酶活性在 D4 组达到最高值，D1 组显著低于其他各组（$P<0.05$）；D1、D6 组超氧化物歧化酶活性显著低于其他各组（$P<0.05$），D1、D6 组间无显著性差异（$P>0.05$），D5 组达到最高值。丙二醛含量随着饲料中维生素 D_3 含量的增加呈现下降趋势；过氧化氢酶活性及总抗氧化能力受饲料中维生素 D_3 含量的影响不显著（$P>0.05$）（表 2-24）。

表 2-24 维生素 D_3 对刺参幼参肠道抗氧化酶活性的影响

免疫指标	组别					
	D1	D2	D3	D4	D5	D6
碱性磷酸酶活性/（U/g prot）	2.58±0.02^a	3.06±0.01^b	3.90±0.00^c	5.16±0.00^d	3.88±0.12^c	3.80±0.05^c
超氧化物歧化酶活性/（U/mg prot）	186.46±3.31^a	228.52±8.06^b	229.71±2.06^b	231.73±5.96^b	284.40±6.71^c	221.83±4.61^a
丙二醛含量/（nmol/mg prot）	4.48±0.17^c	4.24±0.03^c	3.92±0.12^b	3.88±0.03^b	2.95±0.03^a	2.91±0.07^a
过氧化氢酶活性/（U/mg prot）	41.32±0.27	42.67±0.66	41.80±0.69	41.26±1.29	40.82±1.27	41.31±0.58
总抗氧化能力/（U/mg prot）	0.33±0.00	0.33±0.02	0.33±0.01	0.34±0.03	0.33±0.01	0.32±0.02

注：同行数据无字母或上标相同字母表示差异不显著（$P>0.05$），上标不同字母表示差异显著（$P<0.05$）。

2.1.5.3 讨论

1. 维生素 D_3 对刺参幼参生长性能的影响

关于维生素 D_3 促进动物生长的研究报道已有很多，维生素 D_3 可从激素、细胞及分子水平上影响动物的生长和 Ca 结合蛋白的形成（Maalouf et al.，2008），饲料中适量添加维生素 D_3 可促进动物生长，但过量添加会抑制动物的生长，甚至产生毒性。本实验中，随着饲料中维生素 D_3 含量的增加，各组刺参幼参存活率没有显著性差异，增重率和特定生长率呈现先上升后下降的趋势，D4 组达到最高值；当饲料中维生素 D_3 含量为 3000IU/kg 时，刺参幼参增重率和特定生长率明显下降，这与对中华绒螯蟹（孙新瑾，2009）、凡纳滨对虾（He et al.，1992）、皱纹盘鲍（周歧存和麦康森，2004）的研究结果一致。刺参幼参体壁 Ca 含量比较丰富，适量的维生素 D_3 能够促进甲状腺激素的分泌，进而提高刺参幼参的生长速度（张璐等，2016）；而高水平的维生素 D_3 对维生素 A 和维生素 E 的吸收有拮抗作用，这在一定程度上降低动物的生长速度（陈娟等，2010）。随着饲料中维生素 D_3 含量的增加，刺参幼参的肠体比呈现先增大后减小的趋势，说明适量的维生素 D_3 能够促进肠道的发育，提高刺参幼参的生长性能。以增重率为评价指标，经一元二次回归分析得出，刺参幼参对饲料中维生素 D_3 的最适需求量为 1587.5IU/kg。

2. 维生素 D_3 对刺参幼参体成分的影响

养殖动物的体成分随着配合饲料组成的不同而发生变化，包括粗蛋白、粗脂肪、水分、粗灰分等。本实验中，刺参幼参体壁水分含量不受饲料中维生素 D_3 含量的影响，这与蒋明等（2009）对草鱼的研究结果一致，表明不同含量的维生素 D_3 对刺参幼参体壁干物质的积累程度无显著影响；刺参幼参的生长性能及体壁粗蛋白含量与饲料蛋白质水平有关（王庆吉等，2014；赵斌等，2016），而本实验中各组刺参幼参体壁粗蛋白含量差异不显著，表明当饲料中蛋白质含量一致时，体壁粗蛋白含量并不能真实地反映刺参幼参的生长情况。随着维生素 D_3 含量的增加，各组刺参幼参体壁粗脂肪含量无显著性差异，这与对虹鳟（Barnett et al.，1982）及皱纹盘鲍（周歧存和麦康森，2004）的研究结果不一致，刺参自身对脂肪的需求量与鱼类相比较低，因此维生素 D_3 对刺参的脂肪代谢可能影响不大。羟脯氨酸是胶原蛋白的主要成分之一，是反映胶原蛋白数量的常见指标。本实验中，随着饲料中维生素 D_3 含量的增加，刺参幼参体壁中羟脯氨酸含量呈现先下降后上升的趋势，D4 组为最低值。骨骼的主要成分为骨盐（钙盐和磷酸盐）和有机基质（I型胶原），从刺参体壁中提取的胶原蛋白的氨基酸组成与 I 型胶原蛋白的氨基酸组成相似（崔凤霞等，2006），当摄食饲料中维生素 D_3 缺乏和过多时，会导致刺参体壁骨盐减少、I 型胶原含量上升，从而使体壁中羟脯氨酸含量升高；刺参体壁中维生素 D_3 的沉积量随着饲料中维生素 D_3 含量的增加呈现先上升后下降的趋势，说明维生素 D_3 在促进动物生长的同时增加了自身在机体内的沉积量。

维生素 D_3 对骨骼矿化以及 Ca、P 的沉积作用是浓度依赖型，在适宜浓度范围内，维生素 D_3 含量的增加能够促进骨骼矿化和 Ca、P 沉积，过高或过低均会导致骨骼矿化异常。本实验中，饲料中维生素 D_3 含量显著影响了刺参幼参体壁粗灰分及 Ca、P 含量，D4、D5 组刺参幼参体壁粗灰分、Ca、P 含量显著高于 D1

组，D6 组粗灰分及 Ca 含量呈现下降趋势，但 P 含量趋于稳定。刺参生活在含 Ca 丰富的海水中，可以从海水中吸收 Ca 满足机体的生理代谢需要。饲料中适量的维生素 D_3 能够促进矿物质在刺参体壁中的积累，粗灰分及 Ca 含量上升；但过量地补充维生素 D_3 则抑制矿物质沉积（Graff et al.，2015），粗灰分及 Ca 含量下降；而海水中 P 含量较低，刺参体壁中的 P 主要来源于饲料，因此饲料中维生素 D_3 含量过高时，同样能够促进 P 的吸收。Mg 在细胞代谢及骨骼发育过程中起着重要的作用，在体内 Mg 与 Ca 的代谢密切相关。本实验中，刺参幼参体壁 Mg 含量随着饲料中维生素 D_3 含量的增加而上升，D1 组显著低于其他组，Ca 与 Mg 结构相似，在一定范围内，维生素 D_3 能够促进 Mg 的吸收。Fe 是各种动物必需的矿物元素，是动物营养中最重要的矿物质之一，是生物机体生长发育不可缺少的条件（张佳明，2007），随着维生素 D_3 含量的增加，刺参幼参体壁 Fe 含量呈现先上升后下降的趋势，说明机体在维生素 D_3 促进 Fe^{2+} 吸收的同时受其他阳离子的影响。Mn 是骨正常形成所必需的元素，机体对 Mn 的吸收利用受很多因素的影响（王秋梅，2007）。本实验中，随着维生素 D_3 含量的增加，各组刺参幼参体壁 Mn 含量无显著性差异，表明机体对 Mn 的吸收不受维生素 D_3 含量的影响，这与对肉鸡的研究结果（张淑云和王安，2010）不一致，可能与养殖对象的种类有关。

3. 维生素 D_3 对刺参幼参肠道抗氧化酶活性的影响

刺参缺乏特异性的免疫系统，只有体壁和体内防御机制共同组成的非特异性免疫体系（白阳等，2016）。刺参体内的免疫主要分为细胞免疫和体液免疫，二者之间密切相关，相辅相成。刺参肠道免疫酶活性及抗氧化酶活性的高低可以间接地反映机体的健康状况（常杰等，2011）。

碱性磷酸酶是一种含 Zn 对底物专一性较低的磷酸单酯水解酶，在体内直接参与磷酸基团的转移和代谢，与机体的 Ca、P 代谢有关（Haussler et al.，1970）。本实验中，随着饲料中维生素 D_3 含量的增加，刺参幼参肠道碱性磷酸酶活性呈现先上升后下降的趋势，D4 组达到最高值，与对斑节对虾的研究结果（Shiau and Hwang，1994）一致。碱性磷酸酶存在于刺参幼参骨、肠道等组织中，在维生素 D_3 生理剂量范围内，促进肠道对 Ca、P 的吸收，碱性磷酸酶活性上升，当维生素 D_3 含量过高时，抑制肠道对 Ca、P 的吸收，碱性磷酸酶活性下降。实验结果显示，维生素 D_3 与碱性磷酸酶活性存在良好的相关性。

抗氧化物酶系统被认为是免疫系统的一部分，是细胞抵御氧化损伤的主要武器。由总超氧化物歧化酶、过氧化氢酶和总抗氧化能力等组成的抗氧化系统可以清除过多的自由基，保护机体免受活性氧基团的损伤。超氧化物歧化酶和过氧化氢酶的活性的高低间接反映机体清除氧自由基的能力。本实验中，随着饲料中维生素 D_3 含量的增加，刺参幼参肠道超氧化物歧化酶活性呈现先上升后下降的趋势,并在 D5 组达到最高值,表明维生素 D_3 对超氧化物歧化酶的生成有较强的诱导能力,增强机体的抗氧化能力。过氧化氢酶具有重要的生理功能,在细胞内与产生 H_2O_2 的需氧脱氢酶类同时存在,可以催化过氧化氢分解为水和氧气,和超氧化物歧化酶协同完成细胞内的抗氧化作用。本实验中,刺参幼参肠道过氧化氢酶活性受饲料中维生素 D_3 含量的影响不显著,这与对黄颡鱼的研究结果(段鸣鸣等,2014)不一致。刺参是无脊椎动物,缺少脊椎动物所具有的特异性免疫器官,与超氧化物歧化酶相比,较高含量的维生素 D_3 才能最大限度地提高过氧化氢酶活性。总抗氧化能力可以反映机体的抗应激能力(Rengpipat et al., 2000),随着维生素 D_3 含量的增加,各组刺参幼参肠道总抗氧化能力无显著性差异,而张淑云和王安(2010)发现,随着饲粮中维生素 D_3 含量的增加,肉鸡血清和肝脏中总抗氧化能力呈现上升趋势,这与本实验的结果不一致,可能与动物种类及实验条件不同有关。丙二醛是脂质过氧化反应链式终止阶段产生的小分子产物,其含量可以间接反映自由基的产生情况和机体组织细胞的脂质过氧化程度(Onderci et al., 2003)。本实验中,随着饲料中维生素 D_3 含量的增加,刺参幼参肠道中丙二醛含量呈现下降趋势,添加量为 3000IU/kg(D6 组)时丙二醛含量最低,这与对肉鸡(陈娟等,2010)和蛋雏鸭(解俊美和王安,2012)的研究结果一致,说明维生素 D_3 在提高刺参幼参抗氧化能力方面具有一定的积极作用。

2.1.5.4 结论

(1)以增重率为评价指标,经一元二次回归分析可知,初始体质量为 15.43g 的刺参幼参对饲料中维生素 D_3 的最适需求量为 1587.5IU/kg。

(2)饲料中添加适量的维生素 D_3 能够提高刺参幼参的生长性能,对刺参幼参体组成产生一定的影响。

(3)饲料中维生素 D_3 含量对刺参幼参肠道蛋白酶、淀粉酶及脂肪酶的活性无显著性影响,但是适量的维生素 D_3 能够提高机体的抗氧化能力。

2.2 刺参对氨基酸和脂肪酸的需求

2.2.1 刺参幼参对亮氨酸最适需求量的研究

亮氨酸是水生动物的必需氨基酸之一,分子式为 $C_6H_{13}NO_2$,其与异亮氨酸和缬氨酸被统称为支链氨基酸。亮氨酸是支链氨基酸中唯一的生酮氨基酸,在机体合成蛋白质、能量代谢、葡萄糖平衡等方面具有重要作用,是一种功能性氨基酸(Lynch and Adams, 2014)。亮氨酸的主要氧化部位在肌肉,是合成机体蛋白质的原料,可以抑制蛋白质降解和促进肌肉蛋白合成(杨霞等,2014)。饲料中添加的游离氨基酸与蛋白态氨基酸吸收不同步是导致鱼虾类对游离氨基酸利用效果不佳的主要原因,对晶体氨基酸进行包膜缓释处理可改善晶体氨基酸的作用效果(韩秀杰等,2019)。研究表明,亮氨酸能够增强团头鲂(Liang et al.,2018)、尼罗罗非鱼(Gan et al.,2016)、大黄鱼(Li et al.,2010)、三疣梭子蟹(Huo et al.,2017)、斑节对虾(Millamena et al.,1999)等水产动物的生长性能、抗氧化能力及免疫能力,并确定了其对亮氨酸的需求量。

刺参被列为"海八珍"之一,具有很高的食用及药用价值,是我国渤海、黄海海域重要的海水养殖品种。对刺参幼参支链氨基酸的营养需求研究中,韩秀杰等(2019)报道了刺参幼参对异亮氨酸和缬氨酸的最适需求量,而亮氨酸作为支链氨基酸中唯一的生酮氨基酸,其对刺参生长性能、消化能力及免疫能力影响的研究尚未见报道。研究刺参幼参对亮氨酸的最适需求量,对配制营养均衡、环境友好的配合饲料有重要的指导意义。因此,本实验通过在饲料中添加不同含量的亮氨酸,研究其对刺参幼参的生长性能、消化酶活性及抗氧化能力的影响,从而为刺参配合饲料的合理配制提供依据(刘财礼等,2022)。

2.2.1.1 材料与方法

1. 实验饲料

本实验以鱼粉、海藻粉和谷朊粉为主要蛋白源,以鱼油和大豆卵磷脂为主要脂肪源,设计粗蛋白含量约为20.00%、粗脂肪含量约为2.70%的基础饲料配方,在基础饲料中分别添加0、0.80%、1.60%、2.40%、3.20%和4.00%的包膜亮氨酸,配成亮氨酸含量分别为1.29%、1.63%、1.98%、2.22%、2.58%和2.97%的6组等氮等脂的实验饲料,分别命名为D1(对照组)、D2、D3、D4、D5和D6组(表2-25,表2-26)。

表2-25 实验饲料配方及营养组成(干物质)

	项目	组别					
		D1	D2	D3	D4	D5	D6
原料组成	鱼粉/%	7.00	7.00	7.00	7.00	7.00	7.00
	海藻粉/%	30.00	30.00	30.00	30.00	30.00	30.00
	花生粕/%	5.00	5.00	5.00	5.00	5.00	5.00
	谷朊粉/%	7.00	7.00	7.00	7.00	7.00	7.00
	包膜甘氨酸/%	4.80	4.00	3.20	2.40	1.60	0.80
	包膜亮氨酸/%	0.00	0.80	1.60	2.40	3.20	4.00
	维生素预混料/%	1.00	1.00	1.00	1.00	1.00	1.00
	矿物质预混料/%	1.00	1.00	1.00	1.00	1.00	1.00
	抗氧化剂/%	0.10	0.10	0.10	0.10	0.10	0.10

续表

项目		组别					
		D1	D2	D3	D4	D5	D6
原料组成	鱼油/%	1.00	1.00	1.00	1.00	1.00	1.00
	大豆卵磷脂/%	1.00	1.00	1.00	1.00	1.00	1.00
	海泥/%	42.10	42.10	42.10	42.10	42.10	42.10
营养组成	粗蛋白/%	20.46	20.19	20.43	20.42	20.32	20.17
	粗脂肪/%	2.75	2.73	2.63	2.72	2.68	2.66
	粗灰分/%	53.87	53.81	55.11	54.48	55.41	54.52
	能量/（kJ/g）	7.90	7.95	7.96	8.02	8.13	8.21

注：①每千克维生素预混料含维生素 A 7500.00IU、维生素 D 1500.00 IU、维生素 E 60.00mg、维生素 K_3 18.00mg、维生素 B_1 12.00mg、维生素 B_2 12.00mg、维生素 B_{12} 0.10mg、泛酸 48.00mg、烟酰胺 90.00mg、叶酸 3.70mg、D-生物素 0.20mg、盐酸吡哆醇 60.00mg、维生素 C 310.00mg；②每千克矿物质预混料含锌 35.00mg、锰 21.00mg、铜 8.30mg、铁 23.00mg、钴 1.20mg、碘 1.00mg、硒 0.30mg。

表 2-26 实验饲料氨基酸组成（干物质）（%）

氨基酸	组别					
	D1	D2	D3	D4	D5	D6
亮氨酸	1.29	1.63	1.98	2.22	2.58	2.97
甘氨酸	3.13	2.97	2.52	2.18	1.73	1.33
天冬氨酸	1.33	1.29	1.37	1.28	1.35	1.36
苏氨酸	0.66	0.67	0.69	0.64	0.66	0.67
丝氨酸	0.82	0.85	0.86	0.84	0.84	0.86
谷氨酸	3.86	4.01	4.17	4.00	4.08	4.09
丙氨酸	1.08	0.85	1.02	0.84	1.09	0.86
半胱氨酸	0.64	0.66	0.62	0.70	0.64	0.63
缬氨酸	0.69	0.74	0.75	0.73	0.71	0.75
蛋氨酸	0.27	0.22	0.14	0.32	0.20	0.09
异亮氨酸	0.52	0.58	0.59	0.58	0.56	0.58
酪氨酸	0.89	0.93	0.87	0.90	0.87	0.88
苯丙氨酸	1.19	1.20	1.21	1.20	1.21	1.22
赖氨酸	0.76	0.76	0.79	0.76	0.76	0.77
脯氨酸	1.05	1.14	1.06	1.11	1.10	1.10
组氨酸	0.29	0.31	0.31	0.30	0.31	0.32
精氨酸	0.88	0.92	0.93	0.91	0.91	0.90
总氨基酸	19.34	19.69	19.87	19.50	19.59	19.39

实验所用亮氨酸购自上海麦克林生化科技有限公司（纯度≥99%），β-环糊精购自孟州市华兴生物化工有限责任公司，参考胡友军等（2002）的方法对亮氨酸进行包被，包被后的有效亮氨酸含量约为 48%。实验所用固体原料经超微粉碎后过 200 目标准筛，按配方配比称重，加入鱼油及适量蒸馏水充分混匀，用小型颗粒饲料挤压机制成直径为 0.20cm、长度为 0.30cm 的颗粒，60℃烘干，用小型粉

碎机破碎，筛选粒度为20～100目的颗粒备用。

2. 饲养管理及样品采集

养殖实验在山东省海洋资源与环境研究院东营实验基地循环水养殖系统中进行，实验时间为2019年11月6日至2020年1月4日，共计60d。实验用刺参幼参购自山东安源种业科技有限公司。正式实验前，将刺参幼参置于养殖系统中用对照组饲料驯养2周，停食24h后，挑选450头参刺坚挺粗壮、体质量为（16.40±0.14）g的刺参幼参，随机平均分配到18个圆柱形循环水桶（$h_{桶}$=80cm，$h_{水}$=65cm，$d_{桶}$=60cm）中，每桶放置一个海参养殖筐。每组饲料随机投喂3个桶，每天定时（16:00）投喂1次，初始投喂量为刺参幼参体质量的3%，每天观察刺参幼参的摄食情况，根据摄食情况调整次日投喂量。每3d换水1次，换水量为养殖桶内水位的1/2，换水时用虹吸软管将桶底残饵及粪便吸出，养殖实验进行一个月时更换海参养殖筐1次并彻底清洗循环水桶。养殖期间，控制水流流速约为2L/min，水温为13～17℃，溶氧量大于6mg/L，pH为7.5～8.2，盐度为28，氨氮和亚硝酸盐浓度不超过0.05mg/L，室内保持弱光环境。

养殖实验结束后，停食24h，将每桶的刺参全部捞出称总重并记录刺参数量，计算增重率、特定生长率、存活率等。每桶随机选取10头刺参，放置于白色托盘中，待其身体自然舒展后，测量其体长及体质量，然后解剖，收集体壁及肠道，测量体壁重、肠道重及肠道长，用于计算肠壁比及肠长比。体壁保存于–20℃条件下，肠道保存于–80℃条件下，待测。

3. 检测指标与方法

肠道中淀粉酶、脂肪酶、蛋白酶、酸性磷酸酶、碱性磷酸酶、谷草转氨酶、谷丙转氨酶、过氧化氢酶和超氧化物歧化酶的活性和丙二醛含量、总抗氧化能力的测定均使用南京建成生物工程研究所生产的试剂盒，各种酶活性单位参照试剂盒说明书表示。

4. 数据统计分析

实验数据用SPSS 17.0软件进行单因素方差分析，并用邓肯多重范围检验进行多重比较分析，当$P<0.05$时表示具有显著性差异，统计结果用平均值±标准差表示。采用一元二次回归分析，确定刺参幼参对亮氨酸的最适需求量。

2.2.1.2 结果分析

1. 亮氨酸对刺参幼参生长性能及形体指标的影响

各组刺参幼参之间的存活率无显著性差异（$P>0.05$），存活率为96.00%～

98.67%；随着饲料中亮氨酸含量从 1.29%增加到 1.98%，刺参幼参的增重率和特定生长率均显著升高（$P<0.05$），而随着亮氨酸含量进一步增加，增重率和特定生长率呈现下降的趋势（$P<0.05$），增重率在 D3 组达到最大值 100.84%，D3、D4 和 D5 组增重率显著高于对照组（$P<0.05$），且 3 组之间无显著性差异（$P>0.05$）。各组间肠壁比和肠长比均无显著性差异（$P>0.05$）（表 2-27）。

表 2-27　亮氨酸对刺参幼参生长性能及形体指标的影响

项目	组别					
	D1	D2	D3	D4	D5	D6
初始体质量/g	16.40±0.019	16.40±0.015	16.41±0.009	16.41±0.011	16.40±0.006	16.39±0.012
终末体质量/g	31.05±0.43^a	31.68±0.47^{ab}	32.95±0.66^c	32.48±0.41^{bc}	32.15±0.67^{bc}	30.65±0.43^{ab}
增重率/%	89.32±2.77^a	93.18±2.90^{ab}	100.84±4.01^c	97.91±2.39^{bc}	96.08±3.97^{bc}	86.99±2.97^{ab}
特定生长率/(%/d)	1.06±0.03^a	1.10±0.02^{ab}	1.16±0.04^c	1.14±0.02^{bc}	1.12±0.04^{bc}	1.04±0.02^{ab}
肠壁比/%	5.73±0.39	5.65±0.50	5.83±0.35	5.77±0.42	5.81±0.42	5.78±0.54
肠长比/%	3.49±0.15	3.47±0.15	3.53±0.18	3.50±0.20	3.53±0.18	3.53±0.21
存活率/%	96.00±4.00	97.33±4.62	97.33±4.62	96.00±6.93	98.67±2.31	96.00±4.00

注：同行数据无字母或上标相同字母表示差异不显著（$P>0.05$），上标不同字母表示差异显著（$P<0.05$）。

以增重率为评价指标，经一元二次回归分析得出，初始体质量为 16.40g 的刺参幼参对饲料中亮氨酸的最适需求量为 2.11%（10.37%饲料粗蛋白）（图 2-7）。

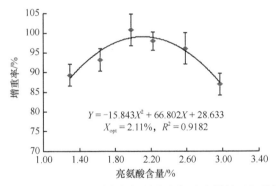

图 2-7　刺参幼参增重率与饲料中亮氨酸含量的回归分析

2. 亮氨酸对刺参幼参体壁基本营养成分及氨基酸含量的影响

随着饲料中亮氨酸含量从 1.29%增加到 1.98%，刺参幼参体壁的粗脂肪含量显著升高（$P<0.05$），在 D3 组达到最大值 5.50%，且显著高于其他组（$P<0.05$），随着亮氨酸含量进一步增加，刺参幼参体壁粗脂肪含量显著降低（$P<0.05$）；各组间水分、粗蛋白和粗灰分含量均无显著性差异（$P>0.05$）（表 2-28）。

表 2-28　亮氨酸对刺参幼参体壁基本营养成分的影响（%）

项目	组别					
	D1	D2	D3	D4	D5	D6
水分	91.88±0.39	92.16±0.28	91.54±0.36	92.25±0.14	92.13±0.48	91.86±0.44
粗蛋白	44.61±1.22	44.89±0.82	45.10±0.95	44.61±0.25	44.22±1.35	44.23±0.78
粗脂肪	4.49±0.28^a	4.89±0.37^b	5.50±0.15^c	4.18±0.03^a	4.35±0.12^a	4.21±0.13^a
粗灰分	32.36±0.52	32.39±0.89	30.49±1.58	32.20±0.36	32.72±1.77	32.67±1.32

注：粗灰分、粗蛋白和粗脂肪的含量为干基含量；同行数据无字母或上标相同字母表示差异不显著（$P>0.05$），上标不同字母表示差异显著（$P<0.05$）。

随着饲料中亮氨酸含量的增加，刺参幼参体壁亮氨酸和蛋氨酸含量显著升高（$P<0.05$），体壁酪氨酸、丙氨酸和脯氨酸含量呈现先升高后降低的趋势（$P<0.05$），而苯丙氨酸和甘氨酸呈现先降低后升高的趋势（$P<0.05$）；各组间体壁总氨基酸含量无显著性差异（$P>0.05$）（表 2-29）。

表 2-29　亮氨酸对刺参幼参体壁氨基酸含量的影响（%）

项目	组别					
	D1	D2	D3	D4	D5	D6
精氨酸	2.91±0.05^{ab}	2.99±0.07^b	2.92±0.04^b	2.95±0.06^b	3.00±0.02^b	2.82±0.05^a
组氨酸	0.62±0.02^{ab}	0.62±0.04^{ab}	0.60±0.06^a	0.65±0.01^b	0.62±0.01^{ab}	0.60±0.02^a
苏氨酸	2.37±0.04	2.39±0.06	2.45±0.09	2.38±0.05	2.42±0.03	2.38±0.07
异亮氨酸	1.28±0.05^{ab}	1.31±0.01^{ab}	1.30±0.02^{ab}	1.31±0.04^{ab}	1.33±0.02^b	1.26±0.06^a
亮氨酸	2.27±0.02^a	2.30±0.03^{ab}	2.31±0.02^{ab}	2.32±0.01^b	2.31±0.04^{ab}	2.32±0.02^b
赖氨酸	1.92±0.06	1.96±0.07	1.91±0.07	1.99±0.01	1.93±0.05	1.91±0.05
缬氨酸	1.89±0.07	1.96±0.02	1.93±0.06	1.96±0.02	1.97±0.05	1.88±0.06
蛋氨酸	0.76±0.01^a	0.85±0.06^b	1.08±0.04^{cd}	1.02±0.03^c	1.02±0.05^c	1.11±0.01^d
半胱氨酸	1.55±0.01	1.61±0.05	1.58±0.09	1.63±0.03	1.65±0.09	1.58±0.04
苯丙氨酸	2.52±0.07^d	1.61±0.03^a	2.04±0.03^b	2.03±0.03^b	1.61±0.06^a	2.28±0.09^c
谷氨酸	6.25±0.05^{bc}	6.32±0.14^c	6.09±0.10^a	6.23±0.02^b	6.26±0.02^{bc}	6.24±0.03^b
甘氨酸	5.33±0.04^{bc}	5.46±0.14^c	5.11±0.12^a	5.26±0.05^{ab}	5.33±0.06^{bc}	5.34±0.10^{bc}
酪氨酸	2.14±0.08^{ab}	2.12±0.04^{ab}	2.23±0.08^b	2.26±0.05^b	2.15±0.09^{ab}	2.06±0.08^a
天冬氨酸	4.36±0.08^a	4.46±0.11^b	4.55±0.08^b	4.46±0.08^b	4.47±0.05^{ab}	4.40±0.10^{ab}
丙氨酸	2.53±0.06^a	2.63±0.06^{bc}	2.68±0.07^c	2.59±0.05^{abc}	2.56±0.02^{ab}	2.51±0.04^a
脯氨酸	2.89±0.08^c	3.01±0.06^d	2.76±0.07^{ab}	2.83±0.01^{bc}	2.84±0.03^{bc}	2.70±0.06^a
丝氨酸	2.28±0.04^a	2.32±0.02^a	2.52±0.07^b	2.26±0.05^a	2.25±0.03^a	2.33±0.09^a
总氨基酸	43.87±0.78	43.95±0.57	44.06±0.84	44.14±0.37	43.70±0.48	43.72±0.33

注：同行数据无字母或上标相同字母表示差异不显著（$P>0.05$），上标不同字母表示差异显著（$P<0.05$）。

3. 亮氨酸对刺参幼参肠道消化酶及抗氧化能力的影响

饲料中亮氨酸含量在 2.22% 及以下时，刺参幼参肠道淀粉酶活性呈现平稳的趋势，当亮氨酸含量超过 2.22% 时，淀粉酶活性显著降低（$P<0.05$），D2、D3 和 D4 组淀粉酶活性显著高于 D6 组（$P<0.05$）；随着饲料中亮氨酸含量从 1.29% 增

加到 1.98%，刺参幼参肠道脂肪酶和蛋白酶的活性显著升高（$P<0.05$），饲料中亮氨酸含量进一步增加时，脂肪酶和蛋白酶的活性均显著降低（$P<0.05$），D3 组脂肪酶活性显著高于其他组（$P<0.05$），D2、D3 和 D4 组蛋白酶活性显著高于其他组（$P<0.05$）（表 2-30）。

表 2-30 亮氨酸对刺参幼参肠道抗氧化能力的影响

项目	组别					
	D1	D2	D3	D4	D5	D6
淀粉酶活性/（U/mg prot）	1.04±0.05[ab]	1.14±0.07[b]	1.14±0.11[b]	1.14±0.10[b]	1.06±0.11[ab]	0.96±0.07[a]
脂肪酶活性/（U/g prot）	1.05±0.05[a]	1.20±0.10[b]	1.95±0.06[e]	1.35±0.04[d]	1.27±0.07[bc]	1.29±0.03[cd]
蛋白酶活性/（U/mg prot）	187.97±17.77[a]	210.01±18.27[b]	223.89±15.97[b]	216.06±17.36[b]	187.16±19.25[a]	168.61±8.67[a]
酸性磷酸酶活性/（U/g prot）	93.81±1.22[abc]	95.09±8.51[bc]	100.93±10.35[c]	99.22±8.03[bc]	91.24±5.45[ab]	85.62±5.06[a]
碱性磷酸酶活性/（U/g prot）	349.87±15.00[b]	359.82±42.99[b]	379.53±23.44[b]	362.98±27.28[b]	308.62±26.92[a]	302.77±16.05[a]
谷草转氨酶活性/（U/g prot）	0.80±0.06[a]	0.92±0.07[b]	1.11±0.08[c]	1.16±0.15[c]	1.12±0.13[c]	1.06±0.11[c]
谷丙转氨酶活性/（U/g prot）	2.55±0.15[a]	3.19±0.16[c]	3.80±0.32[d]	2.97±0.18[bc]	3.08±0.18[c]	2.80±0.18[b]
总抗氧化能力/（nmol/mg prot）	75.99±1.47[a]	81.83±1.53[ab]	101.19±4.23[d]	95.30±7.98[cd]	89.47±3.72[bc]	88.01±5.37[bc]
过氧化氢酶活性/（U/g prot）	500.05±22.80[bc]	497.71±26.85[bc]	492.34±22.80[b]	524.61±24.16[c]	447.00±22.87[a]	434.02±21.95[a]
丙二醛含量/（nmol/mg prot）	3.38±0.14[d]	2.40±0.17[b]	1.82±0.18[a]	1.98±0.18[a]	2.36±0.10[b]	3.09±0.14[c]
超氧化物歧化酶活性/（U/mg prot）	15.43±0.71[a]	16.02±1.27[a]	16.24±0.71[ab]	17.54±1.41[b]	16.22±1.22[ab]	16.05±1.14[a]

注：同行数据无字母或上标相同字母表示差异不显著（$P>0.05$），上标不同字母表示差异显著（$P<0.05$）。

饲料中亮氨酸含量在 2.22%以下时，刺参幼参肠道酸性磷酸酶和碱性磷酸酶的活性各组间无显著性差异（$P>0.05$），当亮氨酸含量超过 2.22%时，酸性磷酸酶和碱性磷酸酶的活性呈降低趋势，酸性磷酸酶和碱性磷酸酶的活性均在 D3 组达到最大值，D3 组酸性磷酸酶活性显著高于 D5、D6 组（$P<0.05$），D5、D6 组碱性磷酸酶活性显著低于其他组（$P<0.05$）。随着饲料中亮氨酸含量从 1.29%增加到 1.98%，谷草转氨酶、谷丙转氨酶的活性和总抗氧化能力显著升高（$P<0.05$），亮氨酸含量超过 1.98%后，谷草转氨酶活性趋于平稳，而谷丙转氨酶活性和总抗氧化能力呈降低趋势（$P<0.05$），各实验组谷草转氨酶活性显著高于对照组（$P<0.05$），D3 组谷丙转氨酶活性显著高于其他组（$P<0.05$），D3 组总抗氧化能力显著高于 D1、D2、D5 和 D6 组（$P<0.05$）。饲料中亮氨酸含量为 2.22%时，过氧化氢酶和超氧化物歧化酶的活性达到最大值，D4 组过氧化氢酶活性显

著高于 D3、D5 和 D6 组（$P<0.05$），D4 组超氧化物歧化酶活性显著高于 D1、D2 和 D6 组（$P<0.05$）。随着饲料中亮氨酸含量从 1.29%增加到 1.98%，丙二醛含量显著降低（$P<0.05$），饲料中亮氨酸含量超过 2.22%后，丙二醛含量显著升高（$P<0.05$），D3 和 D4 组丙二醛含量显著低于其他组（$P<0.05$）。

2.2.1.3 讨论

1. 亮氨酸对刺参幼参生长性能的影响

亮氨酸是水产动物的必需氨基酸，许多研究表明，饲料中添加适量亮氨酸可提高水产动物的生长性能（Abidi and Khan，2007）。本实验中，随着饲料中亮氨酸含量从 1.29%增加到 1.98%，刺参幼参的增重率和特定生长率均显著升高，当饲料中亮氨酸含量达到 1.98%时，增重率达到最大值 100.84%。这是因为饲料中缺乏亮氨酸，影响氨基酸平衡，体内过多的其他氨基酸无法被有效利用而被机体代谢，抑制了机体蛋白质的合成，不利于水产动物的生长，饲料中添加适量的亮氨酸能改善氨基酸平衡，促进水产动物生长。而随着饲料中亮氨酸含量进一步增加，刺参幼参的增重率和特定生长率呈现下降的趋势，刺参幼参生长受到抑制，对吉富罗非鱼（石亚庆等，2014）、卵形鲳鲹（Tan et al.，2016）、花鲈（路凯，2015）、三疣梭子蟹（Huo et al.，2017）和斑节对虾（Millamena et al.，1999）等水产动物的研究也得到了类似的结果，饲料中添加适量的亮氨酸有利于其生长，而过量的亮氨酸会抑制其生长。其原因可能有：①亮氨酸与异亮氨酸、缬氨酸统称为支链氨基酸，支链氨基酸之间存在拮抗作用，过量的亮氨酸在小肠壁处与异亮氨酸和缬氨酸竞争载体，影响了异亮氨酸和缬氨酸的吸收和转运，降低了氨基酸利用率；②亮氨酸过量影响氨基酸平衡，过量的亮氨酸提高了支链氨基酸转氨酶和脱氢酶的活性，使机体消耗能量将多余的亮氨酸代谢及排出，同时这两种酶也增强了异亮氨酸和缬氨酸的代谢，引起异亮氨酸和缬氨酸缺乏，影响机体生长（黄爱霞等，2018）；③亮氨酸是生酮氨基酸，过量的亮氨酸在代谢过程中产生了较多的酮类和其他有害代谢物，从而抑制了水产动物的生长（Abidi and Khan，2007）。但是，对虹鳟（Choo et al.，1991）、草鱼（黄爱霞等，2018）和异育银鲫（李桂梅，2009）的研究显示，过量的亮氨酸并没有对其生长性能产生显著的抑制作用，这可能是不同物种对亮氨酸的耐受力不同导致的。因此，饲料中亮氨酸含量对不同水产动物生长性能影响的作用机制有待进一步研究。

以增重率为评价指标，经一元二次回归分析得出，初始体质量为 16.40g 的刺参幼参对饲料中亮氨酸的最适需求量为 2.11%，占饲料粗蛋白的 10.37%，该数值（占饲料粗蛋白的比例）远高于一般鱼虾蟹类水产动物对亮氨酸的需求量（张圆圆和王连生，2020），如大黄鱼（6.79%）（Li et al.，2010）、斑节对虾（4.3%）（Millamena et al., 1999)、凡纳滨对虾（5.71%）(张德瑞等，2016)、三疣梭子蟹（5.14%）(胡友军等，2002)、中华绒螯蟹（5.88%）(杨霞等，2014) 等，该结果与刺参幼参对精氨酸（王际英等，2015)、赖氨酸（王吉桥等，2009) 和缬氨酸（韩秀杰等，2019)等氨基酸需求量的研究结果类似。其原因可能是：①刺参对饲料粗蛋白的需求量低于一般鱼虾蟹类水产动物（朱伟等，2005；李素红等，2012)，而在刺参体壁的 10种必需氨基酸中，亮氨酸是含量最高的 3 种氨基酸之一，仅次于精氨酸含量，与苏氨酸含量相近，因此补充大量的亮氨酸才能满足刺参的快速增长（王际英等，2015)；②不同物种对亮氨酸需求量的差异可能与实验对象、生长阶段、评价指标、蛋白质来源和饲养管理等因素有关；③刺参与一般鱼虾蟹类相比，具有独特的摄食习性——舔食性，摄食活动持续时间长，饲料与水体接触面积大，且实验饲料在水中停留时间远远长于鱼虾蟹类饲料，实验所用晶体亮氨酸，存在一定的溶失现象。

2. 亮氨酸对刺参幼参体壁成分的影响

已有研究证明，亮氨酸可以激活 mTOR 通路，促进氨基酸合成蛋白质，并降低机体蛋白质的分解代谢，从而起到蛋白质沉积的作用，提高水产动物机体蛋白质的含量（Guertin et al.，2006；Yoshizawa，2004)。在对中华绒螯蟹（杨霞等，2014)、吉富罗非鱼（石亚庆等，2014) 和卵形鲳鲹（Tan et al.，2016) 的研究中，饲料中添加适量亮氨酸都显著提高了全鱼和肌肉的粗蛋白含量。但也有研究得出了不同的结论，如对凡纳滨对虾（王用黎，2013)、花鲈（李燕，2010）的研究表明，饲料中亮氨酸含量对机体粗蛋白含量无显著影响，这与本实验的结果一致，表明刺参蛋白质的合成不仅仅受亮氨酸水平的影响，刺参机体还具有维持体壁蛋白质恒定的趋势。在对不同水产动物的亮氨酸研究中，出现以上不同实验结果的原因可能是实验对象、生长阶段、生长速率等因素存在差异。支链氨基酸对于机体的脂肪代谢和沉积具有重要的调控作用，能够显著影响机体的脂肪代谢和沉积，机体缺乏亮氨酸会激活脂肪组织解偶联蛋白 1 和脂肪内激素敏感性脂肪酶，促进脂肪脂解产热，导致机体脂肪含量减少（Cheng et al.，2010)，而亮氨酸过量则影响氨基酸平衡，导致机体过度消耗能量，也会减少机体脂肪的沉积（韩秀杰等，2019)。本研究也证明了这一点，亮氨酸缺乏或过量，显著降低刺参体壁粗脂肪含量，适量的亮氨酸具有提高刺参体壁粗脂肪含量的作用。

随着饲料中亮氨酸含量的增加，刺参体壁必需氨基酸中蛋氨酸含量的增长最为显著。另外，饲料中添加适量的亮氨酸，显著提高刺参体壁酪氨酸、丙氨酸和脯氨酸的含量，降低苯丙氨酸和甘氨酸的含量。饲料中亮氨酸含量不超过 2.58%时，刺参体壁精氨酸含量各组间无显著性差异，当亮氨酸含量超过 2.58%后，精氨酸含量显著降低，说明只有饲料中亮氨酸含量变化比较明显时，部分氨基酸之间才会表现出明显的协同或拮抗作用。各组刺参体壁的总氨基酸含量无显著性差异，说明刺参机体对自身的氨基酸模式具有一定的调节能力，有维持机体蛋白质恒定的作用。

3. 亮氨酸对刺参幼参肠道消化酶活性及抗氧化能力的影响

有研究表明，亮氨酸在代谢过程中能够为谷氨酸和谷氨酰胺提供碳源和氮源（Chen et al., 2009），而谷氨酰胺能够促进蛋白酶、脂肪酶的分泌（叶元土等, 2007），且亮氨酸代谢产物 β-羟基-β-甲基戊二烯二酰 CoA 是胆固醇的前体物质，而胆固醇可保持细胞膜的流动性，从而有利于维持肠上皮细胞的完整性，进而保证消化酶的分泌（王镜岩等, 2007）。亮氨酸缺乏显著降低青鱼肠道 α-淀粉酶、胰蛋白酶、糜蛋白酶和弹性蛋白酶的活性（Wu et al., 2017）；饲料中添加适量亮氨酸，吉富罗非鱼胃蛋白酶、肠蛋白酶、肠脂肪酶和肠 Na^+-K^+-ATP 等酶的活性均得到提高（石亚庆等, 2014），这些研究结果也证明了饲料中的亮氨酸可以影响水产动物消化酶的活性。本实验中，随着饲料中亮氨酸含量从 1.29%增加到 1.98%，刺参幼参肠道脂肪酶和蛋白酶的活性显著升高，饲料中亮氨酸含量进一步增加时，脂肪酶和蛋白酶的活性均呈降低趋势，脂肪酶和蛋白酶的活性均在 D3 组达到最大值，表明饲料中添加适量亮氨酸能提高改善刺参肠道的消化能力，有利于蛋白质和脂肪的消化吸收利用。同时，亮氨酸经过包膜处理，延缓了其在水中的溶失及在消化道中的吸收，有利于外源添加晶体氨基酸与蛋白态氨基酸的同步吸收，提高其吸收利用能力，从而促进刺参幼参的生长（韩秀杰等, 2019）。饲料中亮氨酸含量在 2.22%及以下时，刺参幼参肠道淀粉酶活性呈现平稳的趋势，当亮氨酸含量超过 2.22%时，淀粉酶活性呈降低趋势，表明亮氨酸对刺参幼参淀粉代谢影响较小，只有亮氨酸含量过高时，才会抑制刺参幼参的淀粉代谢，这可能是饲料中糖类物质含量较少及刺参对糖类物质利用较弱造成的。

酸性磷酸酶和碱性磷酸酶是重要的水解酶，能催化磷酸单酯的水解及磷酸基团的转移反应，在免疫反应中发挥作用，对动物的生存具有重要的意义，一直以来都是动物免疫学中的一个重要指标（刘青和赵恒寿, 2007）。本实验中，饲料中亮氨酸含量在2.22%以下时，刺参肠道酸性磷酸酶和碱性磷酸酶的活性各组间无显著性差异，当亮氨酸含量超过 2.22%时，酸性磷酸酶和碱性磷酸酶的活性呈降低趋势，酸性磷酸酶和碱性磷酸酶的活性均在 D3 组达到最大值，说明只有亮氨酸含量变化明显时，才会显著影响刺参幼参肠道酸性磷酸酶和碱性磷酸酶的活性，过量的亮氨酸能抑制刺参幼参的非特异性免疫能力。谷草转氨酶和谷丙转氨酶是氨基酸代谢中起重要作用的两种酶，其活性与氨基酸代谢强弱有关（李桂梅, 2009）。肝脏是一般鱼类重要营养物质的合成代谢器官，所以在一般鱼类机体中，肝脏中的谷草转氨酶和谷丙转氨酶活性较高。刺参作为一种低营养等级的舔食性底栖生物，不具备肝脏等功能性器官（臧元奇, 2012），因此其营养物质的合成代谢与鱼类存在较大差异。本实验中，随着饲料中亮氨酸含量从 1.29%增加到 1.98%，刺参幼参肠道谷草转氨酶和谷丙转氨酶的活性显著升高，亮氨酸含量超过 1.98%后，谷草转氨酶活性趋于平稳，而谷丙转氨酶活性显著降低，且各实验组谷草转氨酶和谷丙转氨酶的活性均显著高于对照组，这表明刺参幼参肠道参与了营养物质的合成代谢，同时表明饲料中添加适量的亮氨酸能显著增强刺参的营养物质代谢能力。

生物机体正常代谢会产生大量的活性氧，活性氧族如超氧阴离子自由基、过氧化氢、羟自由基和单线态氧都是有氧代谢的副产物，线粒体是活性氧的一个重要来源，线粒体消耗的氧中有 2%被转化成氧自由基（谢玉英，2009）。活性氧会对动物机体造成氧化损伤，如对核酸和蛋白质造成氧化损伤，甚至会损伤细胞膜上的多不饱和脂肪酸，使其发生脂质过氧化反应，破坏细胞膜的完整性。水生动物的抗氧化能力主要通过机体的自身免疫反应实现，总抗氧化能力、超氧化物歧化酶和过氧化氢酶水平是生物机体抗氧化的重要的指标，在机体抗氧化中起着关键的作用（Bagnyukova et al.，2003）。本实验中，当饲料中亮氨酸含量为 2.22%时，过氧化氢酶和超氧化物歧化酶的活性达到最大值，D4 组过氧化氢酶活性显著高于 D3、D5 和 D6 组，D4 组超氧化物歧化酶活性显著高于 D1、D2 和 D6 组；总抗氧化能力随着饲料中亮氨酸含量的增加呈现先升高后降低的趋势，D3 组总抗氧化能力显著高于 D1、D2、D5 和 D6 组。这表明在饲料中添加适量的亮氨酸能提高刺参幼参肠道抗氧化酶的活性，从而提高刺参幼参清除氧自由基和抗氧化的能力。生物膜中含有大量的多不饱和脂肪酸，极易受氧自由基攻击而形成脂质过氧化产物丙二醛，所以水产动物体内的丙二醛含量能间接反映机体的受损伤程度和抗氧化能力（Martinez-Álvarez et al.，2002）。本实验中，随着饲料中亮氨酸含量从 1.29%增加到 1.98%，丙二醛含量显著降低，当饲料中亮氨酸含量超过 2.22%后，丙二醛含量显著升高，D3 和 D4 组丙二醛含量显著低于其他组，这表明饲料中添加适量的亮氨酸能减轻活性氧对刺参幼参机体的损伤，对团头鲂（Liang et al.，2018）、卵形鲳鲹（Tan et al.，2016）和三疣梭子蟹（Huo et al.，2017）的研究也得到了类似的结果。

2.2.1.4　结论

综上所述，在本实验条件下，以增重率为评价指标，体质量为 16.40g 的刺参幼参对饲料中亮氨酸的最适需求量为 2.11%（10.37%饲料粗蛋白）。饲料中添加适量的亮氨酸能有效增强刺参幼参对营养物质的消化能力，提高抗氧化及免疫能力，进而促进刺参幼参的生长。

2.2.2　刺参幼参对缬氨酸最适需求量的研究

缬氨酸是组成蛋白质的 20 种氨基酸之一，化学名称为 2-氨基-3-甲基丁酸，是人体必需氨基酸之一和生糖氨基酸。缬氨酸在生物体正常生命活动中,尤其是在神经系统的正常运转中起重要作用,缬氨酸缺乏会导致哺乳动物神经功能的衰退,它与异亮氨酸和亮氨酸一起促进身体正常生长,修复组织,调节血糖,并提供需要的能量(廖国周和柴仕名,2005)。缬氨酸还可以帮助肝脏清除多余的氮,并将身体需要的氮运输到各个部位。缬氨酸也是水生动物的必需氨基酸之一,对水生动物的生长性能、蛋白质利用等有重要的影响。目前,国内外已经开展了一些鱼类和虾类缬氨酸营养需求的研究,确定了鲤、草鱼(罗莉等,2010)、尼罗罗非鱼(Santiago and Lovell, 1988)、斑点叉尾鮰(Wilson et al., 1980)、虹鳟、印鲮(Ahmed and Khan, 2006)和南亚野鲮(Abidi et al., 2004)等多种水生动物的缬氨酸需求量。

目前,对刺参幼参氨基酸营养需求的研究较少,仅精氨酸(王际英等,2015)、谷氨酰胺(Yu et al., 2016)和赖氨酸(王吉桥等,2009)的需求量已有研究确定,对缬氨酸的需求量则尚未见报道。研究刺参幼参对缬氨酸的营养需求,对配制营养均衡、廉价高效、低污染的配合饲料有重要的指导意义(李桂梅等,2010)。因此,本实验通过在饲料中添加不同含量的缬氨酸,研究其对刺参幼参的生长性能、免疫能力及消化酶活性的影响,从而为刺参幼参对缬氨酸需求的研究提供依据。

2.2.2.1 材料与方法

1. 实验设计与饲料配制

本实验以白鱼粉、小麦粉和海藻粉为主要蛋白源,鱼油为主要脂肪源配制基础饲料,在基础饲料中分别添加0、0.80%、1.60%、2.40%、3.20%和4.00%的包膜缬氨酸,配成缬氨酸含量为0.61%、1.14%、1.46%、1.73%、2.17%和2.64%的6组等氮等能的实验饲料,分别命名为D1、D2、D3、D4、D5和D6组。实验饲料配方、营养组成及氨基酸组成分别见表2-31和表2-32。

表2-31 实验饲料配方及营养组成(干物质)

项目		组别					
		D1	D2	D3	D4	D5	D6
原料组成	白鱼粉/%	16.00	16.00	16.00	16.00	16.00	16.00
	小麦粉/%	8.00	8.00	8.00	8.00	8.00	8.00
	海藻粉/%	34.00	34.00	34.00	34.00	34.00	34.00
	包膜甘氨酸/%	2.00	1.60	1.20	0.80	0.40	0.00
	包膜缬氨酸/%	0.00	0.80	1.60	2.40	3.20	4.00
	多维/%	0.50	0.50	0.50	0.50	0.50	0.50
	多矿/%	0.50	0.50	0.50	0.50	0.50	0.50
	鱼油/%	0.50	0.50	0.50	0.50	0.50	0.50

续表

项目		组别					
		D1	D2	D3	D4	D5	D6
原料组成	抗氧化剂/%	0.10	0.10	0.10	0.10	0.10	0.10
	大豆卵磷脂/%	0.50	0.50	0.50	0.50	0.50	0.50
	海泥/%	37.90	37.50	37.10	36.70	36.30	35.90
营养组成	粗蛋白/%	19.22	19.44	19.39	19.40	19.47	19.45
	粗脂肪/%	1.75	1.78	1.72	1.74	1.70	1.72
	粗灰分/%	48.69	48.73	49.09	49.50	49.86	49.80
	能量（kJ/g）	11.37	11.72	11.86	11.98	12.00	12.02

注：①β-环糊精购自孟州市华兴生物化工有限责任公司；②多维（mg/kg 饲料）为维生素 A 38.00mg、维生素 D_3 13.20mg、维生素 K_3 10.00mg、维生素 B_1 115.00mg、维生素 B_2 380.00mg、盐酸吡哆醇 88.00mg、泛酸 368.00mg、烟酸 1030.00mg、生物素 10.00mg、叶酸 20.00mg、维生素 B_{12} 1.30mg、肌醇 4000.00mg、抗坏血酸 500.00mg；③多矿（mg/kg 饲料）为 NaCl 100.00mg、KCl 3020.50mg、$KAl(SO_4)_2$ 11.30mg、$ZnSO_4·7H_2O$ 363.00mg、$CuSO_4·5H_2O$ 8.00mg、$MgSO_4·7H_2O$ 3568.00mg、$MnSO_4·4H_2O$ 65.10mg、Na_2SeO_3 2.30mg、$CoCl_2$ 28.00mg、KI 7.50mg、NaF 4.00mg、$NaH_2PO_4·2H_2O$ 25 558.00mg、乳酸钙 15 978.00mg、$C_6H_5O_7Fe·5H_2O$ 1523.00mg。

表 2-32 实验饲料氨基酸组成（干物质）（%）

氨基酸	组别					
	D1	D2	D3	D4	D5	D6
精氨酸	0.98	1.01	0.90	0.96	1.00	1.01
组氨酸	0.24	0.28	0.24	0.26	0.27	0.28
异亮氨酸	0.51	0.54	0.50	0.49	0.53	0.56
亮氨酸	1.36	1.28	1.37	1.51	1.48	1.32
赖氨酸	0.84	0.95	0.82	0.89	0.93	0.97
蛋氨酸	0.42	0.44	0.42	0.44	0.46	0.39
苯丙氨酸	0.72	0.77	0.73	0.76	0.77	0.78
苏氨酸	0.70	0.81	0.68	0.75	0.76	0.82
缬氨酸	0.61	1.14	1.46	1.73	2.17	2.64
天冬氨酸	1.62	1.74	1.66	1.65	1.72	1.76
丝氨酸	0.96	0.94	0.90	0.93	0.95	0.94
甘氨酸	2.35	2.20	1.81	1.72	1.59	1.34
丙氨酸	1.19	1.18	1.14	1.18	1.21	1.21
半胱氨酸	0.62	0.56	0.61	0.62	0.54	0.61
酪氨酸	0.46	0.44	0.45	0.49	0.51	0.46
谷氨酸	3.24	3.11	2.88	3.10	3.09	3.10
脯氨酸	0.84	0.92	0.85	0.91	0.94	0.93
总氨基酸	17.65	18.32	17.40	18.40	18.90	19.11

实验所用缬氨酸购自上海麦克林生化科技有限公司（纯度＞99.9%），参考胡友军等（2002）的方法进行包被，包被后缬氨酸含量约为40%。固体原料经超微粉碎后过200目标准筛，按配比称重，加入新鲜鱼油及适量蒸馏水混匀，用小型饲料挤压机制成直径为0.3cm、厚度为0.05cm的片状饲料，60℃烘干。

2. 饲养管理及样品采集

养殖实验在山东省海洋资源与环境研究院东营实验基地循环水养殖系统中进行，为期8周。实验所用刺参幼参购自山东安源种业科技有限公司。实验开始前，将实验参放养于养殖系统中驯养2周，驯养期间投喂基础饲料。正式实验开始之前，控食24h，挑选体质健壮、规格相同、初始质量为（9.20±0.12）g的720头刺参幼参随机分到18个圆柱塑料养殖桶（$h_{桶}$=80cm，$h_{水}$=65cm，$d_{桶}$=60cm）中，每桶放置2个海参养殖筐。实验分为6个处理，每个处理3个重复，每种饲料随机投喂3个桶。每天定时（16:00）投喂1次，投喂量为刺参幼参初始体质量的3%，每天观察刺参幼参的摄食情况，调整次日投喂量。实验期间，每天换水1/3，换水时用虹吸法将残饵及粪便吸出，每隔两天清理一次，控制水中溶氧量大于5mg/L，水温为17.0~19.0℃，盐度为27，pH为7.6~8.2，氨氮和亚硝酸盐浓度不超过0.05mg/L，养殖实验在弱光环境中进行。

养殖实验结束后，控食24h。记录每桶刺参幼参的数量并称重，计算存活率、增重率和特定生长率等，每桶随机取10头实验参，测量体质量、体长后解剖，测量肠道重、肠道长、体壁重，计算肠壁比和肠长比，并采集体壁和肠道，分别置于–20℃和–80℃保存，待测。

3. 样品分析

饲料及刺参幼参体壁氨基酸测定参照《实验动物 配合饲料 氨基酸的测定》（GB/T 14924.10—2008），采用全自动氨基酸测定仪（Hitachi L-8900，Japan）测定。

肠道中超氧化物歧化酶、谷丙转氨酶、谷草转氨酶、酸性磷酸酶、碱性磷酸酶、过氧化氢酶、蛋白酶、脂肪酶和淀粉酶的活性和总抗氧化能力、丙二醛含量的测定均采用南京建成生物工程研究所生产的试剂盒，酸性磷酸酶和碱性磷酸酶的活性单位转换为国际单位，其他酶的活性单位参照试剂盒说明书。

4. 测定指标

实验中分别统计刺参幼参存活率、特定生长率、增重率、肠壁比、肠长比。

5. 数据统计分析

实验所得数据采用Microsoft Excel 2007和SPSS 17.0软件进行单因素方差分

析，差异显著时（$P<0.05$）采用邓肯多重范围检验进行多重比较分析。统计结果以平均值±标准差表示。

2.2.2.2 结果分析

1. 缬氨酸对刺参幼参生长性能及形体指标的影响

缬氨酸对刺参幼参生长性能及形体指标的影响见表 2-33。缬氨酸对刺参幼参的存活率无显著影响（$P>0.05$）；增重率和特定生长率随饲料中缬氨酸含量的升高呈现先上升后下降的趋势，D4 组增重率显著高于其他各组（$P<0.05$），D1 组特定生长率显著低于其他各组（$P<0.05$）；缬氨酸对肠壁比和肠长比均无显著影响（$P>0.05$）。

表 2-33 缬氨酸对刺参幼参生长性能及形体指标的影响

项目	组别					
	D1	D2	D3	D4	D5	D6
初始体质量/g	9.14±0.19	9.26±0.16	9.20±0.07	9.19±0.13	9.11±0.15	9.24±0.06
终末体质量/g	17.32±1.48	18.45±3.65	19.87±2.42	21.86±3.77	21.18±2.21	21.35±0.74
增重率/%	89.50±2.10^a	99.23±2.32^b	116.00±3.53^c	137.90±3.51^e	132.47±3.11^d	131.10±2.90^d
特定生长率/（%/d）	1.10±0.03^a	1.19±0.01^b	1.33±0.01^c	1.49±0.02^e	1.45±0.01^d	1.44±0.01^d
肠壁比/%	4.37±0.12	4.67±0.51	4.37±0.26	4.65±0.61	4.44±0.45	4.67±0.26
肠长比/%	3.52±0.08	3.58±0.05	3.59±0.02	3.60±0.06	3.56±0.08	3.59±0.02
存活率/%	95.83±1.44	91.25±1.77	96.67±3.82	95.83±3.82	90.83±2.89	90.00±0.00

注：同行数据无字母或上标相同字母表示差异不显著（$P>0.05$），上标不同字母表示差异显著（$P<0.05$）。

以增重率为评价指标，经折线回归分析得，体质量为 9.20g 的刺参幼参对饲料中缬氨酸的最适需求量为 1.79%（9.18%饲料粗蛋白）（图 2-8）。

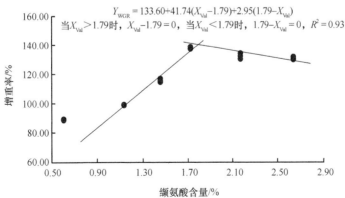

图 2-8 刺参幼参增重率与饲料中缬氨酸含量的折线回归分析

2. 缬氨酸对刺参幼参体壁基本营养成分及氨基酸含量的影响

缬氨酸对刺参幼参体壁基本营养成分的影响见表 2-34。刺参幼参体壁组成中，水分含量为 90.76%~91.46%，各组间无显著性差异（$P>0.05$）。随着饲料中缬氨酸含量的增加，粗蛋白含量呈现先上升后下降的趋势，在 D4 组达到最大值（$P<0.05$）；粗脂肪含量呈现先上升后下降的趋势，在 D3 组达到最大值（$P<0.05$）；粗灰分随着缬氨酸含量的增加呈现先下降后升高的趋势，D4 组达到最低值（$P<0.05$）。

表 2-34　缬氨酸对刺参幼参体壁基本营养成分的影响（%）

项目	组别					
	D1	D2	D3	D4	D5	D6
水分	91.18±0.33	91.14±0.16	91.46±0.08	91.06±0.50	91.27±0.26	90.76±0.31
粗蛋白	43.66±0.43^a	44.61±0.27^b	44.77±0.41^b	45.67±0.23^c	44.75±0.47^b	44.57±0.18^b
粗脂肪	3.32±0.16^a	4.27±0.18^{cd}	4.61±0.04^d	4.50±0.20^d	3.72±0.15^{ab}	3.88±0.07^{bc}
粗灰分	33.03±0.22^b	32.30±0.19^{ab}	32.87±0.23^{ab}	32.04±0.27^a	32.38±0.09^{ab}	32.70±0.16^{ab}

注：粗蛋白、粗脂肪和粗灰分的含量为干基含量；同行数据无字母或上标相同字母表示差异不显著（$P>0.05$），上标不同字母表示差异显著（$P<0.05$）。

缬氨酸对刺参幼参体壁氨基酸组成的影响见表 2-35。随着饲料中缬氨酸含量从 0.61%增加到 2.64%，刺参幼参体壁中组氨酸、缬氨酸、赖氨酸、苯丙氨酸、苏氨酸、天冬氨酸、必需氨基酸及总氨基酸的含量显著升高（$P<0.05$），亮氨酸、异亮氨酸和酪氨酸的含量显著降低（$P<0.05$），但是精氨酸、蛋氨酸、丝氨酸、甘氨酸、丙氨酸、半胱氨酸、谷氨酸及脯氨酸的含量无显著性差异（$P>0.05$），缬氨酸添加量 1.46%组精氨酸含量显著低于 2.64%组（$P<0.05$）。

表 2-35　缬氨酸对刺参幼参体壁氨基酸组成的影响（%）

氨基酸	组别					
	D1	D2	D3	D4	D5	D6
精氨酸	3.13±0.05^{ab}	3.12±0.14^{ab}	3.05±0.02^a	3.18±0.04^{ab}	3.20±0.11^{ab}	3.24±0.12^b
组氨酸	0.59±0.02^{ab}	0.60±0.02^{abc}	0.58±0.01^a	0.61±0.01^{bc}	0.62±0.01^c	0.64±0.02^d
异亮氨酸	1.32±0.03^c	1.26±0.05^b	1.25±0.06^b	1.22±0.04^{ab}	1.19±0.03^a	1.21±0.06^{ab}
亮氨酸	2.66±0.04^c	2.57±0.05^b	2.51±0.03^a	2.46±0.05^a	2.49±0.05^a	2.50±0.03^a
赖氨酸	1.89±0.04^a	1.92±0.07^a	1.89±0.03^a	1.95±0.06^{ab}	1.96±0.08^{ab}	2.05±0.10^b
蛋氨酸	0.54±0.07	0.56±0.04	0.55±0.06	0.56±0.07	0.56±0.05	0.55±0.06
苯丙氨酸	1.36±0.03^a	1.37±0.05^{ab}	1.36±0.02^a	1.41±0.06^{ab}	1.40±0.06^{ab}	1.44±0.04^b
苏氨酸	2.37±0.03^a	2.40±0.09^{ab}	2.34±0.02^a	2.43±0.08^{ab}	2.43±0.09^{ab}	2.52±0.10^b
缬氨酸	1.58±0.02^a	1.63±0.05^{abc}	1.61±0.04^{ab}	1.71±0.08^c	1.71±0.09^c	1.69±0.01^{bc}
天冬氨酸	4.50±0.02^a	4.64±0.16^{ab}	4.50±0.03^a	4.59±0.09^{ab}	4.66±0.12^{ab}	4.74±0.18^b
丝氨酸	2.27±0.03	2.32±0.09	2.23±0.03	2.30±0.05	2.31±0.06	2.34±0.10
甘氨酸	5.27±0.07	5.30±0.16	5.12±0.02	5.32±0.13	5.32±0.18	5.31±0.21
丙氨酸	2.70±0.03^b	2.68±0.06^{ab}	2.63±0.02^a	2.71±0.02^b	2.72±0.03^b	2.72±0.06^b

续表

氨基酸	组别					
	D1	D2	D3	D4	D5	D6
半胱氨酸	1.31±0.09	1.29±0.04	1.31±0.11	1.30±0.10	1.30±0.14	1.31±0.07
酪氨酸	1.18±0.03^{cd}	1.19±0.05^{cd}	1.23±0.02^{d}	1.14±0.06^{bc}	1.11±0.04^{b}	0.90±0.07^{a}
谷氨酸	7.45±0.10^{ab}	7.51±0.32^{ab}	7.28±0.07^{a}	7.53±0.05^{ab}	7.58±0.22^{ab}	7.68±0.27^{b}
脯氨酸	3.02±0.06	3.00±0.17	2.96±0.04	3.08±0.03	3.08±0.11	3.12±0.14
必需氨基酸	11.72±0.14^{a}	11.68±0.12^{a}	11.51±0.13^{a}	11.74±0.11^{ab}	11.74±0.09^{ab}	11.96±0.13^{b}
总氨基酸	43.14±0.46^{b}	43.33±0.24^{bc}	42.40±0.35^{a}	43.50±0.34^{bc}	43.64±0.38^{bc}	43.96±0.33^{c}

注：同行数据无字母或上标相同字母表示差异不显著（$P>0.05$），上标不同字母表示差异显著（$P<0.05$）。

3. 缬氨酸对刺参幼参肠道抗氧化能力及消化酶活性的影响

缬氨酸对刺参幼参肠道抗氧化能力及消化酶活性的影响见表 2-36。随着饲料中缬氨酸含量的增加，刺参幼参肠道中总抗氧化能力和超氧化物歧化酶、过氧化氢酶、谷丙转氨酶、谷草转氨酶、酸性磷酸酶及碱性磷酸酶的活性均有显著升高（$P<0.05$），其中 D1 组的总抗氧化能力和超氧化物歧化酶、谷草转氨酶的活性显著低于其他各组（$P<0.05$），总抗氧化能力和谷草转氨酶活性在 D3 组达到最大值，超氧化物歧化酶活性在 D5 组达到最大值，酸性磷酸酶和碱性磷酸酶的活性均呈现先上升后下降的趋势，分别在 D4、D5 组达到最大值。饲料中添加缬氨酸显著降低了丙二醛含量，D1 组显著高于其他各组（$P<0.05$）。随着饲料中缬氨酸含量增加，刺参幼参肠道中蛋白酶和脂肪酶的活性也有显著升高（$P<0.05$），分别在 D3、D4 组达到最大值（$P<0.05$）；淀粉酶活性不受饲料中缬氨酸含量的影响（$P>0.05$）。

表 2-36 缬氨酸对刺参幼参肠道抗氧化能力及消化酶活性的影响

项目	组别					
	D1	D2	D3	D4	D5	D6
总抗氧化能力/（U/g prot）	0.21±0.01^{a}	0.26±0.01^{b}	0.26±0.02^{b}	0.26±0.01^{b}	0.24±0.02^{b}	0.26±0.02^{b}
超氧化物歧化酶活性/（U/mg prot）	174.43±6.08^{a}	188.62±6.39^{b}	188.85±5.8^{b}	191.23±6.81^{b}	191.83±4.27^{b}	192.34±3.45^{b}
丙二醛含量/（nmol/mg prot）	0.70±0.02^{d}	0.52±0.02^{ab}	0.53±0.02^{ab}	0.50±0.03^{a}	0.61±0.02^{c}	0.56±0.02^{b}
过氧化氢酶活性/（U/mg prot）	27.62±2.46^{a}	29.74±2.41^{ab}	34.39±1.57^{c}	32.84±1.48^{bc}	33.90±1.59^{c}	30.99±1.75^{bc}
谷丙转氨酶活性/（U/g prot）	3.48±0.08^{a}	3.82±0.17^{b}	3.82±0.10^{b}	3.99±0.17^{b}	3.49±0.16^{a}	3.33±0.16^{a}
谷草转氨酶活性/（U/g prot）	2.35±0.12^{a}	2.78±0.12^{b}	2.87±0.14^{b}	2.72±0.12^{b}	2.83±0.13^{b}	2.81±0.11^{b}
酸性磷酸酶活性/（U/g prot）	1.50±0.08^{a}	1.58±0.10^{ab}	1.59±0.05^{ab}	1.78±0.13^{c}	1.74±0.13^{bc}	1.51±0.11^{a}

续表

项目	组别					
	D1	D2	D3	D4	D5	D6
碱性磷酸酶活性/（U/g prot）	2.06±0.05^a	2.12±0.05^{ab}	2.18±0.06^{bc}	2.24±0.04^c	2.25±0.03^c	2.21±0.06^{bc}
蛋白酶活性/（U/mg prot）	1083.39±27.73^a	1184.10±41.80^c	1243.83±58.23^d	1186.01±22.10^c	1126.45±27.81^{ab}	1152.09±26.48^{bc}
脂肪酶活性/（U/g prot）	27.55±2.08^a	28.30±2.28^a	32.24±1.56^{bc}	33.13±1.64^c	32.91±2.18^{bc}	29.44±1.36^{ab}
淀粉酶活性/（U/mg prot）	4.09±0.31	4.24±0.21	4.24±0.46	4.42±0.18	4.42±0.39	4.40±0.22

注：同行数据无字母或上标相同字母表示差异不显著（$P>0.05$），上标不同字母表示差异显著（$P<0.05$）。

2.2.2.3 讨论

1. 缬氨酸对刺参幼参生长性能的影响

本实验中，刺参幼参的存活率不受饲料中缬氨酸含量的影响，饲料中添加缬氨酸的实验组的增重率及特定生长率与对照组相比均有显著提高，证明饲料中添加适宜含量的缬氨酸能够促进刺参幼参的生长。本研究表明，刺参幼参的生长性能随饲料中缬氨酸含量的增加呈现先升高后降低的趋势，表明饲料中缬氨酸含量过低或过高均会抑制刺参幼参的生长。产生这种生长抑制特征的主要原因有：①饲料中缬氨酸含量过低或过高，均会破坏氨基酸代谢平衡，导致机体代谢紊乱，从而使生长受到抑制（周歧存等，2015）；②缬氨酸参与机体的许多生理代谢过程，如作为蛋白质合成的原料抑制其分解、参与机体组织修复、维持机体氮代谢等，饲料中缬氨酸含量过低，机体蛋白质合成受限，从而对生长产生抑制作用；③饲料中缬氨酸含量过高，就会破坏氨基酸平衡，饲料中氨基酸含量越高，机体氨基酸分解代谢就越旺盛，而氨基酸的分解需要消耗能量，从而降低蛋白质和饲料的利用率，对生长产生抑制作用。以增重率为评价指标，采用折线回归分析得出，刺参幼参对缬氨酸的最适需求量为1.79%，占饲料粗蛋白的9.18%。该数值（占饲料粗蛋白的比例）远高于鲤、真赤鲷、尼罗罗非鱼（Santiago and Lovell，1988）等对缬氨酸的需求量，这可能与实验动物的种属及规格大小等有关。

2. 缬氨酸对刺参幼参体成分的影响

在构成机体蛋白质的20种氨基酸中，支链氨基酸是唯一不局限于肝脏可在全身组织特别是肌肉组织进行分解代谢的必需氨基酸。各支链氨基酸结构相似，它们在代谢途径中的转氨基和脱氢作用可共用支链氨基酸转氨酶和脱氢酶，因此它们在代谢过程中会出现拮抗作用，且拮抗程度在各动物间存在差异，最终影响动物的代谢及生长。本实验中，随着饲料中缬氨酸含量的增加，过量的缬氨酸抑制了体壁中亮氨酸和异亮氨酸的沉积，导致二者含量显著降低。当饲料中某种必需氨基酸含量不足时，就会表现为机体生长缓慢，且饲料和蛋白质利用率低，随着饲料中该必需氨基酸含量的增加，机体生长逐渐加快，当该必需氨基酸含量超过一定水平时，其生长速率又呈现下降趋势，饲料和蛋白质利用率也降低（Hughes et al., 1983; Ravi and Devaraj, 1991; Kaushik, 1998）。本实验中，刺参幼参的生长性能和饲料利用率等指标也出现类似的变化，饲料中缬氨酸含量过低或过高均会增加机体代谢负担，抑制生长。对哺乳动物的研究发现，支链氨基酸对于脂肪代谢具有重要的调控作用，能够显著影响体内脂肪沉积（Guo and Cavener, 2007; Zhang et al., 2007; Cheng et al., 2010）。饲料缺乏或有过量的支链氨基酸均能够减少小鼠体内脂肪沉积（Nishimura et al., 2010; Chen et al., 2012a; Hasek et al., 2013）。本实验所得结果与以上研究结果相似，即饲料中缺乏缬氨酸能够抑制机体脂肪积累，而缬氨酸过量则导致脂肪含量减少，推测这是由氨基酸不平衡导致过度的能量消耗引起的。

本实验中，随着饲料中缬氨酸含量从 0.61%增加到 2.64%，刺参幼参体壁中组氨酸、缬氨酸、赖氨酸、苯丙氨酸、苏氨酸及天冬氨酸的含量显著提高，亮氨酸、异亮氨酸及酪氨酸含量显著降低，但在低添加组中大多无显著性差异，说明只有缬氨酸含量达到一定值时，上述几种氨基酸之间才会表现出显著的协同或拮抗作用。

蛋白质中各种必需氨基酸的构成比例称为氨基酸模式，即根据蛋白质中必需氨基酸的含量，以含量最低的氨基酸为 1 计算出的其他氨基酸的相应比值。尽管刺参对不同氨基酸绝对需求量的差异比较大，但就某一阶段而言，对各种氨基酸的需求量相对于某一种氨基酸（蛋氨酸）的比例具有一定规律性。对刺参而言，其氨基酸模式相对恒定，受品种、年龄等因素的影响较小；随着刺参的生长，改变的只是各种氨基酸的绝对需求量，而相互比例基本保持不变。本实验中，刺参幼参体壁中各实验组内几种必需氨基酸与蛋氨酸的比值均无显著性差异，该研究结果显示，不同氨基酸对促进蛋白质的合成应有适当的比例与浓度，氨基酸的平衡性对于提高蛋白质含量和合成效率具有重要意义，这也反映了氨基酸的吸收利用是按照一定模式进行的。目前，尚无关于刺参体壁氨基酸模式的报道，但通过其他研究中刺参体壁氨基酸含量的数据（王际英等，2015），可得出与本实验较为相近的结果。

3. 缬氨酸对刺参幼参肠道抗氧化能力及消化酶活性的影响

水生动物的抗氧化作用主要通过机体的自身免疫反应实现（Lin and Shiau, 2005）。研究发现，水生动物机体的抗氧化系统包括由超氧化物歧化酶、过氧化氢酶、谷胱甘肽过氧化物酶（GPx）等组成的酶性抗氧化系统和由谷胱甘肽、维生素C和维生素E组成的非酶性抗氧化系统（Valko et al.，2007）。本实验中，随着饲料中缬氨酸含量的增加，刺参幼参肠道中超氧化物歧化酶活性和总抗氧化能力均显著提高，但各添加组之间无显著性差异，说明饲料中添加适宜含量的缬氨酸能够提高肠道中超氧化物歧化酶活性和总抗氧化能力，从而增强刺参幼参机体清除氧自由基和抗氧化能力。刺参体内具有一定的抗氧化保护系统，但是其机体组织仍易受到氧自由基的攻击，生物膜中含有大量的多不饱和脂肪酸，极易受氧自由基攻击而形成脂质过氧化产物丙二醛，因此丙二醛含量常用来反映机体脂质发生过氧化的程度。本实验中，对照组刺参幼参肠道中丙二醛含量显著高于其他各组，当缬氨酸的添加水平达到 1.73%时，刺参幼参肠道中丙二醛含量显著降低且达到最小值，随着缬氨酸含量的进一步增加，刺参幼参肠道中丙二醛含量又显著升高。这说明饲料中添加适宜含量的缬氨酸可以降低刺参的脂肪氧化损伤，提高抗氧化能力。

谷丙转氨酶和谷草转氨酶是机体最重要的两种转氨酶，广泛分布于机体的各组织器官，通常分布于细胞膜、细胞质和线粒体中，谷丙转氨酶在肝脏中含量最高，谷草转氨酶在心脏中含量最高，肝脏次之（王震等，2016）。这两种酶常参与机体内蛋白质和氨基酸代谢，其活性代表体内氨基酸代谢强弱（Coloso et al.，1999；任和和占秀安，2006）。本实验中，随着饲料中缬氨酸含量的增加，刺参幼参肠道中谷丙转氨酶和谷草转氨酶的活性有显著提高，但各添加组之间大多差异不大，说明随着缬氨酸的添加，氨基酸代谢反应增强，但具体机制需要进一步研究。酸性磷酸酶和碱性磷酸酶可以催化磷酸单酯水解及磷酸基团转移，是生物磷代谢的重要酶类，与物质代谢关系密切（Kaushik et al.，2013；Topham，2013），它们还是动物体内重要的解毒体系，并在脂肪、葡萄糖、钙以及磷的消化、吸收和转运过程中起重要作用（Rosalki et al.，1970）。本实验中，随着饲料中缬氨酸含量的增加，刺参幼参肠道中酸性磷酸酶和碱性磷酸酶的活性均有显著提高，说明缬氨酸能够提高酸性磷酸酶和碱性磷酸酶的活性，从而促进脂肪、钙及磷等物质的吸收。

肠道对食物蛋白质水解产物的吸收、转运是营养学研究的重要内容之一。虽然目前有许多研究资料表明，动物肠道除了对蛋白质的最终水解产物氨基酸进行有效的吸收，还对次级水解产物如二肽、三肽等小肽进行有效的吸收和转移，且与氨基酸的吸收、转运存在不同的生理机制和通道，但是对氨基酸的吸收、转运机制及其影响因素的研究依然还有许多空白（叶元土等，2003）。消化酶活性能够反映刺参对不同营养成分的消化能力，刺参营养学是以刺参消化生理为基础的，因此测定刺参体内消化酶的活性及性质，对于加强刺参消化生理的研究具有重要意义（姜令绪等，2007）。本实验中，饲料中添加缬氨酸显著提高了刺参幼参肠道蛋白酶和脂肪酶的活性，且分别在 D3 和 D4 组达到最大值，说明缬氨酸能够提高蛋白酶和脂肪酶的活性，从而促进肠道发育，改善肠道抗氧化状态，提高机体的抗氧化能力和免疫能力。缬氨酸经过包膜处理，减少了在水中的溶失，延缓了在消化道中的吸收速度，促进了外源添加氨基酸与蛋白态氨基酸的同步吸收，改善了其吸收性，从而促进刺参幼参的生长。刺参幼参肠道淀粉酶活性不受饲料中缬氨酸含量的影响，各实验组差异不大，说明饲料中添加缬氨酸对淀粉代谢的影响比较小。

2.2.2.4 结论

在本实验条件下，根据实验结果可得出以下结论：①饲料中添加适宜水平的缬氨酸可以促进刺参幼参的生长，以增重率为评价指标，得出体质量为9.20g的刺参幼参对缬氨酸的最适需求量为1.79%（9.18%饲料蛋白）；②饲料中添加适宜水平的缬氨酸可以提高刺参幼参的生长性能，改善体壁的常规成分及调控氨基酸含量；③饲料中添加适宜水平的缬氨酸可以促进刺参幼参肠道发育，提高抗氧化酶和消化酶的活性，改善肠道的抗氧化状态，降低肠道的氧化损伤，从而提高机体的抗氧化能力及免疫能力。

2.2.3 刺参幼参对精氨酸最适需求量的研究

精氨酸是水生动物的必需氨基酸之一，广泛参与生物体细胞分裂、伤口复原、激素分泌及机体免疫功能的调节等生理过程，还是一氧化氮、尿素、鸟氨酸及肌丁胺的直接前体，是合成肌肉素的重要元素，且被用于合成聚胺、瓜氨酸及谷氨酰胺（Kaushik et al.，1988）。此外，精氨酸还是合成胶原蛋白的重要原料（Zimmermann and Rothenberg，2006）。已有的研究表明，精氨酸可以提高牙鲆（Alam et al.，2002）、黑棘鲷（Zhou et al.，2011a）、杂交鲈（Cheng et al.，2012）、青石斑鱼（Zhou et al.，2012a）等水生动物的生长性能及免疫能力，减少大西洋鲑（Oehme et al.，2010）对环境的应激。

刺参体壁中含有丰富的精氨酸（王永辉等，2010），被誉为"精氨酸大富翁"，精氨酸在刺参生命生理过程中起着重要的作用。目前海参精氨酸方面的研究仅见于精氨酸激酶折叠与结构（郭勤，2005），而精氨酸对刺参生长性能、免疫能力及消化酶活性的影响尚未见报道。因此，本实验通过在饲料中添加不同含量梯度的精氨酸，研究其对刺参幼参生长性能、免疫能力及消化酶活性的影响，以期为刺参对饲料中精氨酸需求的研究提供参考。

2.2.3.1 材料与方法

1. 实验设计与饲料配制

实验分为5个处理，每个处理3个重复，每个重复放养9.10g左右的刺参幼参30头。实验所用精氨酸购自江苏金维氨生物工程有限公司（USP35医药级，纯度>99.9%）。参考胡友军等（2002）、王冠（2005）的方法对精氨酸进行包被，包被后精氨酸含量为39.67%。分别添加0%、1%、2%、3%和4%的包膜精氨酸，用海泥进行调平，配制粗蛋白含量约为20%的5种实验饲料，精氨酸含量分别为0.32%、0.73%、1.16%、1.61%和1.99%，分别命名为D1、D2、D3、D4和D5组（表2-37）。固体原料经超微粉碎后过200目标准筛，按配比称重，加入新鲜鱼油及适量蒸馏水混匀，用小型颗粒饲料挤压机制成直径为0.3cm的颗粒，60℃烘干，用小型粉碎机破碎，筛选粒度为80～100目的颗粒备用。

表2-37 实验饲料配方及营养组成

项目		组别				
		D1	D2	D3	D4	D5
原料组成	海泥/%	35	34	33	32	31
	包膜精氨酸/%	0	1	2	3	4
	鱼油/%	1	1	1	1	1
	磷脂/%	1	1	1	1	1
	维生素预混料/%	2	2	2	2	2
	矿物质预混料/%	1	1	1	1	1
	其他/%	60	60	60	60	60
营养组成	粗蛋白/%	20.52	20.96	21.38	21.83	22.29
	粗脂肪/%	3.97	4.01	3.94	4.03	4.07
	粗灰分/%	43.44	42.89	41.37	40.81	40.26
	能量/（kJ/g）	10.23	10.33	10.43	10.52	10.62
	精氨酸/%	0.32	0.73	1.16	1.61	1.99

注：①维生素预混料及矿物质预混料配方参考文献（王际英等，2014）；②其他原料包括鱼粉10%、大豆浓缩蛋白10%、豆粕15%、虾粉5%、面粉10%。

2. 实验用刺参及实验管理

养殖实验在山东省海洋资源与环境研究院东营实验基地循环水养殖系统中进行，实验用刺参购自山东安源种业科技有限公司。实验开始前，将1000头体质量为9g左右的刺参幼参放养于养殖系统中，使其适应养殖环境。21d后，挑选伸展状态好、肉刺尖挺的幼参450头，随机平均置于15个循环水桶（$h_{桶}$=80cm，$h_{水}$=60cm，$d_{桶}$=65cm）中，每桶放置海参养殖筐1个，控制水流流速约为2L/min。每天定时（16:00）投喂1次，投喂量为刺参幼参初始体质量的3%，饲料投喂时加入适量水润湿，泼洒投喂，每天观察刺参幼参的摄食情况，调整次日投喂量。实验期间，每隔2d换水1次，换水时用虹吸法将残饵及粪便吸出。每隔15d更换海参养殖筐1次。养殖期间，控制水温为19～21℃，盐度为30，溶氧量大于7mg/L，氨氮浓度低于0.05mg/L。实验时间为2013年8月2日至9月30日，共60d。

3. 样品采集与计算

养殖实验结束后，停饲 48h。将刺参全部捞出计数，放置在洁净的泡沫板上，待其自然舒展后称总重，然后随机选取 15 头，测量体长和体质量后解剖，采集体壁、肠道和体腔液，测量体壁重、肠道重和肠道长，计算增重率、特定生长率、饲料系数、蛋白质效率、存活率、脏壁比、肠壁比及肠长比。体腔液用超声波细胞破碎仪破碎后过 0.45μm 滤膜，肠道用 0.75% 生理盐水冲洗干净，与体壁一起置于-20℃保存，待测。分别统计刺参幼参的特定生长率、增重率、存活率、饲料系数、蛋白质效率、脏壁比、肠长比、肠壁比。

4. 样品测定

待体壁酸水解后，用安捷伦 1200 系列高效液相色谱法（HPLC）测定氨基酸，用南京建成生物工程研究所试剂盒测定羟脯氨酸。

体腔液中超氧化物歧化酶、总抗氧化能力、一氧化氮及一氧化氮合酶均采用南京建成生物工程研究所试剂盒测定。

体腔液中碱性磷酸酶、谷丙转氨酶及谷草转氨酶采用生化分析仪（Hitachi, 7020）测定，试剂购自北京利德曼生化股份有限公司。

肠道粗酶液用吴永恒等（2012）的方法提取。蛋白酶活性采用福林酚方法测定，在 pH 为 8.0、50℃条件下，每分钟分解酪蛋白生成 1μg 酪氨酸的肠道蛋白的量为 1 个酶活性单位（U）；纤维素酶活性采用沈文英等（2004）的方法测定，在 pH 为 7.0、30℃条件下，每分钟催化纤维素生成 1mg 葡萄糖的肠道蛋白的量为 1 个酶活性单位（U）（姜令绪等，2007）；淀粉酶及脂肪酶活性采用南京建成生物工程研究所生产的试剂盒测定。

5. 数据统计

实验所得数据采用 Microsoft Excel 2007 及 SPSS 11.0 软件进行单因素方差分析，结果以平均值±标准差表示，差异显著（$P<0.05$）时用邓肯多重范围检验进行多重比较分析。

2.2.3.2 结果分析

1. 精氨酸对刺参幼参生长性能及形体指标的影响

精氨酸含量对各组刺参幼参的存活率无显著影响，但随着精氨酸含量的增加，实验组刺参幼参的增重量、增重率、特定生长率及蛋白质效率均呈现先升后降的趋势，并在 D4 组达到最大值，显著大于其他各组（$P<0.05$）（表 2-38）。D3、D4 及 D5 组刺参幼参的脏壁比、肠壁比及肠长比显著小于 D1（$P<0.05$），但 3

组之间差异不显著（$P>0.05$）。

表 2-38　精氨酸对刺参幼参生长性能及形体指标的影响（$n=15$）

项目	组别				
	D1	D2	D3	D4	D5
初始体质量/g	9.12±0.08	9.05±0.06	9.11±0.08	9.10±0.04	9.11±0.09
终末体质量/g	13.80±0.30^a	16.12±0.28^b	19.51±0.51^c	21.41±0.58^d	18.85±0.69^c
增重量/g	4.68±0.23^a	7.07±0.25^b	10.40±0.48^c	12.31±0.55^d	9.74±0.60^c
增重率/%	51.32±2.11^a	78.08±2.78^b	114.08±4.83^c	135.3±5.64^d	106.88±5.53^c
特定生长率/(%/d)	0.69±0.03^a	0.96±0.03^b	1.27±0.04^c	1.42±0.04^d	1.21±0.05^c
投喂量/g	900.00	900.00	900.00	874.00	874.00
饲料系数	6.42±0.30^d	4.24±0.16^c	2.88±0.14^b	2.38±0.12^a	3.00±0.19^b
蛋白质效率/%	78.94±3.86^a	117.25±4.25^b	168.54±7.66^c	200.95±8.95^d	155.48±8.54^e
脏壁比/%	18.43±0.93^b	17.4±0.7^{ab}	16.5±0.98^a	15.83±0.7^a	16.57±0.71^a
肠壁比/%	4.64±0.38^b	4.44±0.27^b	3.59±0.35^a	3.08±0.48^a	3.67±0.31^a
肠长比/%	4.24±0.14^b	3.96±0.19^{ab}	3.87±0.19^a	3.64±0.12^a	3.70±0.19^a
存活率/%	86.67±8.81	81.11±8.38	82.22±7.69	88.89±7.70	83.33±8.81

注：同行数据无字母或上标相同字母表示差异不显著（$P>0.05$），上标不同字母表示差异显著（$P<0.05$）。

分别以增重率和饲料系数为评价指标，经二次曲线回归拟合，体质量为 9.1g 的刺参幼参对饲料中精氨酸的最适需求量分别为 1.55%（7.10%饲料粗蛋白）和 1.54%（7.05%饲料粗蛋白）（图 2-9）。

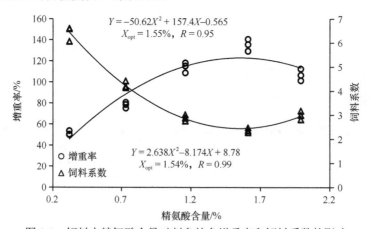

图 2-9　饲料中精氨酸含量对刺参幼参增重率和饲料系数的影响

2. 精氨酸对刺参幼参体壁基本成分及氨基酸含量的影响

随着饲料中精氨酸含量的增加，刺参幼参体壁粗蛋白含量显著升高（$P<0.05$），粗灰分含量显著降低（$P<0.05$），水分及粗脂肪的含量差异不显著（$P>0.05$）（表 2-39）。

表 2-39 精氨酸对刺参幼参体壁基本成分的影响（%）

项目	组别				
	D1	D2	D3	D4	D5
水分	91.81±0.63	91.82±0.57	91.93±0.77	92.41±0.48	92.06±0.21
粗蛋白	43.13±0.64^a	44.51±0.41^b	45.97±0.04^c	46.33±0.23^{cd}	46.84±0.28^d
粗脂肪	2.06±0.45	2.28±0.56	2.42±0.71	2.03±0.43	2.44±0.17
粗灰分	34.78±1.84^b	34.25±0.87^{ab}	33.07±1.57^a	33.06±1.54^a	33.63±0.79^{ab}

注：粗蛋白、粗脂肪、粗灰分的含量为干基含量；同行数据无字母或上标相同字母表示差异不显著（$P>0.05$），上标不同字母表示差异显著（$P<0.05$）。

随着饲料中精氨酸含量的增加，刺参幼参体壁天冬氨酸、谷氨酸、丝氨酸、苏氨酸、精氨酸、甘氨酸、亮氨酸、组氨酸、酪氨酸及羟脯氨酸的含量均有显著升高（$P<0.05$），赖氨酸含量有显著降低（$P<0.05$），但是丙氨酸、脯氨酸、缬氨酸、蛋氨酸、异亮氨酸、苯丙氨酸及半胱氨酸的含量均无显著性差异（$P>0.05$）。随着饲料中精氨酸含量的增加，刺参幼参体壁必需氨基酸及总氨基酸的含量显著升高（$P<0.05$）（表 2-40）。

表 2-40 精氨酸对刺参幼参体壁氨基酸含量的影响（%）

氨基酸	组别				
	D1	D2	D3	D4	D5
天冬氨酸	35.46±0.52^a	38.09±1.52^b	37.90±1.13^b	41.61±1.53^c	40.89±0.82^c
谷氨酸	66.60±2.89^a	66.54±0.97^a	68.46±4.97^{ab}	68.76±0.73^b	70.94±0.99^c
丝氨酸	17.18±1.91^{ab}	18.57±0.96^b	16.21±0.46^a	17.27±1.16^{ab}	18.72±0.32^b
苏氨酸	16.17±1.01^a	17.45±0.86^a	16.57±3.60^a	19.21±1.22^b	17.37±0.29^a
精氨酸	33.02±1.25^a	34.76±0.96^{ab}	35.88±1.36^{bc}	37.71±1.12^c	37.22±1.45^c
甘氨酸	41.16±2.52^a	42.24±3.46^{ab}	44.31±1.31^{ab}	46.23±1.59^b	44.16±1.41^{ab}
丙氨酸	21.09±1.00	21.48±0.72	22.16±0.94	21.69±1.21	22.73±0.78
脯氨酸	12.33±0.60	13.23±1.26	12.84±0.64	12.85±1.54	13.56±0.31
缬氨酸	18.05±0.72	19.43±1.59	19.23±1.94	20.25±1.11	20.76±2.23
蛋氨酸	6.60±0.08	6.96±0.91	6.49±0.44	6.09±0.96	6.49±0.93
异亮氨酸	12.78±0.29	14.16±1.00	14.28±0.41	14.24±1.24	14.63±0.43
亮氨酸	19.29±0.21^a	20.34±0.45^{ab}	21.15±0.70^b	20.21±1.43^{ab}	21.19±1.06^b
苯丙氨酸	12.70±0.15	13.85±0.93	13.87±0.22	14.26±1.85	14.27±0.52
半胱氨酸	4.06±0.32	3.93±0.72	3.97±1.09	5.08±0.85	3.98±0.51
赖氨酸	18.76±0.62^c	18.31±0.43^{bc}	17.72±0.76^b	15.54±1.14^{cd}	16.43±0.99^a
组氨酸	4.12±0.52^a	5.24±0.76^{ab}	5.09±0.78^{ab}	6.56±1.14^b	5.99±0.55^b
酪氨酸	10.78±0.32^a	10.93±0.81^a	10.72±1.57^a	11.33±0.95^{ab}	13.02±0.81^b
羟脯氨酸	25.64±1.75^a	30.27±1.85^b	36.61±1.94^c	39.48±1.27^c	40.55±2.04^c
必需氨基酸	141.49±3.29^a	149.55±3.42^b	149.28±1.50^b	151.41±5.20^c	153.35±3.27^c
总氨基酸	375.79±10.09^a	394.84±5.32^b	402.48±4.95^b	415.73±8.42^c	421.89±6.94^c

注：同行数据无字母或上标相同字母表示差异不显著（$P>0.05$），上标不同字母表示差异显著（$P<0.05$）。

3. 精氨酸对刺参幼参体腔液抗氧化能力、精氨酸代谢产物及代谢酶活力的影响

随着饲料中精氨酸含量的增加，体腔液中总抗氧化能力、一氧化氮含量和超氧化物歧化酶、一氧化氮合成酶、碱性磷酸酶、谷丙转氨酶及谷草转氨酶的活性均有显著升高（$P<0.05$）（表 2-41）。其中，D3、D4、D5 组超氧化物歧化酶活性显著高于 D1 组（$P<0.05$），但与 D2 组无显著性差异（$P>0.05$）；D4、D5 组总抗氧化能力显著高于其他 3 组，D2、D3 组显著高于 D1 组（$P<0.05$）；一氧化氮合成酶活性及一氧化氮含量随着饲料中精氨酸含量的增加而逐渐升高（$P<0.05$），在 D5 组达到最大值；碱性磷酸酶活性随着饲料中精氨酸含量的增加呈现先上升后平稳的趋势，D2、D3、D4、D5 组活性显著高于 D1 组；D1 组谷丙转氨酶及谷草转氨酶的活性显著低于 D4、D5 组（$P<0.05$），但与 D2、D3 组之间无显著性差异（$P<0.05$）。

表 2-41 精氨酸对刺参幼参体腔液抗氧化能力、精氨酸代谢产物及代谢酶活力的影响

项目	饲料精氨酸水平/%				
	D1	D2	D3	D4	D5
超氧化物歧化酶/（U/ml）	62.82±2.84^a	65.61±3.79^{ab}	69.01±2.57^b	69.63±1.91^b	69.94±3.65^b
总抗氧化能力/（U/ml）	6.66±0.49^a	13.44±1.27^b	14.18±1.35^b	17.51±1.20^c	17.64±1.43^c
一氧化氮含量/（μmol/ml）	5.95±0.08^a	6.57±0.16^b	7.75±0.60^c	9.47±0.32^d	10.95±0.22^e
一氧化氮合成酶/（U/ml）	21.59±4.82^a	36.93±2.18^b	54.55±3.21^c	69.31±8.04^d	95.55±4.27^e
碱性磷酸酶/（U/ml）	2.34±0.58^a	4.33±0.58^b	5.67±1.15^{bc}	6.67±1.15^c	6.33±1.15^c
谷丙转氨酶/（U/ml）	4.20±0.36^a	4.63±0.32^{ab}	4.77±0.31^{ab}	5.13±0.64^b	5.37±0.47^b
谷草转氨酶/（U/ml）	4.43±0.58^a	5.10±0.62^{ab}	5.47±0.57^{ab}	6.23±0.76^b	5.90±0.32^b

注：同行数据无字母或上标相同字母表示差异不显著（$P>0.05$），上标不同字母表示差异显著（$P<0.05$）。

4. 精氨酸对刺参幼参肠道消化酶活性的影响

随着饲料中精氨酸含量增加，刺参幼参肠道蛋白酶活性显著降低（$P<0.05$），纤维素酶活性显著升高（$P<0.05$），而淀粉酶及脂肪酶的活性不受饲料中精氨酸含量的影响（$P>0.05$）（表 2-42）。刺参幼参肠道蛋白酶活性与饲料中精氨酸含量呈显著的二次负相关关系（$Y_{酶活}=-3.228X^2_{精氨酸}+2.865X_{精氨酸}+26.24$，$R^2=0.934$）；D3、D4 和 D5 组纤维素酶活性显著高于 D1 和 D2 组（$P<0.05$），D1、D2 组之间差异不显著（$P>0.05$）。

表 2-42 精氨酸对刺参幼参肠道消化酶活性的影响

消化酶	组别				
	D1	D2	D3	D4	D5
蛋白酶/（U/mg prot）	26.36±1.85^b	27.89±2.51^b	24.19±2.49^b	22.63±1.84^b	19.27±2.95^a
淀粉酶/（U/mg prot）	6.63±0.26	6.83±0.38	6.76±0.41	7.17±0.55	6.98±0.36
脂肪酶/（U/mg prot）	0.74±0.10	0.86±0.09	0.78±0.05	0.87±0.13	0.81±0.08
纤维素酶/（U/mg prot）	0.75±0.57^a	1.29±0.42^a	1.58±0.29^b	1.85±0.44^b	1.96±0.73^b

注：同行数据无字母或上标相同字母表示差异不显著（$P>0.05$），上标不同字母表示差异显著（$P<0.05$）。

2.2.3.3 讨论

1. 精氨酸对刺参幼参生长性能及体成分的影响

关于精氨酸促进动物生长研究的报道已有很多，精氨酸可从激素、细胞及分子水平上影响动物生长及组织蛋白的合成（Luzzana et al., 1998；Kwak et al., 1999）。本实验中，低水平精氨酸组的刺参幼参没有出现生长抑制或病理现象，随着饲料中精氨酸含量的增加，刺参幼参的生长性能得到了显著提高，这与对一些鱼类的研究类似（Alam et al., 2002；周凡，2011；周恒永，2011），但随着饲料中精氨酸含量的进一步增加，不同的水生动物表现出不同的生长响应，中华绒螯蟹（叶金云等，2010）、大菱鲆（魏玉婷，2010）、军曹鱼（赵红霞等，2007）等表现为生长性能降低，而银大麻哈鱼（Luzzana et al., 1998）、斑点叉尾鮰（Buentello and Gatlin Ⅲ，2000）、红鼓鱼（Cheng et al., 2011）等表现为维持生长性能。这主要是由 3 种原因导致：①饲料中精氨酸水平，目前针对精氨酸需求的研究多为单因素梯度实验，梯度的大小可能影响实验动物的生长表征，从氨基酸平衡角度分析，饲料中过高的精氨酸含量会导致动物对氨基酸利用的紊乱（周凡，2011；Berge et al., 1999；Iaccarino et al., 2009），由此导致动物生长受阻；②养殖环境，实验所采用的基础饲料（Fournier et al., 2003）、养殖条件、养殖动物不同生长阶段（Griffin et al., 1994；姚永峰等，2014）、养殖管理等也是造成研究结果存在差异的原因；③养殖动物的生理功能，某些鱼类体内存在与哺乳动物类似的精氨酸合成途径和机制（Buentello and Gatlin Ⅲ，2000），可以将瓜氨酸和鸟氨酸转化为精氨酸，而刺参是比鱼类低等的生物，其体内很可能不存在精氨酸合成体系，因此其机体对饲料中精氨酸含量的反应更为明显（万军利等，2006），这有待进一步研究。

以增重率或饲料系数为评价指标，刺参幼参对精氨酸的最适需求量分别为 1.55%和 1.54%，相当于分别占饲料粗蛋白的 7.10%和 7.05%。该数值（占饲料粗蛋白的比例）远高于鱼类对精氨酸的需求（Luo et al., 2007；Murillo-Gurrea et al., 2001）。刺参对饲料粗蛋白的需求量远低于一般海水鱼类（朱伟等，2005；李素红等，2012），而其体壁 10 种必需氨基酸中精氨酸含量最高，刺参幼参快速生长需要补充大量精氨酸，因此其对精氨酸的需求量远高于海水鱼类。在营养需求研究中，通常采用生长速率指标或营养物质利用率指标评价其最适需求量，但后一类指标在很大程度上依赖于对养殖动物的管理（Mai et al., 1995）。在刺参养殖实验中，摄食量是非常难以控制的，因此对其饲料系数的计算只能以投喂量为依据，由此导致饲料转化率高于实际数值。本实验中，以饲料系数为评价指标经二次回归得出的精氨酸最适需求量虽然与以增重率为评价指标得出的数值较为接近，但只能作为参考。

饲料中精氨酸含量增加，提高了刺参幼参体壁粗蛋白的沉积，这与对中华绒鳌蟹（叶金云等，2010）和海水鱼（魏玉婷，2010；赵红霞等，2007；Buentello and Gatlin Ⅲ，2000；Cheng et al.，2011）的研究一致。随着精氨酸含量的增加，饲料氨基酸模式得到改善，从而提高了养殖动物对饲料粗蛋白的利用。随着精氨酸含量的继续增加，氨基酸模式遭到破坏，尤其是精氨酸与赖氨酸之间发生了拮抗作用（Kaushik et al.，1988），降低了体壁粗蛋白的沉积。随着饲料中精氨酸含量的增加，刺参幼参体壁必需氨基酸中精氨酸含量增加最显著，各实验组苏氨酸、亮氨酸、赖氨酸、组氨酸的含量虽然显著高于对照组，但各实验组之间无显著性差异。此外，赖氨酸含量只有高精氨酸组才有显著降低，说明只有在某种氨基酸含量变化明显时，氨基酸之间的协同或拮抗作用才比较明显，由此推测水生动物最优氨基酸模式可能为一个范围。饲料中精氨酸含量的增加，显著增加了体壁羟脯氨酸和必需氨基酸的含量，说明体壁中胶原蛋白的含量有了显著提高。胶原蛋白和必需氨基酸含量的增加，提高了刺参的可食性和营养价值。

2. 精氨酸对刺参幼参体腔液酶活性的影响

刺参体内免疫反应主要分为细胞免疫和体液免疫，二者之间密切相关，相辅相成（常杰等，2011），刺参体腔液中主要免疫或抗氧化酶活性可以反映刺参机体的生理状况。超氧化物歧化酶和总抗氧化能力是机体 2 个重要的抗氧化指标，在清除氧自由基、防止生物分子损伤方面具有重要作用。本实验中，随着饲料中精氨酸含量的增加，刺参幼参体腔液中超氧化物歧化酶活性和总抗氧化能力有显著提高，但各添加组之间差异不大，说明饲料中添加适量的精氨酸能提高刺参幼参机体抗氧化能力，这与对团头鲂（Ren et al.，2013）和凡纳滨对虾（Zhou et al.，2012b）的研究一致。本实验中，一氧化氮合成酶及一氧化氮含量随着饲料中精氨酸含量的升高而显著升高，一氧化氮合成酶是刺参敏感免疫学指标（常杰，2010），而一氧化氮的生成依赖于一氧化氮合成酶，一氧化氮是软体动物体内重要的免疫调节因子和神经信使分子（王晓安等，2003）。对杂色鲍的研究表明，一氧化氮含量与超氧化物歧化酶、一氧化氮合成酶、碱性磷酸酶的活性等免疫指标存在良好的相关性（王广军等，2008）。因此，可以推测转氨酶是催化氨基酸与酮酸之间氨基转移的一类酶，普遍存在于动物心肌、脑、肝、肾等组织中，谷丙转氨酶和谷草转氨酶通常被用来评价动物肝脏细胞受损的状况。对哺乳动物和鱼类的研究表明，肝脏是重要营养物质合成代谢器官。刺参作为一种低营养等级的沉食性底栖生物，不具备肝脏等功能性器官（臧元奇，2012；于东祥等，2010），因此其营养物质的合成代谢与鱼类存在较大差异。本实验中，刺参幼参体腔液中谷丙转氨酶和谷草转氨酶的活性随着饲料中精氨酸的含量而显著上升，说明体腔液中氨基酸代谢活性有了显著升高，由此推测体腔液可能参与了刺参幼参营养物质代谢，但具体途径有待进一步研究。

3. 精氨酸对刺参幼参肠道消化酶活性的影响

目前，关于刺参消化酶的报道较多，研究结果不尽相同，报道的蛋白酶活性差异也很大。例如，付雪艳（2004）认为前肠具有较高的酸性蛋白酶及丝氨酸蛋白酶活性，后肠碱性蛋白酶活性最高；而姜令绪等（2007）的研究表明，在偏碱性范围内，中肠蛋白酶活性最高。造成这种结果的原因除了对消化道的分节存在差异，还与实验刺参的不同生长阶段（周玮等，2010）及测定条件（姜令绪等，2007）密切相关。本实验中，实验刺参终末体质量较小，为 13~20g，肠道质量为 0.4~1.5g，前中后肠分节不明显，故取整个肠道进行消化酶的测定。

本实验中，饲料中添加精氨酸降低了刺参幼参肠道蛋白酶活性，且蛋白酶活性与饲料中精氨酸含量呈显著二次负相关关系（$Y_{酶活}=-3.228X^2_{精氨酸}+2.865X_{精氨酸}+26.24$，$R^2=0.934$）。目前，关于精氨酸含量与机体蛋白酶活性之间的关系尚未见报道，但付雪艳（2004）、李淑霞（2010）的研究证明，刺参肠道蛋白酶在 pH 为 8.0 的条件下，活性最大的可能为丝氨酸蛋白酶中的胶原酶。胶原酶作用于蛋白质中 L-精氨酸、L-赖氨酸、甘氨酸和 L-瓜氨酸残基。饲料中精氨酸含量增加相当于增加了酶产物浓度，从而抑制了酶活性。此外，精氨酸含量增加可能破坏了饲料的氨基酸平衡，从而导致蛋白质消化利用率降低，这与王吉桥等（2009）在饲料中添加赖氨酸对刺参消化酶活性的影响研究一致。刺参肠道内纤维素酶活性较低且周年变化不明显，受肠道中微生物数量和种类的影响较大（王吉桥等，2007b）。本实验中，刺参幼参肠道纤维素酶活性上升的原因可能是精氨酸含量增加影响了刺参幼参的肠道微生物，从而使纤维素酶活性发生了改变。本实验中，刺参幼参肠道中淀粉酶和脂肪酶的活性不受饲料中精氨酸含量的影响，这与王吉桥等（2009）对刺参的研究类似，与对罗氏沼虾（董云伟等，2001）、大菱鲆（陈超和陈京华，2012）、宝石鲈（邵庆均等，2004）等的研究也类似，说明添加晶体氨基酸对脂肪及淀粉的代谢影响甚微。

2.2.4 饲料中花生四烯酸含量对刺参幼参生长性能、抗氧化能力及脂肪酸代谢的影响

近年来，n-6 高不饱和脂肪酸，尤其是花生四烯酸，在海洋动物中的作用得到了越来越多的关注。已经证实，花生四烯酸是多种生物活性化合物的前体，统称为类二十烷酸，主要包括前列腺素、血栓素和白三烯（Johnson et al.，1983）。这些生物活性物质在动物体内具有重要的作用，能够调节一系列重要的生理代谢过程。同时，花生四烯酸不仅对海水动物的生长、存活、抗应激及免疫等起到重要的调节功能（Atalah et al.，2011；王成强等，2016；Tian et al.，2014），还对海水动物的脂质代谢及性类固醇合成具有一定的调控作用（Luo et al.，2012；张圆琴

等，2017）。朱伟等（2005）的研究表明，刺参稚参饲料中粗脂肪的适宜添加量为3%～5%；廖明玲（2014）通过养殖实验确定，饲喂 n-3 高不饱和脂肪酸水平为0.22%～0.38%的饲料能够使刺参具有最佳的生长性能和更高的营养价值；左然涛等（2017）报道，饲料中添加 0.60%的二十二碳六烯酸时，刺参成参的生长速率最快。而花生四烯酸作为一种海水鱼类的必需脂肪酸，在其他海水动物生长中也有一定的生理作用，但其在海参生长中的作用尚未见相关报道。因此，本实验通过研究花生四烯酸对刺参幼参生长、抗氧化及脂肪酸代谢的影响，旨在探究花生四烯酸在刺参饲料中的适宜添加水平，也初步阐述花生四烯酸对刺参脂肪酸代谢的影响，不仅可以更好地完善刺参营养学参数数据库，还可以为研究刺参脂肪酸代谢提供一定的理论依据（王成强等，2018）。

2.2.4.1 材料与方法

1. 实验饲料

基础饲料以鱼粉和发酵豆粕为主要蛋白源，以小麦粉为主要糖源。通过在基础饲料中分别添加 0、0.50%、1.00%、1.50%、2.00%和 2.50%的花生四烯酸-纯化油（花生四烯酸含量约占总脂肪酸的 48.70%），用硬脂酸甘油三酯进行调平，制成 6 组等氮等脂的饲料，分别命名为 D1（对照组）、D2、D3、D4、D5 和 D6 组（表 2-43）。经过气相色谱分析，6 组饲料中花生四烯酸的含量分别为 0.02%、0.17%、0.36%、0.51%、0.59%和 0.98%（饲料干重）。实验饲料中脂肪酸组成见表 2-44。

表 2-43 实验饲料配方及营养组成（%）

项目		组别					
		D1	D2	D3	D4	D5	D6
原料组成	鱼粉	8.00	8.00	8.00	8.00	8.00	8.00
	发酵豆粕	17.00	17.00	17.00	17.00	17.00	17.00
	小麦粉	14.00	14.00	14.00	14.00	14.00	14.00
	海藻粉	20.00	20.00	20.00	20.00	20.00	20.00
	海泥	35.50	35.50	35.50	35.50	35.50	35.50
	预混料	2.00	2.00	2.00	2.00	2.00	2.00
	大豆卵磷脂	1.00	1.00	1.00	1.00	1.00	1.00
	花生四烯酸-纯化油	0.00	0.50	1.00	1.50	2.00	2.50
	硬脂酸甘油三酯	2.50	2.00	1.50	1.00	0.50	0
营养组成	粗蛋白	19.90	20.18	19.70	19.94	19.87	19.69
	粗脂肪	4.93	4.95	4.97	4.98	5.01	5.02

续表

项目		组别					
		D1	D2	D3	D4	D5	D6
营养组成	粗灰分	45.92	45.90	45.71	45.60	45.63	45.38
	花生四烯酸	0.02	0.17	0.36	0.51	0.59	0.98

注：①预混料包括1%矿物质预混料和1%维生素预混料，均购自山东升索渔用饲料研究中心，其中每千克矿物质预混料含锌35.0mg、锰21.0mg、铜8.3mg、铁23.0mg、钴1.2mg、碘1.0mg、硒0.3mg，每千克维生素预混料含维生素A 7500.0IU、维生素D 1500.0IU、维生素E 60.0mg、维生素K_3 18.0mg、维生素B_1 12.0mg、维生素B_2 12.0mg、维生素B_{12} 0.1mg、泛酸48.0mg、烟酰胺90.0mg、叶酸3.7mg、D-生物素0.2mg、肌醇60.0mg、维生素C 310.0mg；②花生四烯酸-纯化油：花生四烯酸占总脂肪酸的比例为48.70%，购自嘉必优生物技术（武汉）股份有限公司。

表2-44 实验饲料主要脂肪酸组成（%）

脂肪酸种类	花生四烯酸-纯化油	组别					
		D1	D2	D3	D4	D5	D6
C18:2n-6	6.80	12.69	14.18	15.98	17.27	16.28	20.17
C20:4n-6	48.70	0.36	4.07	8.69	12.98	17.12	24.33
C18:3n-3	—	1.35	1.37	1.52	1.51	1.39	1.68
C20:5n-3	—	0.99	1.01	1.04	1.11	1.01	1.20
C22:5n-3	—	0.26	0.28	0.32	0.42	0.34	0.50
C22:6n-3	—	0.97	1.01	1.03	1.03	0.95	1.12

首先将所有原料粉碎过筛，之后将各种原料均匀混合，再将花生四烯酸-纯化油与混合好的原料充分混匀，然后加入适量的水制成面团，再用自动制粒机进行制粒，将制好的饲料放置于45℃左右的烘箱中烘干，之后保存于阴凉干燥处，备用。

2. 养殖管理

实验地点是山东省海洋资源与环境研究院东营实验基地，养殖周期为56d，养殖方式为循环水养殖，实验参为该基地当年繁育的同一批参苗。首先将所有实验参放置于养殖桶中，用基础饲料暂养15d，使其适应实验饲料与养殖条件。暂养结束后，挑选体色健康、大小均一的刺参幼参随机分到18个养殖桶（$h_桶$=80cm，$h_水$=50cm，$d_桶$=60cm）中，刺参幼参初始体质量为（10.78±0.06）g，每个养殖桶放置40头刺参，每个实验组3个重复，每个桶内放置2个布满波纹板的海参专用筐。每天在规定时间（7:00）投喂，投喂量为刺参体质量的2%，具体投喂量根据刺参的摄食情况进行调整。实验期间，水温控制在17~19℃，盐度为24~26，溶氧量大于7mg/L，氨氮和亚硝酸氮浓度均低于0.05mg/L，每2天换水一次，同时进行吸污。

3. 样品采集与分析

56d 的养殖实验结束时，将实验参禁食 48h，然后对每个养殖桶内的刺参进行计数和称重。之后，从每个养殖桶中随机取出 10 头刺参，置于冰盘上进行解剖取样，分别取体腔液、肠道及体壁，将肠道和体壁样品及时保存于冰箱（-80℃）中，同时对体腔液进行离心（3000r/min，4℃，10min），取上清液分装在离心管中，之后也将其保存于冰箱（-80℃）中。

4. 样品常规及生化分析

饲料和刺参体壁脂肪酸含量测定方法参考气相色谱法（Mourente et al.，1999），并稍作修改。取 100mg 左右冷冻干燥后磨碎的样品，置于 15ml 的顶空进样玻璃瓶中，加入 1 mol/L KOH-甲醇溶液 3ml，放在 75℃水浴中加热 20min，冷却至室温后，加入 2 mol/L HCl-甲醇溶液 3ml，放在 75℃水浴中加热 20min，冷却之后加入 1.5ml 正己烷（色谱级），振荡萃取，静置分层。小心吸取上层正己烷和脂肪酸甲酯的混合物，用微量进样器吸取 1μl 注入气象色谱仪（HP5890Π，美国）中，采用火焰电离检测器进行检测。最后，根据标准脂肪酸出峰时间确定样品中脂肪酸种类，通过峰面积归一法进行测定。

肠道中总抗氧化能力、丙二醛含量和过氧化氢酶、超氧化物歧化酶、脂肪酸合成酶、肉毒碱棕榈酰转移酶-1 及乙酰辅酶 A 羧化酶的活性均利用南京建成生物工程研究所生产的相应试剂盒测得。

5. 计算公式及统计分析

统计刺参幼参的特定生长率、增重率、存活率和饲料效率。

2.2.4.2 结果分析

1. 花生四烯酸对刺参幼参生长性能的影响

表 2-45 显示了花生四烯酸对刺参幼参生长性能的影响。实验结果表明，随着饲料中花生四烯酸含量增加，刺参幼参的存活率无显著性差异（$P>0.05$），为 87.50%~94.17%；当饲料中花生四烯酸含量增加时，刺参幼参增重率呈现先上升后降低的趋势，D3、D4 组刺参幼参增重率显著高于其他组（$P<0.05$），且这两组间无显著性差异（$P>0.05$）。刺参幼参的特定生长率与增重率具有相同的变化趋势。当花生四烯酸含量从 0.02%增加到 0.51%时，刺参幼参的饲料效率从 57.93%升高到 111.64%，在 D4 组出现最高值，且显著高于其他组（$P<0.05$），随着花生四烯酸含量的进一步增加，饲料效率呈现显著下降的趋势（$P<0.05$）。通过对刺参幼参的增重率和饲料中花生四烯酸含量进行二次回归分析（图 2-10），得到二次

曲线方程 $Y=-71.831X^2+73.387X+30.194$ ($R^2=0.6968$)，刺参幼参增重率最高时对应的花生四烯酸含量为 0.51%。

表 2-45 花生四烯酸对刺参幼参生长性能的影响（$n=3$）

指标	组别					
	D1	D2	D3	D4	D5	D6
初始体质量/g	10.78±0.07	10.78±0.13	10.79±0.08	10.79±0.09	10.81±0.05	10.79±0.05
终末体质量/g	14.03±0.07^c	15.24±0.21^b	16.11±0.17^a	16.76±0.20^a	15.06±0.17^b	14.50±0.16^{bc}
存活率/%	89.17±2.20	89.17±0.83	94.17±3.63	93.33±1.67	87.50±1.44	90.83±1.44
增重率/%	30.16±1.45^c	41.37±0.66^b	49.40±2.42^a	55.39±2.15^a	39.39±1.55^b	34.37±0.98^{bc}
特定生长率/(%/d)	0.47±0.02^c	0.62±0.01^b	0.72±0.03^a	0.79±0.02^a	0.59±0.02^b	0.53±0.02^{bc}
饲料效率/%	57.93±2.56^d	79.58±2.42^c	100.48±7.03^b	111.64±5.67^a	74.42±2.08^c	67.39±2.33^d

注：同行数据无字母或上标相同字母表示差异不显著（$P>0.05$），上标不同字母表示差异显著（$P<0.05$）。

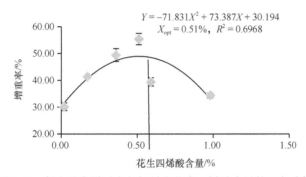

图 2-10 刺参幼参增重率与饲料中花生四烯酸含量的回归分析

2. 花生四烯酸对刺参幼参体壁化学组成的影响

由表 2-46 可见，花生四烯酸对刺参幼参体壁的水分、粗蛋白以及粗灰分的含量均无显著性影响（$P>0.05$），其中粗蛋白含量为 46.15%~46.89%（干重），粗灰分含量为 31.06%~32.46%；随着饲料中花生四烯酸含量的增加，刺参体壁中粗脂肪含量呈现先降低后升高的趋势，D4 组含量达到最低值（含量为 3.71%），且显著低于 D1、D2 和 D6 组（$P<0.05$），与 D3 和 D5 组无显著性差异（$P>0.05$）。

表 2-46 花生四烯酸对刺参幼参体壁化学组成的影响（%）

项目	组别					
	D1	D2	D3	D4	D5	D6
水分	90.69±0.17	90.94±0.06	91.02±0.10	91.07±0.23	90.88±0.21	90.58±0.40
粗蛋白	46.67±0.52	46.83±0.62	46.89±0.68	46.65±0.63	46.15±0.92	46.59±1.24
粗脂肪	4.66±0.10^{ab}	4.61±0.15^{ab}	3.99±0.15^{bc}	3.71±0.14^c	4.29±0.26^{abc}	4.88±0.11^a
粗灰分	31.37±0.67	31.77±0.30	32.46±0.22	32.34±0.41	32.04±0.48	31.06±0.73

注：同行数据无字母或上标相同字母表示差异不显著（$P>0.05$），上标不同字母表示差异显著（$P<0.05$）。

3. 花生四烯酸对刺参幼参体壁脂肪酸组成的影响

表 2-47 显示了花生四烯酸对刺参幼参体壁脂肪酸组成的影响。实验结果表明，随着饲料中花生四烯酸含量的增加，刺参幼参体壁中花生四烯酸和 n-6 多不饱和脂肪酸的含量呈现显著上升趋势，而二十碳五烯酸、二十二碳六烯酸和 n-3 多不饱和脂肪酸的含量显著降低（$P<0.05$）。同时，当饲料中花生四烯酸含量增加时，刺参幼参体壁中多不饱和脂肪酸含量从 35.08% 显著升高至 38.00%，而 $\sum$n-3 多不饱和脂肪酸/$\sum$n-6 多不饱和脂肪酸却从 0.48 显著减小至 0.31（$P<0.05$）。

表 2-47　花生四烯酸对刺参幼参体壁脂肪酸组成的影响

脂肪酸组成及比值	组别					
	D1	D2	D3	D4	D5	D6
C14:0/%	1.10±0.04	1.06±0.03	1.09±0.03	1.08±0.08	1.06±0.08	0.95±0.01
C16:0/%	7.11±0.51	6.19±0.12	6.52±0.21	6.08±0.10	6.45±0.34	6.86±0.49
C18:0/%	4.19±0.12	4.20±0.04	4.30±0.14	4.44±0.14	4.28±0.03	4.52±0.15
C20:0/%	1.30±0.05	1.35±0.04	1.32±0.03	1.39±0.02	1.32±0.02	1.40±0.01
$\sum$饱和脂肪酸/%	13.70±0.63	12.80±0.19	13.23±0.27	13.00±0.16	13.11±0.38	13.73±0.66
C16:1n-7/%	3.17±0.06^c	3.09±0.04^c	3.33±0.02^{bc}	3.49±0.11^{ab}	3.31±0.02^{bc}	3.69±0.10^a
C18:1n-9/%	6.43±0.21^a	5.31±0.26^b	5.51±0.22^{ab}	4.89±0.16^b	5.50±0.24^{ab}	4.98±0.12^b
C18:1n-7/%	4.45±0.21^a	3.95±0.04^{ab}	3.88±0.11^{ab}	3.81±0.06^b	3.82±0.10^b	3.91±0.18^{ab}
C20:1n-7/%	3.04±0.14^a	2.89±0.04^{ab}	2.79±0.06^{ab}	2.56±0.15^b	2.75±0.07^{ab}	2.68±0.08^{ab}
$\sum$单不饱和脂肪酸/%	17.09±0.41^a	15.25±0.28^b	15.50±0.06^b	14.74±0.33^b	15.38±0.29^b	15.26±0.24^b
C18:2n-6/%	8.23±0.37^a	6.84±0.35^{ab}	5.73±0.26^{bc}	5.10±0.19^{cd}	5.47±0.40^{bcd}	4.07±0.19^d
C18:3n-6/%	3.50±0.17	3.49±0.05	3.34±0.04	3.69±0.27	3.16±0.14	3.24±0.04
C20:4n-6/%	12.00±0.26^e	17.60±0.15^d	19.51±0.31^c	20.05±0.45^{bc}	21.05±0.35^{ab}	21.70±0.25^a
$\sum$n-6 多不饱和脂肪酸/%	23.73±0.22^b	27.92±0.29^a	28.58±0.55^a	28.84±0.38^a	29.68±0.55^a	29.01±0.46^a
C18:3n-3/%	1.43±0.06^a	1.28±0.02^{ab}	1.22±0.01^b	1.22±0.01^b	1.15±0.02^b	1.21±0.05^b
C20:5n-3/%	4.61±0.16^a	3.43±0.10^b	3.42±0.12^b	3.44±0.13^b	3.10±0.03^b	3.23±0.13^b
C22:5n-3/%	2.31±0.11	2.43±0.02	2.32±0.04	2.40±0.06	2.15±0.06	2.18±0.09
C22:6n-3/%	3.00±0.02^a	2.57±0.04^b	2.55±0.07^b	2.53±0.10^b	2.17±0.08^c	2.37±0.03^{bc}
$\sum$n-3 多不饱和脂肪酸/%	11.35±0.20^a	9.72±0.18^b	9.51±0.19^b	9.59±0.20^b	8.57±0.17^c	8.99±0.12^{bc}
$\sum$多不饱和脂肪酸/%	35.08±0.40^b	37.64±0.12^a	38.09±0.56^a	38.43±0.18^a	38.26±0.71^a	38.00±0.39^a
$\sum$饱和脂肪酸/$\sum$多不饱和脂肪酸	0.39±0.02	0.34±0.01	0.35±0.01	0.34±0.01	0.34±0.02	0.36±0.02
$\sum$n-3 多不饱和脂肪酸/$\sum$n-6 多不饱和脂肪酸	0.48±0.01^a	0.35±0.01^b	0.33±0.01^b	0.33±0.01^b	0.29±0.01^c	0.31±0.01^{bc}

注：同行数据无字母或上标相同字母表示差异不显著（$P>0.05$），上标不同字母表示差异显著（$P<0.05$）。

另外，随着饲料中花生四烯酸含量的增加，刺参幼参体壁中单不饱和脂肪酸和 C18:2n-6 的含量呈现显著降低趋势（$P<0.05$）。当饲料中花生四烯酸的含量从 0.02% 增加至 0.98% 时，刺参幼参体壁中单不饱和脂肪酸和 C18:2n-6 的含量分别从 17.09% 降至 15.26% 和从 8.23% 降至 4.07%（$P<0.05$）。而不同实验组间刺参幼参体壁中饱和脂肪酸（C14:0、C16:0、C18:0 和 C20:0）的含量并无显著性差异（$P>0.05$）。

4. 花生四烯酸对刺参幼参肠道抗氧化能力的影响

表 2-48 显示，随着饲料中花生四烯酸含量的增加，刺参幼参肠道中超氧化物歧化酶活性呈现先升高后降低的趋势，D4 组达到最高值，且显著高于 D1、D2 和 D6 组（$P<0.05$），同 D3 和 D5 组无显著性差异（$P>0.05$）；刺参幼参肠道总抗氧化能力和过氧化氢酶活性呈现同超氧化物歧化酶相似的变化趋势；同时，当饲料中花生四烯酸含量从 0.02% 增加到 0.51% 时，刺参幼参肠道中丙二醛含量无显著性变化（$P>0.05$），而当花生四烯酸含量进一步增加至 0.98% 时，肠道中丙二醛含量呈现显著上升趋势，且在 D6 组达到最高值，显著高于 D1、D2、D3 和 D4 组（$P<0.05$），同 D5 组无显著性差异（$P>0.05$）。

表 2-48 花生四烯酸对刺参幼参肠道抗氧化能力的影响

指标	组别					
	D1	D2	D3	D4	D5	D6
超氧化物歧化酶活性/（U/mg prot）	46.64 ± 0.56^c	50.46 ± 1.55^{bc}	55.18 ± 0.70^a	55.23 ± 1.17^a	51.55 ± 0.93^{ab}	49.44 ± 0.47^{bc}
丙二醛含量/（nmol/mg prot）	3.28 ± 0.07^b	3.49 ± 0.09^b	3.37 ± 0.09^b	3.22 ± 0.07^b	5.43 ± 0.16^a	5.56 ± 0.23^a
总抗氧化能力/（U/mg prot）	1.14 ± 0.04^d	1.34 ± 0.08^{cd}	1.77 ± 0.04^{ab}	1.86 ± 0.05^a	1.58 ± 0.09^{abc}	1.52 ± 0.06^{bc}
过氧化氢酶活性/（U/mg prot）	89.34 ± 1.91^{cd}	98.35 ± 3.35^{bcd}	109.52 ± 3.56^{ab}	115.60 ± 4.75^a	103.16 ± 3.76^{abc}	85.94 ± 3.71^d

注：同行数据无字母或上标相同字母表示差异不显著（$P>0.05$），上标不同字母表示差异显著（$P<0.05$）。

5. 花生四烯酸对刺参幼参肠道脂肪酸合成酶、乙酰辅酶 A 羧化酶和肉毒碱棕榈酰转移酶-1 活性的影响

由表 2-49 可见，当饲料中花生四烯酸含量从 0.02% 增加到 0.59% 时，刺参幼参肠道中脂肪酸合成酶活性无显著性差异（$P>0.05$），而当饲料中花生四烯酸含量进一步增加至 0.98% 时，脂肪酸合成酶活性显著下降，显著低于其他各处理组（$P<0.05$）；刺参幼参肠道中乙酰辅酶 A 羧化酶活性呈现同脂肪酸合成酶相似的变化趋势；另外，刺参幼参肠道中肉毒碱棕榈酰转移酶-1 活性随饲料中花生四烯酸含量直接呈现先升高后降低的趋势，D3、D4 和 D5 组肉毒碱棕榈酰转移酶-1 活性显著高于 D1、D2 和 D6 组（$P<0.05$），且这三组间无显著性差异，同时 D1、D2 和 D6 组的肉毒碱棕榈酰转移酶-1 活性也无显著性差异（$P>0.05$）。

表 2-49　花生四烯酸对刺参幼参肠道脂肪酸合成酶、乙酰辅酶 A 羧化酶和肉毒碱棕榈酰转移酶-1 酶活性的影响　　　　（单位：ng/ml）

指标	组别					
	D1	D2	D3	D4	D5	D6
脂肪酸合成酶	4.32±0.05[a]	4.38±0.05[a]	4.64±0.11[a]	4.31±0.09[a]	4.53±0.08[a]	2.94±0.07[b]
乙酰辅酶 A 羧化酶	25.34±0.07[a]	24.58±0.31[ab]	25.37±0.61[a]	25.51±0.69[a]	25.45±0.51[a]	23.15±0.25[b]
肉毒碱棕榈酰转移酶-1	22.23±0.11[b]	22.72±0.33[b]	25.32±0.37[a]	25.00±0.19[a]	24.61±0.23[a]	23.01±0.20[b]

注：同行数据无字母或上标相同字母表示差异不显著（$P>0.05$），上标不同字母表示差异显著（$P<0.05$）。

2.2.4.3　讨论

1. 花生四烯酸对刺参幼参生长性能的影响

近年来，随着对花生四烯酸研究的深入，众多研究已表明，花生四烯酸对海水动物的生长和存活具有重要的作用。当饲料中的花生四烯酸含量为 0.5%～1.0%（干物质）时，大菱鲆幼鱼具有较佳的生长效率和存活率（Castell et al.，1994）。对鲈的研究表明，基于生长性能分析，大规格鲈鱼对花生四烯酸的最适需求量为 0.37%（王成强等，2016）。同样，对牙鲆（刘镜格等，2005）、大黄鱼（谢奉军，2011）等的研究均表明，适量的花生四烯酸对其生长与存活具有一定的促进作用。同时，也有研究报道，饲料中花生四烯酸含量过高会抑制鱼类的生长。Xu 等（2010）发现，在鲈鱼幼鱼饲料中添加过量的花生四烯酸（0.56%～2.12%，干物质）时，严重抑制其生长，并引发相关炎症。Ishizaki 等（1998）对黄尾鱼幼鱼的研究表明，当用 4%花生四烯酸含量的饲料饲喂黄尾鱼幼鱼时，对其生长有负面影响。本研究结果显示，刺参幼参的增重率和饲料效率均随着饲料中花生四烯酸含量变化呈现一定的变化趋势，以增重率为评定指标，刺参幼参饲料中花生四烯酸的最适添加量为 0.51%（饲料干物质），这表明适量的花生四烯酸对刺参幼参的生长具有一定的促进作用。同时研究也表明，当饲料中花生四烯酸含量高于 0.51%（饲料干物质）时，刺参幼参的增重率与饲料效率均受到下降，这同之前对海水鱼的研究结果一致。产生这种生长抑制的主要原因有以下两点：一是由于花生四烯酸和二十碳五烯酸的代谢产物有竞争作用，当饲料中花生四烯酸含量较高时，刺参幼参体内二十碳五烯酸的生物转化会受到抑制（Furuita et al.，2003），从而影响动物体的生长；二是较高含量的花生四烯酸会导致机体产生炎症反应（左然涛等，2015），机体的免疫系统受到影响，对生长产生抑制作用。另外，也有研究证实，花生四烯酸对部分海水动物的生长性能无显著影响（Salini et al.，2016）。这种差异可能是养殖种类、饲养环境与养殖方式不同造成的。

相关研究已经证实，花生四烯酸是多种生物活性化合物的前体，统称为类二十烷酸，其响应于激素刺激和炎症反应而从磷脂释放（Josephkiii et al.，2011）。

这些生物活性物质在动物体内具有重要的作用，能够调节一系列重要的生理代谢，包括脂质蛋白的代谢、心血管系统、白细胞功能、血小板激活、生殖功能和神经系统的控制等（Sargent et al.，1995；Tang et al.，1996）。这也说明花生四烯酸之所以能够对海水动物生长和存活起到一定的调控作用，主要是因为其衍生物的存在。同时也说明为何花生四烯酸能对生物体的抗应激能力起到一定的增强作用，已有研究报道，在饲料中添加适量的花生四烯酸能够显著提高海水仔稚幼鱼的存活率（谢奉军，2011；Lund et al.，2007）。而有研究表明，刺参的存活率并未随花生四烯酸含量的变化发生变化，均维持在 87.50%~94.17%，不同处理组间无显著性差异（$P>0.05$），这同对鱼类的研究结果类似（王成强等，2016；Bessonart et al.，1999）。造成这一差异的原因可能是幼体阶段刺参自身免疫系统较弱，故需要较高含量的花生四烯酸来增强其抗应激能力；而当发育之后，由于其自身具有较强的抗应激能力，因此花生四烯酸对其作用较小。结合本实验结果可以推测，刺参本身可能具有一定的抗应激性能，故不同含量的花生四烯酸饲料对其存活率无显著影响，其相关调控机制有待深入研究。

2. 花生四烯酸对刺参幼参体壁生化组成及脂肪酸组成的影响

众多研究表明，饲料中不同含量的花生四烯酸会影响动物体的常规生化组成，这可能是由于不同含量的花生四烯酸能够对动物体内相关代谢酶活性产生不同的影响。本研究结果显示，随着饲料中花生四烯酸含量增加，刺参幼参体壁粗脂肪含量呈现先降低后升高的趋势，花生四烯酸含量为 0.51%的饲料组粗脂肪含量最低，而当花生四烯酸含量为 0.02%或 0.98%时，其粗脂肪含量均显著高于花生四烯酸含量为 0.51%的饲料组，这表明适量的花生四烯酸可能促进刺参幼参体内脂肪细胞的活化，从而促进脂肪代谢，进而降低刺参幼参体壁脂肪水平。而高含量的花生四烯酸会导致饲料中脂肪酸营养不平衡或饲料中有机酸过高，使得有机体不能很好地摄取与吸收，从而在体壁中造成沉积，导致体壁脂肪含量较高。这同 Xu 等（2010）、王成强等（2016）的研究结果类似。而对大菱鲆幼鱼的研究表明，饲料中高含量的花生四烯酸导致鱼体和肌肉中粗脂肪的含量显著降低，这可能是因为花生四烯酸通过抑制载脂蛋白 B-100 分子分泌，从而抑制极低密度脂蛋白胆固醇的组装来调控肝脏脂肪转运，同时高含量的花生四烯酸能够抑制脂肪酸合成酶基因表达，进而影响脂肪在鱼体和肌肉中的沉积（艾庆辉等，2016）。导致这种差异的原因可能是饲料中花生四烯酸的实际含量存在较大差异，以及不同实验对象对花生四烯酸的需求及利用能力不同。

对海水鱼的研究表明，饲料中脂肪酸组成能够显著影响鱼体和组织脂肪酸组成（Trushenski et al.，2011；Zuo et al.，2012）。本研究中，随着饲料中花生四烯酸含量的增加，刺参幼参体壁中花生四烯酸含量呈现显著升高的趋势（$P<0.05$），两者

呈正相关关系，另外值得注意的是，不同处理组刺参幼参体壁的花生四烯酸含量均是该组中含量最高的脂肪酸，这也暗示花生四烯酸在刺参幼参体壁中具有较强的富集能力，这与Yu等（2015a）、Wen等（2016b）对刺参的研究结果一致。而刺参幼参体壁中的二十碳五烯酸含量却同饲料中花生四烯酸含量呈负相关关系，这可能是由于花生四烯酸和二十碳五烯酸的代谢产物有竞争作用，从而抑制了刺参参体内二十碳五烯酸的生物转化，这在其他研究中也得到了一致的结果（Furuita et al., 2003）。另外，虽然饲料中C16:0、C18:0和饱和脂肪酸的含量呈现下降趋势，但是刺参幼参体壁中三者的含量却基本保持恒定，同时饲料中C18:2n-6含量与刺参幼参体壁中花生四烯酸含量变化趋势一致，这可能是因为刺参幼参本身具有潜在的高不饱和脂肪酸生物合成通路。Yu等（2015a）对刺参的研究也表明，随着饲料中鼠尾藻［花生四烯酸含量为12%（干物质）］被玉米粉和豆粕替代的比例增加，刺参体壁中花生四烯酸含量并未大幅度降低，因此推测刺参体内可能存在合成n-6高不饱和脂肪酸的潜在通路，其能够将C18:2n-6经过延伸和去饱和作用生成花生四烯酸，同时该研究组还推测，刺参也可能具有α-亚麻酸（ALA）合成二十碳五烯酸的潜在能力。Hasegawa等（2014）在研究中也得出了类似的结论。

多不饱和脂肪酸的合成过程涉及多种酶类，其中Δ6脂肪酸去饱和酶和多不饱和脂肪酸延伸酶5是脂肪酸合成过程中关键的两种酶。Liu等（2017b）通过同源克隆和cDNA末端快速扩增（RACE）技术获得刺参的Δ6脂肪酸去饱和酶基因，并测定了该基因在刺参不同组织中的表达，结果显示，Δ6脂肪酸去饱和酶在刺参的肠道、精巢、体壁、呼吸树中均有表达，且肠道中表达量最高，同时研究得出，刺参具有将亚麻酸和亚油酸合成γ-亚麻酸和十八烷酸的能力，但是进一步合成二十碳五烯酸、花生四烯酸和二十二碳六烯酸的能力可能受到一定的限制。同时，克隆获得海参中多不饱和脂肪酸延伸酶5基因，并测得其在海参的体壁、性腺、呼吸树、肠道中均有表达。同时，经过酵母培养基的细胞实验证明，多不饱和脂肪酸延伸酶5具有将18:3n-6和20:5n-3分别延伸到20:3n-6和22:5n-3的能力，证明了刺参具有合成多不饱和脂肪酸的能力。同时，研究也表明，刺参的延伸能力随着碳链的增长而减弱，其高不饱和脂肪酸的合成能力可能受饲料中不饱和脂肪酸组成的影响，其中的原因有待进一步研究（Li et al., 2016）。本研究中，刺参幼参体壁中亚麻酸含量与二十碳五烯酸、二十二碳六烯酸的含量呈现相同的变化趋势，从而也进一步暗示刺参具有合成高不饱和脂肪酸的能力，但是其合成能力的强弱还有待进一步论证阐述。

3. 花生四烯酸对刺参幼参肠道抗氧化能力的影响

机体在代谢过程中会产生大量的活性氧，如超氧化物阴离子和过氧化氢，活性氧会对动物机体产生一系列的负面影响，如机体的氧化应激反应、代谢相关酶的失活,甚至会破坏细胞的完整性(Winston and Di Giulio,1991)。超氧化物歧化酶和过氧化氢酶是机体内主要的抗氧化酶,可以清除内部活性氧,以避免脂肪酸氧化的发生,减轻活性氧的毒性作用,从而能够保护生物免受氧化损伤,在机体的抗氧化中起着关键作用(Pöortner,2002;Bagnyukova et al.,2003)。因此,超氧化物歧化酶、过氧化氢酶的活性和总抗氧化能力可以反映机体抵抗氧化应激的能力,并间接体现水生动物免疫力水平。在本实验中,花生四烯酸含量为0.36%和0.51%的饲料组刺参幼参肠道超氧化物歧化酶和过氧化氢酶的活性及总抗氧化能力均显著高于对照组($P<0.05$),当饲料中花生四烯酸含量进一步增加到0.98%时均呈现下降趋势,这表明在饲料中添加适量的花生四烯酸能够提高抗氧化酶的活性和抗氧化能力。这些结果同之前的研究结果相似。Xu等(2010)发现,当饲料中花生四烯酸含量为0.56%时,鲈鱼幼鱼血清中超氧化物歧化酶活性显著高于对照组。同时,对大规格鲈鱼的研究也表明,当饲料中花生四烯酸含量为0.37%~0.60%时,实验鱼的血清和肝脏中超氧化物歧化酶活性显著提高,鱼体具有较强的抗氧化能力(王成强等,2016)。另外,对牡蛎(Hurtado et al.,2009)、大菱鲆(谭青,2017)的相关研究也得到了类似的结论。

丙二醛是自由基引发脂质过氧化反应后得到的最终分解产物,其含量可能间接反映水生动物的损伤程度和抗氧化能力,其含量越高说明机体受到的损害越严重(Martinez-Álvarez et al.,2002)。同时有研究指出,脂质过氧化的发生同其不饱和脂肪酸水平有密切的关系。在本研究中,饲料中花生四烯酸含量为0.59%、0.98%时,刺参幼参肠道丙二醛含量明显高于其他各处理组($P<0.05$),这说明当花生四烯酸含量过高时,刺参幼参的脂质过氧化反应较强,机体的抗氧化能力受到损害。从而也进一步说明,当饲料中高不饱和脂肪酸含量过高时,容易引起脂质过氧化反应,从而导致机体抗氧化系统遭到破坏。这同 Wen 等(2016b)对刺参、Zuo 等(2012)对大黄鱼、Xu 等(2010)对鲈的研究结果类似。

4. 花生四烯酸对刺参幼参肠道脂肪酸合成酶、乙酰辅酶 A 羧化酶和肉毒碱棕榈酰转移酶-1 活性的影响

乙酰辅酶 A 羧化酶是脂肪酸合成起始过程中乙酰辅酶 A 转化为丙二酰辅酶 A 的关键酶。脂肪酸合成酶能够催化丙二酰辅酶 A 进行反复延伸形成 16 碳和 18 碳饱和脂肪酸,继而在软脂酰-ACP 硫酯酶作用下产生软脂酸。由此可知,脂肪酸合成酶和乙酰辅酶 A 羧化酶是脂肪酸合成过程中的关键酶,其活性的大小会直接影响脂肪酸合成的进度。近年来,相关研究报道,饲料中多不饱和脂肪酸能够在一定程度上调控脂肪酸合成和氧化过程中相关酶的活性与表达。王成强等(2016)发现,随着饲料中花生四烯酸含量升高,鲈肝脏脂肪酸合成酶的活性与表达水平呈现显著下降趋势;左然涛(2013)的研究表明,大黄鱼肝脏、肌肉、肾脏中脂肪酸合成酶的活性均随着亚麻酸/亚油酸比例的增加而降低；同时，时皎皎等（2012）也指出，日粮中多不饱和脂肪酸含量较高时会抑制小鼠肝脏中脂肪酸合成酶和乙酰辅酶A羧化酶的活性。这可能是因为多不饱和脂肪酸可以在一定程度上抑制6-磷酸脱氢酶和苹果酸酶的活性，限制还原型辅酶Ⅱ的生成，从而使得脂肪酸合成酶的活性受到抑制。因此，饲料中多不饱和脂肪酸能够抑制机体内脂肪酸的合成，同时n-3多不饱和脂肪酸的抑制效果要好于n-6多不饱和脂肪酸。本实验结果显示，刺参幼参肠道中脂肪酸合成酶和乙酰辅酶A羧化酶的活性均随着饲料中花生四烯酸含量的增加而显著降低，因而饲料中脂肪酸组成（特别是高不饱和脂肪酸）能够对机体的脂肪酸合成代谢起到一定的调控作用，高不饱和脂肪酸含量较高时会抑制脂肪酸合成相关酶的活性。

肉毒碱棕榈酰转移酶-1是调节线粒体β氧化的关键点，其活性高低可以影响脂肪酸的氧化分解速率。本实验结果显示，当饲料中花生四烯酸含量增加时，刺参幼参肠道内肉毒碱棕榈酰转移酶-1活性呈现先升高后降低的趋势，过高或过低含量的花生四烯酸均对肉毒碱棕榈酰转移酶-1活性起到抑制作用。相似的结果在对鲈的实验中也有报道，鲈肝脏中肉毒碱棕榈酰转移酶-1基因表达随着饲料中花生四烯酸含量增加也呈现先升高后降低的趋势，高含量的花生四烯酸对肉毒碱棕榈酰转移酶-1活性起到一定的抑制作用（王成强，2016）；同时，对大黄鱼的研究也表明，共轭亚油酸能够对其肝脏中肉毒碱棕榈酰转移酶-1的活性产生一定的影响（左然涛，2013）。这可能暗示，过氧化物酶体增殖物激活受体α基因的激活可能因脂肪酸组成和含量的不同而有差异，过高或过低含量的多不饱和脂肪酸都会使得激活作用较弱或起到抑制作用。同时，有研究表明，饲料中不同脂肪酸组成会对过氧化物酶体增殖物激活受体α的活性产生不同影响，而肉毒碱棕榈酰转移酶-1又是过氧化物酶体增殖物激活受体α的一个靶基因，故其活性会受到饲料中脂肪酸组成的影响。

2.2.4.4 结论

综上可知，饲料中添加适量的花生四烯酸对刺参幼参的生长具有一定的促进作用。基于生长性能与抗氧化能力指标分析，当饲料中花生四烯酸含量为0.36%~0.51%时，刺参幼参具有较佳的生长性能和机体抗氧化能力。对刺参幼参增重率和饲料中花生四烯酸含量的回归分析表明，刺参幼参对饲料中花生四烯酸的最适需求量为0.51%。同时，研究也表明，饲料中花生四烯酸的含量会对刺参幼参肠道中脂肪酸组成及脂质代谢造成一定的影响，相关代谢机制有待进一步深入研究。

第 3 章 刺参配合饲料原料的开发

3.1 海藻在刺参配合饲料中的应用

3.1.1 海藻在刺参养殖中的营养应用研究进展

在自然界,刺参主要以细菌、原生动物、小型软体动物、海藻、甲壳类、海草及海洋动植物碎屑等为食(Moriarty,1982)。20 世纪 70 年代,我国北方沿海的辽宁、山东、河北三省开展刺参人工育苗及增养殖研究,研究了刺参各发育阶段的最适饵料,仅限于直接投喂单胞藻、底栖硅藻、有机碎屑等天然饵料,很少投喂人工饲料。随着刺参养殖业的发展,池塘养殖和海上吊笼养殖成为主要的养殖方式,天然饵料已经不能满足高密度刺参的营养需求,北方池塘养殖多将海藻粉和廉价贝类进行简单加工撒入池塘,南方多采用泡发海带与鲜杂鱼糜混合投喂,然而粗放的投喂方式对水质污染较重。近年来,刺参池塘养殖全价配合饲料的开发和使用明显缩短了养殖周期,增加了经济效益(于世浩等,2009)。配合饲料中使用较多的是鼠尾藻、马尾藻等大型藻类,然而由于过度开发利用,大型藻类资源日渐匮乏,价格日益攀升,已不能满足刺参养殖业快速发展的需要。鱼粉、鲜杂鱼糜、扇贝边粉、虾壳粉、陆生植物粉(豆粕、麸皮、玉米粉、地瓜粉)及海泥等其他原料的配合使用,使饲料的营养水平差异较大,质量良莠不齐,很难满足刺参的营养需求。因此,为了解决刺参优质海藻资源短缺及配合饲料效果不佳等问题,相继开发了鼠尾藻、马尾藻等刺参天然饵料的替代品和人工配合饲料。我国海岸线绵长,海域广阔,海藻生长繁殖条件优越,已报道的海藻种类有 800 多种(Tseng,1983),有直接经济价值的种类达近百种(曾呈奎,1962),其中海带、裙带菜、紫菜、江蓠和麒麟菜等已广泛养殖。根据《2016 中国渔业统计年鉴》,2015 年我国藻类的养殖面积达 13.06 万 km^2,总产量达 208.92 万 t。海藻粉中含有丰富的碘化物、维生素等营养成分,大都以有机态形式存在,比矿物质更容易被动物消化吸收(丛大鹏和咸洪泉,2009)。目前已有不少研究报道,部分大型海藻替代鼠尾藻和马尾藻用于刺参养殖取得了较好的效果。本章综述微藻及大型海藻在刺参不同发育阶段及不同养殖方式中的营养应用研究进展,以期为海藻在刺参养殖中的应用及配合饲料的开发利用提供参考(王晓艳等,2018)。

3.1.1.1 微藻在刺参养殖中的应用

微藻营养丰富，富含蛋白质、脂肪（多不饱和脂肪酸）、多糖、色素等多种营养成分和生物活性物质，是刺参重要的食物来源。微藻是多不饱和脂肪酸的生产者，如硅藻含有丰富的二十碳五烯酸，金藻含有丰富的二十二碳六烯酸，盐藻富含亚油酸和亚麻酸，其含量占总脂肪酸含量的 16.66%～19.79%和 16.71%～19.00%（滕怀丽等，2010）。藻类作为水产动物苗种的开口饵料和饵料生物的营养强化剂，在水产育苗中的地位无可替代，具有重要的经济价值和社会效益（刘梦坛等，2010）。微藻的蛋白质含量可达 40%以上，如小球藻为 50%～60%，螺旋藻为 60%～70%，微藻中的必需氨基酸含量与鱼粉相当，甚至更优，如微藻中天冬氨酸和谷氨酸的含量高达 7.1%～12.9%，而鱼粉中天冬氨酸和谷氨酸的含量仅为 5%～9%（刘梅等，2016；董育红等，2003）。目前全世界发现的微藻种类超过 6 万种，有记载的超过 6000 种，而应用于生产实践的仅有几十种（魏东和俞建中，2014）。

1. 微藻在刺参幼体培育及稚参养殖中的应用

刺参不同发育阶段的营养需求不同。刺参浮游幼体阶段的适口饵料种类有盐藻、角毛藻、小新月菱形藻等，日投饵量在初耳幼体期为 1.5×10^4cell/ml，在中耳幼体期为 2×10^4～3×10^4cell/ml，在大耳幼体期为 4×10^4～5×10^4cell/ml（袁成玉，2005；汤海燕，2010；张杰，2010）。出现五触手原基或 20%～30%的浮游幼体发育至樽形幼体时投放附着基，幼体附着后主要以舟形藻、卵形藻和菱形藻等底栖硅藻为食，需要逐步增加光照强度使附着基上的底栖硅藻得以繁殖。稚参在体长为 0.4mm 时，以附着基上的底栖硅藻为主要饵料。根据底栖硅藻的附着量，适当投喂大型海藻磨碎液，每天 2 次，每次 20×10^4～30×10^4cell/ml；稚参体长达 2mm 以上时，可完全以鼠尾藻磨碎液为饵料，每日投喂 4～6 次，每次 20×10^{-6}～30×10^{-6}g/头；后期稚参体长达 5～6mm 时，可增加人工配合饲料，日投喂量为体质量的 10%左右（汤海燕，2010；张杰，2010）。

刺参对不同单胞藻类的吸收利用率差异较大。Shi 等（2013）认为，决定刺参生长的不是微藻的营养成分，而是其细胞壁能否被刺参消化。根据多年的刺参育苗实践，刺参幼体的最适饵料主要是角毛藻和盐藻。角毛藻个体小，细胞壁薄，悬浮性强，易为刺参幼体消化、吸收，而盐藻无细胞壁，易消化，富含 β-胡萝卜素，且含有 33.4%的多不饱和脂肪酸（成永旭，2005；周率等，2016），二者的适宜繁殖温度为 18～28℃，与耳状幼体培育水温相似。摄食这两种饵料的幼体生长发育迅速、变态率和存活率高。小新月菱形藻、三角褐指藻、中肋骨条藻等也是刺参的适宜饵料，能够高密度大面积培养，但由于具有硅质外壳，刺参对其消化吸收能力不及角毛藻和盐藻，镜检常见未被刺参幼体消化的完整藻体被排出体外，但仍有相当数量的藻体被消化吸收，胃液颜色正常（侯传宝，2005）。耳状幼体对湛江等鞭金藻和球等鞭金藻 3011 等金藻类的消化能力弱，大多数藻体被完整地排出体外。若刺参幼体以金藻类为食，从中耳幼体期开始，其胃部易出现萎缩，甚至出现糜烂现象，因此金藻类饵料不宜长期单独投喂，与角毛藻和盐藻搭配投喂效果较好。投喂其他单胞藻如扁藻、小球藻和拟微绿球藻等效果较差，不宜作为刺参幼体的开口饵料（詹冬梅等，2004）。而陈书秀等（2014）用拟微绿球藻新鲜藻液、拟微绿球藻冷冻浓缩液、拟微绿球藻干粉+干酵母及干酵母对刺参浮游阶段幼体进行了为期 15d 的投喂实验，结果表明，以拟微绿球藻干粉 50%+干酵母 50% 搭配投喂的刺参幼体生长速率快，大耳幼体存活率及樽形幼体的变态率均较高。刺参浮游幼体培育采用 2~3 种饵料混合投喂可获得更好的培育效果，使单一饵料的营养缺陷得以相互补充，更好地满足刺参幼体生长发育的营养需求。

2. 微藻在刺参幼参至成参养殖中的应用

与稚参一样，刺参幼参对微藻的吸收利用情况不仅受微藻营养物质含量和种类的影响，还取决于微藻的细胞壁结构。Shi 等（2013）用鲜筒柱藻、叉鞭金藻、小新月菱形藻、钝顶螺旋藻和干鼠尾藻与海泥混合配制的饲料饲喂刺参幼参，结果表明，筒柱藻营养价值较低，但其弱硅化的细胞壁易被刺参破坏，因此刺参幼参具有最高的特定生长率，而钝顶螺旋藻虽然蛋白质含量最高，但其具有完整的纤维素细胞壁，很难被刺参破坏，因此刺参幼参生长性能最低，能量浪费比例最高。王吉桥等（2013）设置了 3 组实验饲料投喂刺参幼参，分别是海藻类组（小球藻、筒柱藻和角毛藻按 1：1：1 混合）、微生态制剂组和藻类微生态制剂混合组，结果表明，海藻类组的刺参幼参增重率显著高于微生态制剂组和藻类微生态制剂混合组（$P<0.05$），在水温较高且适宜时，刺参养殖水体中藻类和微生态制剂以 2：1 比例混合施用效果较好，在水温较低时，应适当增加藻类比例，减少微生态制剂用量，以更利于刺参幼参的生长。

3.1.1.2 大型海藻在刺参养殖中的应用

鼠尾藻和马尾藻被认为是最适合刺参养殖的大型藻类。研究证实，鼠尾藻中氨基酸较全面，呈味氨基酸丰富，且限制性氨基酸与 FAO 模式相同（吴海歌等，2008）。然而，由于无度的滥采乱用和不能进行商业化生产，这两种藻类已经不能满足刺参养殖生产的需求，目前生产中多使用海带、裙带菜、巨藻、石莼、紫菜和浒苔等替代鼠尾藻和马尾藻，并取得了良好的效果。

1. 大型海藻在刺参幼体培育及稚参养殖中的应用

活微藻饵料的培育周期长、占地广、成本高，受到自然资源、环境因素等条件的限制（王熙涛等，2014），目前生产上多采用大型海藻磨碎液、海藻干粉、配合饲料或几种方式混合投喂稚参。王吉桥等（2008）将新鲜鼠尾藻、配合饲料和干酵母按照一定的比例混合投喂稚参，最终得出一个投喂模式：耳状幼体、樽形幼体、五触手幼体阶段投喂单胞藻类；五触手幼体到稚参阶段投喂鼠尾藻和酵母，外加少量配合饲料；稚参阶段以后，增加配合饲料的使用量，并添加海泥。殷旭旺等（2015）用孔石莼、角叉菜、裙带菜干粉及其混合干粉，以及鲜孔石莼、鲜角叉菜、鲜裙带菜及其混合磨碎液分别与25%海泥制成8种实验饲料饲养体质量为（1.25±0.02）g 的刺参，结果表明，投喂鲜孔石莼饲料的刺参特定生长率和饲料转化率均显著高于其他组（$P<0.05$），摄食率和排粪率最高，且与其他组有显著性差异（$P<0.05$）。朱建新等（2007）用鼠尾藻干粉、鲜海带磨碎液和鲜石莼磨碎液喂养体长为（2.90±0.04）cm、体质量为（0.49±0.02）g 的稚参 40d，验结束时，鲜石莼磨碎液组稚参的存活率（85%）、体长[（5.15±0.25）cm]和体质量[（2.76±0.39）g]都显著高于鼠尾藻干粉组和鲜海带磨碎液组（$P<0.05$），这说明鲜石莼磨碎液具有较好的育参效果。但此方法投喂的饲料营养不均衡，水质污染较严重。20世纪初，暴发性的刺参腐皮综合征在山东造成 10 亿元以上的经济损失，其中一个主要原因就是替代单胞藻的大型海藻磨碎液、酵解残渣或沉降藻膏等死细胞引起水质恶化，病原体滋生。通过细胞工程技术酶解大型海藻可分离出大量的单细胞活饵料（赵瑞祯等，2012）。用海螺酶解获得的游离条斑紫菜叶状体单细胞作为饵料，与底栖硅藻、配合饲料、鼠尾藻磨碎液培育稚参，结果显示，游离紫菜细胞组 30 日龄稚参的存活率高达 90.17%，高于底栖硅藻组，极显著高于配合饲料组及鼠尾藻磨碎液组（$P<0.01$）。游离紫菜细胞组稚参的体长、增重率及特定生长率高于配合饲料组与鼠尾藻磨碎液组，游离紫菜单细胞可以替代底栖硅藻、配合饲料、鼠尾藻磨碎液培育稚参。用大型海藻磨碎液培育刺参幼体的效果较好，但新鲜海藻的采收受季节限制，且易污染水质，使用具有局限性，而用细胞工程技术酶解大型海藻，可以分离出大量的单细胞活饵料，具有工艺简单、生产量大、时间较短、营养丰富和饲料效果好等诸多优点，为刺参育苗期饲料开发开辟了一条崭新的途径，促进了刺参健康养殖业的发展。

2. 大型海藻在刺参幼参至成参养殖中的应用

刺参幼参的饲料类型与稚参基本相同，但用量差异很大。以人工配合饲料为主，以底栖硅藻和大型褐藻磨碎液为辅（钟鸣和胡超群，2016），饲料中配以适量的海泥，以增粗刺参肠道，增加吸收面积，延长食物在刺参消化道中的停留时间（Shi et al.，2015）。大型海藻被广泛应用于刺参幼参养殖，在生产中采用不同配比的海藻干粉与一定量的海泥搭配投喂刺参。Xia 等（2012b）用不同的海藻粉（鼠尾藻粉、马尾藻粉、大叶藻粉、石莼粉、生海带粉和熟海带粉）和海泥以 3∶7 的比例配制成 6 组实验饲料，饲喂（4.0±0.2）g 的刺参幼参 60d，结果显示，石莼粉组和海带粉组的生长性能显著高于其他组，氮排放率显著低于其他组（$P<0.05$），表明用石莼和海带养殖刺参幼参的效果较佳。Yuan 等（2006）用贝类粪便与海藻粉（海带粉、鼠尾藻粉、马尾藻粉）按照不同比例配制成 5 种颗粒饲料饲喂刺参，发现贝类粪便与海藻粉按 3∶1 混合，刺参生长性能最高。Wen 等（2016b）用马尾藻、龙须菜、石莼及三种藻类分别与底栖生物物质按 1∶1 混合，配制成 6 种实验饲料饲喂刺参，结果表明，龙须菜与底栖生物物质 1∶1 混合组刺参增重率显著高于单独添加龙须菜组（$P<0.05$），石莼与底栖生物物质 1∶1 混合组刺参粗蛋白含量显著高于单独添加石莼组（$P<0.05$），添加底栖生物物质的三个组与单独添加藻类组相比，多不饱和脂肪酸含量较高。Wen 等（2016c）用马尾藻、龙须菜、石莼及三种藻类两两 1∶1 混合，配制成 6 种实验饲料饲喂刺参，石莼组刺参的特定生长率与马尾藻组差异不显著（$P>0.05$），而显著高于龙须菜组（$P<0.05$），以脂肪酸组成为判定标准，相对于龙须菜组，石莼组的饲喂效果更好，表明石莼可以替代马尾藻用于人工饲料中，优于龙须菜。然而，Gao 等（2011）研究发现，龙须菜可以替代鼠尾藻满足刺参的营养需求，与 Wen 等（2016c）的研究结果不一致。刺参不同生长阶段的营养需求不同，而石莼不论是在稚幼参还是成参的养殖中，都具有较好的效果，可能是石莼质地松软，极易被刺参消化道破坏。

大叶藻广泛分布于温带浅海，大叶藻草地是包括海参在内的各种海洋生物的重要栖息地。近年来，污水的排放加重了近岸水体富营养化，线形硬毛藻暴发性生长导致绿潮发生（Cooke et al.，2015），严重影响大叶藻海草生态系统的健康（Zhang et al.，2015）。Song 等（2017）用绿潮大型藻类线形硬毛藻和海草大叶藻以一定的配比（100∶0、75∶25、50∶50、25∶75、0∶100）与海泥混合配制成 5 种实验饲料，用鼠尾藻和海泥（4∶6）配制的饲料作为对照组，结果表明，两种藻类的配比显著影响饲料利用率和能量收支，高比例的线形硬毛藻组刺参的能量沉积率显著高于高比例的大叶藻组（$P<0.05$），而且线形硬毛藻组的刺参生长性能与对照组差异不显著（$P>0.05$），这说明线形硬毛藻可以替代鼠尾藻作为刺参的食物来源。郭娜等（2011）的研究表明，绿潮物种浒苔也可以作为刺参的食物来源。

3. 大型海藻在刺参不同养殖方式中的应用

随着刺参养殖业的发展，出现了多种养殖方式，如池塘养殖、海上吊笼养殖、网箱养殖、围堰养殖、室内工厂化养殖、海区底播养殖等，以池塘养殖和海上筏式吊笼养殖为主，以人工配合饲料和大型藻类为主要饵料，具体投喂情况则根据养殖方式、养殖密度和具体环境条件而定。

北方池塘养殖密度较大，一般移植大叶藻和鼠尾藻等大型藻类，为水体增加含氧量和隐蔽处，其碎屑和分泌物可为刺参提供饵料，但受光照和温度的限制，尤其春、秋季水温较低，水中浮游动植物较少，天然饵料难以满足刺参的生长需求，而这时又是刺参生长的适温期，因此，多使用海藻粉及廉价贝类与海泥进行简单加工撒入池塘。由于没有使用较好的加工工艺及黏合剂，饲料原料在水中容易溶失，刺参对其利用率较低（于瑞海等，2012）。虽然配合饲料的使用能有效解决上述问题，缩短半年以上养殖周期，具有可观的经济效益，但目前刺参配合饲料价格较高，难以接受。

海上筏式吊笼养殖一般以干海带和配合饲料为主。为保证饵料在水中的稳定性，通常添加黏合剂，以降低饵料在海水中的散失率。我国北方吊笼养殖一般用体长 3～5cm 的幼参，前期投喂浸泡 2d 后的干海带，7～15d 投喂一次，投喂量为刺参体质量的 5%～10%（李成军和雷帅，2016）。我国福建沿海一般以筏式吊笼养殖为主，刺参饵料主要是浸泡 2～3d 的干海带，与自配的团状饲料搭配投喂。韩承义等（2011）以干海带为主要饵料，补充添加营养物质投喂刺参，结果表明，补充添加鱼糜组的增重率显著高于补充添加虾米糠、麦麸和面粉混合组和投喂纯海带组（$P<0.05$）。王兴春（2014）分别用海带、海带+鲜杂鱼糜、海带+扇贝边粉和粉状配合饲料 4 种饵料投喂刺参，研究了不同饲料对刺参生长和体成分的影响，结果表明，海带+鲜杂鱼糜组刺参的生长性能最高，证明了动物蛋白对刺参生长的重要性。余致远（2015）比较了传统海带饵料和市售的凝胶状海参配合饲料对海上筏式吊笼养殖的 3 种规格刺参（平均体质量分别为 105.57g、73.15g 和 52.89g）生长性能的影响，结果显示，传统海带饵料投喂组小规格刺参的平均个体增重率和特定生长率均显著大于凝胶状海参配合饲料投喂组（$P<0.05$），而两种饲料投喂组大、中规格刺参间的投喂效果差异不显著（$P>0.05$）。海带+鲜杂鱼糜的投喂方式虽然会获得较高的生长性能，但是该方法配方粗略，容易污染水质。不同养殖方式及不同生长阶段刺参的营养需求不同，对饲料的性状要求也不同，生产中需根据具体养殖模式选择适宜的投喂方式，以满足刺参的生长需求。

3.1.1.3 结语

目前，刺参养殖中多使用单一海藻饲喂，营养组成单一，应多种海藻搭配使用，避免单一海藻氨基酸、脂肪酸及其他营养物质不平衡的问题。未来需加强海参消化系统发育学及营养生理学的基础研究，确定不同生长阶段海参的营养需求并制定各阶段配合饲料的营养标准，根据海参摄食特性和消化机能优化选择海参饲料加工工艺，开发低污染、低成本、高效、全价的配合饲料。

3.1.2 5 种海藻在刺参幼参饲料中的应用研究

鼠尾藻和马尾藻被认为是最优质的刺参饵料，由于过度开采已经不能满足产业发展的需要。天然海域中，刺参主要以沉积物中的单细胞藻类、原生动物、细菌、大型海藻碎屑及海泥为食，因为刺参优先选择栖息在富含大型海藻的海域，充足的大型海藻碎屑是海参的主要食物来源，所以大型海藻成为人工配合饲料的主要原料。目前已有不少研究报道，部分大型海藻用于刺参养殖取得了不错的投喂效果（Wen et al.，2016b；殷旭旺等，2015；李猛等，2017；唐薇等，2014）。石莼和海带是中国北方很受欢迎且价格较低的大型海藻，含有丰富的碳水化合物、蛋白质和各种矿物质，目前已被广泛应用于刺参养殖中。铜藻是中国暖温带海域浅海区海藻场的主要连片大型褐藻物种，植株高大，枝叶繁茂，多生长在潮下带浅海岩礁上，成片漂浮于水面，堪称"海中森林"，铜藻不但含有丰富的藻胶、纤维素、半纤维素及矿物质，而且生物活性物质含量相当高，作为工业、食品原料资源具有极大的开发潜力，可惜目前还没有被广泛开发利用（郑海羽等，2008）。海带渣为工业提胶的剩余产品，干海带加水浸泡后切碎，甲醛固色，在加热条件下，利用碳酸氢钠消化后过滤，滤液用于提取海藻酸钠，将不溶性组分干燥后得到海带渣原料，所以海带经过有效成分提取后，仍存在许多粗蛋白、粗纤维和矿物质等营养物质，用来制作刺参饲料可实现海带渣的高值化利用，这一部分资源的利用可使海带干物质的利用率高达45%左右（张俊杰等，2010）。本节在刺参幼参配合饲料中添加海带粉、海带渣粉、铜藻粉及石莼粉，与鼠尾藻的饲喂效果作对比研究，并探讨混合海藻粉的应用效果，以期为大型海藻在刺参人工配合饲料中的应用提供参考（王晓艳等，2019）。

3.1.2.1 材料与方法

1. 实验饲料

以鱼粉、小麦粉、海泥等为主要原料，对照组（D1组）为鼠尾藻粉（购自莱州市海福饲料有限公司），D2、D3、D4、D5和D6组分别用铜藻粉（采集于江苏省盐城市大丰区养殖水域；菌落总数为 20 840CFU/g；大肠杆菌数目为16MPN/100g；沙门氏菌、副溶血弧菌、金黄色葡萄球菌、志贺氏菌均未检出）、海带粉（购自青岛汇福林海洋生物科技有限公司）、海带渣（购自青岛汇福林海洋生物科技有限公司）、石莼粉（购自青岛海之源饲料科技有限公司）及5种海藻粉等比例混合（1:1:1:1:1）替代鼠尾藻粉，配制成6种实验饲料，每种实验饲料包含30%的海藻粉。将所有实验原料逐级混匀，与适量新鲜蒸馏水混合，用螺旋挤压机加工成厚度为2mm的片状饲料，在室温下风干至水分为5%左右，储存于冰箱（-20℃）中备用。实验饲料配方及营养组成见表3-1，实验饲料氨基酸组成见表3-2。

表 3-1　实验饲料配方及营养组成（%）

项目		组别					
		D1	D2	D3	D4	D5	D6
原料组成	鱼粉	12.0	12.0	12.0	12.0	12.0	12.0
	海藻粉	30.0	30.0	30.0	30.0	30.0	30.0
	小麦粉	10.0	10.0	10.0	10.0	10.0	10.0
	微晶纤维素	5.0	5.0	5.0	5.0	5.0	5.0
	海泥	36.4	36.4	36.4	36.4	36.4	36.4
	大豆卵磷脂	1.0	1.0	1.0	1.0	1.0	1.0
	破壁酵母	0.5	0.5	0.5	0.5	0.5	0.5
	黄原胶	0.5	0.5	0.5	0.5	0.5	0.5
	瓜尔豆胶	1.5	1.5	1.5	1.5	1.5	1.5
	抗氧化剂	0.1	0.1	0.1	0.1	0.1	0.1
	维生素预混料	2.0	2.0	2.0	2.0	2.0	2.0
	矿物质预混料	1.0	1.0	1.0	1.0	1.0	1.0
营养组成	粗蛋白	15.26	14.46	16.64	16.83	16.23	15.90
	粗脂肪	1.54	1.91	1.80	1.73	1.41	1.80
	粗灰分	50.03	53.06	47.99	46.32	48.14	49.21
	能量	11.57	10.58	12.30	12.73	12.12	11.94

注：①每千克维生素预混料含维生素 A 38.0mg、维生素 B_1 115.0mg、维生素 B_2 380.0mg、维生素 B_{12} 1.3mg、维生素 D_3 13.2mg、α-生育酚 210.0mg、烟酸 1030.0mg、泛酸 368.0mg、生物素 10.0mg、叶酸 20.0mg、抗坏血酸 500.0mg、盐酸吡哆醇 88.0mg、肌醇 4000.0mg；②每千克矿物质预混料含 KCl 3020.5mg、$KAl(SO_4)_2$ 11.3mg、NaCl 100.0mg、KI 7.5mg、NaF 4.0mg、$ZnSO_4·7H_2O$ 363.0mg、$CuSO_4·5H_2O$ 8.0mg、$MgSO_4·7H_2O$ 3568mg、$MnSO_4·4H_2O$ 65.1mg、Na_2SeO_3 2.3mg、$NaH_2PO_4·2H_2O$ 25 558.0mg、$CoCl_2$ 28.0mg、$C_6H_5O_7Fe·5H_2O$ 1523.0mg、Ca-lactate 15 978.0mg。

表 3-2　实验饲料氨基酸组成及海藻粉营养组成（干物质）（%）

实验饲料氨基酸组成	组别					
	D1	D2	D3	D4	D5	D6
精氨酸	0.77	0.78	0.90	0.89	0.90	0.82
组氨酸	0.26	0.27	0.27	0.28	0.28	0.27
异亮氨酸	0.45	0.47	0.56	0.58	0.57	0.51
亮氨酸	0.97	0.99	1.20	1.21	1.17	1.07
赖氨酸	0.83	0.90	0.91	0.89	0.88	0.85
蛋氨酸	0.33	0.25	0.43	0.43	0.43	0.40
苯丙氨酸	0.59	0.61	0.74	0.74	0.73	0.66
苏氨酸	0.63	0.64	0.78	0.79	0.76	0.70
缬氨酸	0.57	0.59	0.71	0.73	0.72	0.64
天冬氨酸	1.33	1.49	1.67	1.66	1.59	1.51
丝氨酸	0.75	0.76	0.88	0.89	0.86	0.81

续表

实验饲料氨基酸组成	组别					
	D1	D2	D3	D4	D5	D6
甘氨酸	0.92	0.94	1.11	1.10	1.07	1.00
丙氨酸	0.86	0.91	1.07	1.06	1.04	0.96
半胱氨酸	0.19	0.18	0.24	0.25	0.23	0.22
酪氨酸	0.41	0.37	0.41	0.40	0.38	0.40
谷氨酸	3.33	2.78	3.08	3.09	3.00	3.00
脯氨酸	0.74	0.78	0.90	0.90	0.87	0.82
总氨基酸	13.93	13.71	15.86	15.89	15.48	14.64

2. 饲养管理

实验用刺参幼参购于山东安源水产股份有限公司，正式实验前，于实验所用循环水养殖系统中，用对照组（鼠尾藻粉组）饲料投喂 14d，使刺参幼参适应养殖系统的条件及实验饲料。选取 540 头规格统一、体质健康、初始体质量为 (10.02±0.03)g 的刺参幼参随机分配到 18 个玻璃钢水槽内（$h_{桶}$=80cm, $h_{水}$=50cm, $d_{桶}$=60cm），实验分为 6 组，每组 3 个平行，每个平行 30 头刺参幼参，每个桶内放置 2 个聚乙烯波纹板筐作为刺参幼参的遮蔽物。每天定时（16:00）投喂 1 次，初始投喂量为刺参总湿重的 2%，观察摄食情况并及时调整投喂量。每 3d 采用虹吸法吸出残饵及粪便，实验周期为 56d，实验期间每天换水量为 1/3，控制水温为 (18±2)℃，溶氧量大于 6mg/L，pH 为 7.6～8.2，盐度为 28～32，亚硝酸盐浓度低于 0.05mg/L，氨氮浓度低于 0.05mg/L，光照周期为自然光周期。

3. 样品采集

实验中期开始收集刺参粪便，用镊子从每个桶内小心夹取完整粪便，用电热鼓风干燥箱 60℃烘干，用于酸不溶性灰分的测定。饲养结束后禁食 48h，分别对每个桶的刺参进行计数并称量总重，每桶随机选取 10 头刺参，解剖分离内脏、体壁及肠道，称重后保存于冰箱（–80℃）中备用。另外，每个桶随机选取 3 头刺参，沿腹中线的右侧解剖（避免破坏腹肠系膜），剔除呼吸树后，取中肠 0.8cm 左右并固定于波恩氏液（Bouin）中 24h 后，保存于 70%乙醇中，经脱水、透明、浸蜡、包埋后，进行常规石蜡连续切片处理，切片厚度为 7.0μm，然后经苏木精-伊红（HE）染色，用中性树胶封片并编号，在徕卡高清摄像系统下观察并拍照。从冰箱（–80℃）中取出刺参肠道，冰浴解冻匀浆，4℃离心 20min（2500r/min），取上清液（粗酶液）于 4℃保存，用于消化酶和免疫酶活性的测定（24h 内测定）。

4. 计算公式与实验方法

刺参的增重率、特定生长率、存活率、肠体比、脏壁比的计算方法，以及饲料及刺参体壁的水分、粗灰分、粗脂肪、粗蛋白的含量及饲料能量测定方法见附录，酸不溶性灰分根据《饲料中盐酸不溶灰分的测定》（GB/T 23742—2009）测定；肠道粗酶液蛋白质浓度（考马斯亮蓝法），胰蛋白酶、糜蛋白酶、α-淀粉酶、脂肪酶、总超氧化物歧化酶、碱性磷酸酶、酸性磷酸酶的活性，以及丙二醛含量均采用南京建成生物工程研究所生产的试剂盒测定，测定方法均参考试剂盒说明书。

5. 数据分析

所有数据采用 SPSS 18.0 软件进行单因素方差分析，用邓肯多重范围检验进行多重比较分析，$P<0.05$ 为差异显著，$P<0.01$ 为差异极显著，$P>0.05$ 为差异不显著。统计数据以平均值±标准误表示。

3.1.2.2 结果分析

1. 不同海藻粉对刺参幼参生长性能和形体指标的影响

不同海藻粉对刺参幼参的生长性能具有显著影响，如表 3-3 显示，混合海藻粉组与对照组的刺参幼参增重率和特定生长率显著高于其他各组（$P<0.01$），铜藻粉组与海带粉组差异不显著（$P>0.05$），但显著低于石莼粉组（$P<0.01$），海带渣组最低。海带渣组刺参幼参的肠体比和脏壁比均显著小于其他各组（$P<0.05$），其他各组间差异不显著（$P>0.05$）。不同海藻粉对刺参幼参存活率的影响不显著（$P>0.05$）。

表 3-3 不同海藻粉对刺参幼参生长性能和形体指标的影响

项目	组别					
	D1	D2	D3	D4	D5	D6
初始体质量/g	10.01±0.01	10.02±0.02	10.04±0.01	10.04±0.03	10.02±0.03	10.01±0.01
终末体质量/g	18.83±0.23^a	13.17±0.07^c	13.31±0.09^{bc}	11.19±0.11^d	13.72±0.07^b	18.92±0.16^a
增重率/%	88.33±2.33^a	31.29±0.23^c	32.76±0.93^c	11.37±0.57^d	36.97±0.60^b	89.20±1.60^a
特定生长率/(%/d)	1.13±0.02^a	0.49±0.00^c	0.51±0.01^c	0.19±0.01^d	0.56±0.01^b	1.14±0.02^a
肠体比/%	3.34±0.04^{ab}	3.37±0.05^a	3.26±0.05^{ab}	2.37±0.09^c	3.16±0.03^b	3.39±0.05^a
脏壁比/%	7.84±0.38^a	7.12±0.37^a	7.44±0.35^a	5.77±0.27^b	7.49±0.15^a	7.75±0.34^a
存活率/%	91.67±1.67	88.33±1.67	88.33±1.67	83.33±0.00	95.00±5.00	88.89±1.11

注：同行数据无字母或上标相同字母表示差异不显著（$P>0.05$），上标不同字母表示差异显著（$P<0.05$）。

表 3-4 显示，饲料中添加海带粉、海带渣、石莼粉及混合海藻粉均能提高刺参幼参的体壁粗蛋白含量，但与对照组差异不显著（$P>0.05$），不同海藻粉对刺参幼参体壁水分、粗灰分和粗脂肪含量的影响不显著（$P>0.05$）。

表 3-4 不同海藻粉对刺参幼参体壁组成的影响（%）

组成	组别					
	D1	D2	D3	D4	D5	D6
水分	90.80±0.35	91.30±0.08	90.72±0.15	90.58±0.20	91.05±0.21	91.32±0.06
粗蛋白	42.96±0.44	42.71±0.35	43.46±0.07	43.63±0.62	43.63±0.23	43.70±0.09
粗脂肪	3.58±0.26	3.66±0.05	3.58±0.09	3.94±0.14	3.63±0.02	3.77±0.10
粗灰分	36.23±0.04	37.30±0.24	35.92±0.29	36.73±0.16	36.22±0.43	36.44±0.51

2. 不同海藻粉对刺参幼参消化酶活性及酸不溶性灰分含量的影响

鼠尾藻粉组和混合海藻粉组刺参幼参肠道的胰蛋白酶活性显著高于海带粉组、海带渣组和石莼粉组（$P<0.05$），铜藻粉组与对照组差异不显著（$P>0.05$）；海带渣组及石莼粉组刺参幼参肠道的糜蛋白酶活性显著对照组、铜藻粉组和混合藻粉组（$P<0.05$），与海带粉组差异不显著（$P>0.05$）；不同处理组 α-淀粉酶和脂肪酶的活性差异不显著（$P>0.05$）；铜藻粉组刺参幼参粪便的酸不溶性灰分含量极显著高于对照组（$P<0.01$），海带渣组与对照组差异不显著，其他各组均极显著低于对照组（$P<0.01$）（表 3-5）。

表 3-5 不同海藻粉对刺参幼参消化酶活性及酸不溶性灰分含量的影响

项目	组别					
	D1	D2	D3	D4	D5	D6
胰蛋白酶/（U/mg prot）	820.93±21.13^a	798.95±42.57^{ab}	714.61±27.94^{bc}	630.35±24.67^c	720.34±13.07^b	816.15±26.80^a
糜蛋白酶/（U/mg prot）	4.39±0.26^a	4.30±0.14^{ab}	3.75±0.15^{bc}	3.39±0.11^c	3.55±0.14^c	4.29±0.18^{ab}
α-淀粉酶/（U/mg prot）	1.13±0.03	1.24±0.06	1.07±0.06	1.09±0.07	1.07±0.06	1.24±0.07
脂肪酶/（U/g prot）	6.32±0.27	5.60±0.09	6.20±0.35	5.83±0.20	6.53±0.14	6.52±0.20
酸不溶性灰分含量/%	42.24±0.22^b	43.52±0.16^a	39.71±0.15^d	41.85±0.24^b	40.75±0.15^c	40.47±0.13^c

注：同行数据无字母或上标相同字母表示差异不显著（$P>0.05$），上标不同字母表示差异显著（$P<0.05$）。

3. 不同海藻粉对刺参幼参非特异性免疫性能的影响

由表 3-6 可见，不同海藻粉对刺参幼参非特异性免疫具有显著影响，海带渣组刺参幼参肠道的总超氧化物歧化酶活性显著低于其他各组（$P<0.01$），混合海藻粉组显著高于铜藻粉组、海带粉组及石莼粉组（$P<0.01$）；海带渣组刺参幼参肠道的丙二醛含量显著高于其他各组（$P<0.01$），混合海藻粉组与对照组差异不显著（$P>0.05$）；不同海藻粉对刺参幼参肠道的酸性磷酸酶、碱性磷酸酶的活性具有显著影响（$P<0.01$），海带渣组刺参幼参的酸性磷酸酶活性显著低于对照组（$P<0.01$），其余各组与对照组差异不显著（$P>0.05$）；铜藻粉组、海带渣组及石莼粉组刺参幼参肠道的碱性磷酸酶活性显著低于对照组（$P<0.01$），混合海藻粉组与对照组差异不显著（$P>0.05$）。

表 3-6 不同海藻粉对刺参幼参非特异性免疫性能的影响

项目	组别					
	D1	D2	D3	D4	D5	D6
总超氧化物歧化酶活性/（U/mg prot）	20.99±1.07^{ab}	18.81±1.00^{b}	19.14±0.82^{b}	14.96±0.60^{c}	18.37±0.12^{b}	22.65±1.30^{a}
丙二醛含量/（nmol/mg prot）	1.44±0.04^{b}	1.36±0.07^{bc}	1.29±0.06^{bc}	2.01±0.11^{a}	1.17±0.08^{b}	1.31±0.04^{bc}
酸性磷酸酶活性/（U/g prot）	134.22±6.92^{ab}	119.04±3.45^{bc}	129.21±4.45^{ab}	111.19±2.93^{c}	129.57±4.18^{ab}	138.33±8.25^{a}
碱性磷酸酶活性/（U/g prot）	457.58±11.64^{a}	366.27±17.56^{bc}	418.42±14.20^{ab}	352.44±16.66^{c}	385.15±20.42^{bc}	461.99±11.38^{a}

注：同行数据无字母或上标相同字母表示差异不显著（$P>0.05$），上标不同字母表示差异显著（$P<0.05$）。

4. 不同海藻粉对刺参幼参肠道组织结构的影响

图 3-1 为 6 组饲料投喂下刺参幼参的中肠横切组织结构。可见，不同海藻粉对刺参幼参肠道组织结构影响显著，鼠尾藻粉组和混合海藻粉组刺参幼参的肠道皱襞高度明显大于其他各组，且皱襞面积较大，铜藻粉组及海带渣组皱襞高度最小，且皱襞面积最小，海带粉组及石莼粉组皱襞高度居中。

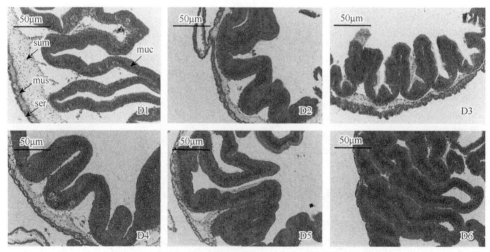

图 3-1 不同海藻粉对刺参幼参肠道组织结构的影响
muc-黏膜层；sum-黏膜下层；mus-肌肉层；ser-浆膜层

3.1.2.3 讨论

海藻为刺参生长所必需，饲料中添加 30%的海藻粉能最大限度提高刺参的生长性能（Yuan et al.，2006），而鼠尾藻是公认的刺参优质饵料，在本实验条件下，铜藻粉、海带粉、海带渣及石莼粉对刺参幼参的增重效果均不如鼠尾藻粉，然而，混合海藻粉却达到了与鼠尾藻粉相同甚至更优的增重效果，这可能是因为多种海藻粉混合营养更全面，弥补了单一海藻粉营养不均衡的问题，与 Gao 等（2011）对刺参的研究结果一致。铜藻的氨基酸组成与鼠尾藻相似（表 3-2），但促生长效果却明显不及鼠尾藻，可能是因为其谷氨酸含量（2.78%）明显低于鼠尾藻粉组（3.33%），而谷氨酸是一种呈味氨基酸（刘长琳等，2015），能增强动物食欲，提高摄食量。本研究中海带渣组刺参幼参的肠体比和脏壁比均显著小于其他各组，可能是因为在提取海带渣的褐藻酸钠过程中，为了保证提取物的色泽而加入的甲醛等固色剂部分残留，影响了刺参幼参的消化功能，而海带渣组的存活率与其他各组差异不显著，表明海带渣对刺参的存活率无不利影响，但不利于肠道的消化吸收。郝继浦（2014）的研究表明，利用海带渣生产刺参饵料，刺参的生长性能显著低于配合饲料组（$P<0.05$），与本研究结果相似。

刺参食道和胃都很短，仅起到运输和机械处理内吞食物的作用，刺参无特化的消化腺（李旭等，2013），前肠和中肠有腺细胞，能分泌消化酶，对食物进行细胞外消化（王吉桥等，2007b）。本研究中，刺参幼参肠道的蛋白酶活性较高，且受不同藻类影响显著，而脂肪酶活性较低，表明刺参参对不同藻类蛋白质的吸收能力不同，对脂肪的耐受性低（Seo and Lee，2011），多项研究也证明了刺参肠道具有较高的蛋白酶活性和较低的脂肪酶活性（唐黎等，2007；赵斌等，2015），与本研究结果一致。铜藻粉组和海带渣组刺参参肠道消化酶活性较低，证明刺参幼参对其消化吸收能力较弱。与此一致，铜藻粉组及海带渣组刺参幼参肠道的皱襞高度明显小于对照组及混合海藻粉组，细胞数目明显减少，而肠绒毛的完整性及高度是营养物质被消化吸收的基本保障，可能是因为消化道内褐藻酸酶活性较低（唐黎等，2007），刺参对铜藻粉和海带渣的吸收能力较弱，绒毛之间的隐窝细胞分裂速度降低，不能及时分化迁移以补充正常脱落的绒毛上皮（王吉桥等，2007b；孙丽娜，2013），消化吸收的表面积减小，从而影响了对营养物质的吸收。此外，酸不溶性灰分是指示饲料消化率的有效指标（Li et al.，2008），本研究中，海带粉组、石莼粉组及混合海藻粉组刺参幼参的酸不溶性灰分含量均显著低于对照组，铜藻粉组显著高于对照组，海带渣组与对照组差异不显著，表明刺参参对海带粉、石莼粉及混合海藻粉的消化吸收率较高，对铜藻粉和海带渣的消化吸收率较低，可能因为较铜藻而言，海带和石莼的质地较软，易被刺参幼参肠道破坏。

超氧化物歧化酶、酸性磷酸酶和碱性磷酸酶的活性是评价刺参机体免疫力的重要指标。超氧化物歧化酶是机体清除活性氧的关键防御酶，其活性直接指示免疫系统的状态及清除自由基的能力（杨宁等，2016；秦搏等，2015；董晓亮等，2013）。丙二醛是自由基作用于脂质发生过氧化的终产物，会引起蛋白质、核酸等生物大分子交联聚合（Wakita et al.，2011）。在本研究中，海带渣组刺参幼参的氧化产物丙二醛含量显著高于其他各组，超氧化物歧化酶活性显著低于其他各组，表明该组刺参幼参体内脂质过氧化反应较为严重，机体清除自由基的能力较弱，混合海藻粉组超氧化物歧化酶活性与对照组差异不显著，但均显著高于其他各组，表明多种海藻粉混合饲喂刺参参可以有效提高机体清除自由基的能力，提高刺参幼参的抗氧化性能。酸性磷酸酶和碱性磷酸酶是溶酶体酶的重要组成部分（郑慧等，2014），同时还是刺参体内参与免疫防御活动的重要水解酶，在刺参抵抗疾病、免疫反应及细胞损伤修复过程中具有一定的生物学意义（李继业，2007）。在本研究中，海带粉和混合海藻粉组的刺参幼参肠道酸性磷酸酶和碱性磷酸酶的活性显著高于海藻渣组，可能是因为海带中含有较多矿物质及多糖成分，能激活免疫相关酶的活性，从而有效提高刺参的免疫性能（李丹彤等，2014）。

3.1.2.4 结论

通过对刺参幼参生长性能、非特异性免疫及消化机能的研究可知，混合海藻粉具有与鼠尾藻粉相似的饲喂效果，因此建议刺参幼参养殖中混合使用多种海藻粉，以保证刺参幼参的营养均衡。

3.1.3 酶解海带粉对刺参幼参生长性能、体组成、消化代谢和抗氧化能力的影响

海藻具有非常重要的经济、营养、药用和生态价值。全球可供人类使用的海藻共有291种（Balbas et al.，2015），主要用于生产食品和水凝胶，提取药物、糖以及生产有机材料、肥料和动物饲料等。海带是一种常见的海产品，属于海藻中的褐藻，具有成本低、产量高、营养丰富等优点。海带作为水产饲料原料，由于产量大并富含糖类、矿物质、维生素、游离氨基酸、脂肪酸等（孙永泰，2015），不仅在自然海域中被作为刺参的天然饵料，也常被用来作为刺参配合饲料中的主要原料来源。但海带中含有纤维素、半纤维素、β-葡聚糖、果胶等非淀粉多糖，难以被动物分泌的内源酶降解，并且其中可溶性非淀粉多糖具有较强的抗营养作用，溶于水后产生较强的黏性，形成食糜黏度增加、营养屏障作用和微生物菌群数量增加等不利影响，以至于降低养殖动物对饲料养分的消化利用率（Bedford and Classen，1992）。因此，将海带的多糖成分进行降解处理是提高其利用率、减少不利影响的有效途径。目前，对海带多糖的降解方法研究较多，如物理降解法、化学降解法和生物降解法。与物理降解法和化学降解法相比，生物降解法具有反应条件温和、专一性强等优势（董学前等，2017；姚骏等，2018）。生物降解法即酶解，是用专一性糖苷酶或非专一性酶通过特异性裂解多糖中的某一糖苷键对多糖进行酶解（郭娜等，2019），将其分解为更容易吸收的小分子营养物质，降低食糜黏度，提高饲料利用率，进而提高养殖动物的生长性能（曹新勇，2013；关莹

等，2021；王美琪等，2023）。

3.1.3.1 材料与方法

1. 酶解海带粉制备

实验用酶制剂购于武汉新华扬生物股份有限公司，包括非淀粉多糖酶（木聚糖酶，8000U/g；β-葡聚糖酶，2000U/g；β-甘露聚糖酶，150U/g）、纤维素酶（500U/g）、中性蛋白酶（100 000U/g）、风味酶（5000U/g）。酶解海带粉是由海带粉经复合酶制剂（非淀粉多糖酶∶纤维素酶∶中性蛋白酶∶风味酶=8∶12∶3∶1）水解制成，复合酶添加量为3%(占料重的比例)，料液比为1∶6，酶解条件为温度50℃、pH=6、反应时间6h。海带粉和酶解海带粉营养组成及还原糖含量见表3-7。

表3-7 海带粉和酶解海带粉营养组成及还原糖含量（干物质）（%）

营养组成	海带粉	酶解海带粉
粗蛋白	9.49	9.85
酸溶蛋白（占粗蛋白的比例）	4.67	8.83
粗脂肪	3.00	3.18
粗灰分	44.05	44.90
粗纤维	9.37	18.82
还原糖	0.016	1.550

2. 实验饲料

本研究以鱼粉、海带粉和酶解海带粉为主要蛋白源，配制粗蛋白含量为12.00%、粗脂肪含量为0.40%、能量为6.20kJ/g的基础饲料，其配方及营养组成、氨基酸组成见表3-8和表3-9。在基础饲料中分别添加0、3.00%、6.00%、9.00%、12.00%和15.00%的酶解海带粉，分别命名为D1（对照组）、D2、D3、D4、D5和D6组，各组均添加同等含量的0.60%牛磺酸以解决海带粉中牛磺酸不足的问题（Zhao et al., 2017），配成6组等氮等能的实验饲料。D1~D6组中牛磺酸的实际含量分别为0.71%、0.72%、0.69%、0.70%、0.68%和0.69%，各组实验饲料添加0.1%氧化钇（Y_2O_3）作为外源指示剂。实验饲料的原料经粉碎后过60目筛，混合，制成粒径为6mm的颗粒。以上各组实验饲料经70℃烘干后于−20℃冰柜中保存备用。

表3-8 实验饲料配方及营养组成（干物质）（%）

项目		组别					
		D1	D2	D3	D4	D5	D6
原料组成	白鱼粉	5.50	5.50	5.50	5.50	5.50	5.50
	海带粉	25.00	22.00	19.00	16.00	13.00	10.00
	酶解海带粉	0.00	3.00	6.00	9.00	12.00	15.00

续表

	项目	组别					
		D1	D2	D3	D4	D5	D6
原料组成	牛磺酸	0.60	0.60	0.60	0.60	0.60	0.60
	虾粉	3.50	3.50	3.50	3.50	3.50	3.50
	豆粕	3.00	3.00	3.00	3.00	3.00	3.00
	酵母粉	0.60	0.60	0.60	0.60	0.60	0.60
	贝壳粉	5.00	5.00	5.00	5.00	5.00	5.00
	卡拉胶	0.50	0.50	0.50	0.50	0.50	0.50
	α-淀粉	5.00	5.00	5.00	5.00	5.00	5.00
	维生素预混料	1.00	1.00	1.00	1.00	1.00	1.00
	矿物质预混料	1.00	1.00	1.00	1.00	1.00	1.00
	氧化钇 Y_2O_3	0.10	0.10	0.10	0.10	0.10	0.10
	海泥	49.20	49.20	49.20	49.20	49.20	49.20
营养组成	粗蛋白	12.26	12.20	12.14	11.99	12.07	11.77
	粗脂肪	0.30	0.31	0.33	0.35	0.36	0.38
	粗灰分	59.98	60.46	61.67	62.02	63.19	63.91
	能量/(kJ/g)	6.25	6.20	6.20	6.18	6.17	6.14

注：①每千克维生素预混料含维生素 A 7500.00IU、维生素 D 1500.00IU、维生素 E 60.00mg、维生素 K_3 18.00mg、维生素 B_1 12.00mg、维生素 B_2 12.00mg、维生素 B_{12} 0.10mg、泛酸 48.00mg、烟酰胺 90.00mg、叶酸 3.70mg、D-生物素 0.20mg、盐酸吡哆醇 60.00mg、维生素 C 310.00mg；②每千克矿物质预混料含锌 35.00mg、锰 21.00mg、铜 8.30mg、铁 23.00mg、钴 1.20mg、碘 1.00mg、硒 0.30mg。

表3-9 实验饲料氨基酸组成（%）

氨基酸		组别					
		D1	D2	D3	D4	D5	D6
必需氨基酸	苏氨酸	0.52	0.53	0.53	0.52	0.53	0.50
	缬氨酸	0.64	0.62	0.65	0.64	0.66	0.60
	蛋氨酸	0.59	0.60	0.60	0.64	0.60	0.59
	异亮氨酸	0.46	0.48	0.49	0.45	0.49	0.45
	亮氨酸	0.90	0.90	0.92	0.91	0.92	0.89
	苯丙氨酸	0.73	0.74	0.75	0.75	0.64	0.70
	赖氨酸	0.66	0.67	0.67	0.66	0.70	0.66
	组氨酸	0.18	0.19	0.19	0.21	0.25	0.19
	精氨酸	0.57	0.58	0.58	0.58	0.59	0.55
非必需氨基酸	天冬氨酸	1.16	1.17	1.19	1.20	1.19	1.19
	丝氨酸	0.60	0.60	0.57	0.58	0.57	0.58
	谷氨酸	1.57	1.62	1.57	1.69	1.63	1.63
	甘氨酸	0.63	0.64	0.63	0.62	0.63	0.61
	丙氨酸	0.65	0.66	0.66	0.68	0.70	0.65
	半胱氨酸	0.36	0.38	0.39	0.39	0.40	0.37

续表

氨基酸		组别					
		D1	D2	D3	D4	D5	D6
非必需氨基酸	酪氨酸	0.57	0.60	0.64	0.66	0.64	0.56
	脯氨酸	0.46	0.45	0.47	0.46	0.48	0.45
总氨基酸		11.27	11.42	11.49	11.62	11.63	11.17

3. 实验用刺参和饲养管理

实验用刺参购于山东安源种业科技有限公司（蓬莱）。饲喂实验在山东省海洋资源与环境研究院东营实验基地循环水养殖系统中进行。所有刺参均饲喂基础饲料 1 周，以使其适应实验饲料和饲养条件。实验开始前禁食 24h。选择初始平均体质量为（11.40±0.04）g 的健康刺参 540 头，随机分配到 18 个圆柱形循环水桶（$h_桶$=80cm，$h_水$=50cm，$d_桶$=60cm），分为 6 个实验组，每组 3 个重复，每个重复 30 头刺参，饲喂期 56d。每天定时（16:00）投饵 1 次，饱食投喂。每 3d 换一次水，并采用虹吸法将残饵和粪便用虹吸软管从桶底吸出，换水量为桶内水位的 1/2。养殖实验进行 1 个月时，彻底清洗水桶。养殖期间水温为 13～17℃，pH 为 7.5～8.2，溶氧量不低于 6mg/L，盐度维持在 28～30。

4. 样品采集

养殖实验期间，每天投喂前将包膜完整的长条状粪便收集起来，–20℃保存，用于后续营养物质表观消化率测定。养殖实验结束后，禁食 24h，然后分别对每个养殖桶中的刺参称重并计数。称重后，随机从每个桶中取 14 头刺参进行解剖，称量体壁重、内脏重和肠重，采集刺参的体壁和肠道，–80℃保存，其中体壁用于后续体成分测定，肠道用于后续酶活性测定。

5. 指标测定

刺参存活率、增重率、特定生长率、脏壁比、肠壁比，以及饲料干物质、营养成分的表观消化率计算方法见附录。

实验饲料、粪便及刺参体壁水分、粗蛋白、粗脂肪、粗灰分的含量和氨基酸含量测定方法见附录。饲料和粪便中总磷含量采用分光光度法[《饲料中总磷的测定 分光光度法》（GB/T 6437—2018）]测定，钇含量采用高频电感耦合等离子体发射光谱法测定，能量采用燃烧法使用量热仪测定。海带粉和酶解海带粉中还原糖通过 3,5-二硝基水杨酸比色法测定（王莉丽等，2020；程柳和李静，2016），以葡萄糖为标准品，3,5-二硝基水杨酸比色法的用量为 3ml，取样品溶液 2ml 于 25ml 具塞刻度试管中，沸水浴加热 6min，冷却后，于 540nm 波长条件下测定还原糖含量。

1）肠道消化酶和代谢酶测定

肠道中淀粉酶、脂肪酶、蛋白酶、ATP 酶（adenosine triphosphatase）、Na^+-K^+-ATP 酶（Na^+-K^+-ATPase）、Ca^{2+}-Mg^{2+}-ATP 酶（Ca^{2+}-Mg^{2+}-ATPase）、葡萄糖激酶（glucokinase，GK）、磷酸烯醇式丙酮酸羧激酶（phospho-enolpyruvate carboxykinase，PEPCK）、谷草转氨酶、谷丙转氨酶的活性均使用南京建成生物工程研究所生产的试剂盒进行测定，各种酶活性单位参照试剂盒说明书表示。

2）肠道抗氧化指标测定

肠道中过氧化氢酶和超氧化物歧化酶的活性、总抗氧化能力以及丙二醛含量均使用南京建成生物工程研究所生产的试剂盒进行测定，各种酶活性单位参照试剂盒说明书表示。

6. 数据统计分析

实验数据采用 SPSS 17.0 软件进行单因素方差分析，当达到差异显著（$P<0.05$），应用邓肯多重范围检验进行多重比较分析。统计结果以平均值±标准差（n=3）表示。

3.1.3.2 结果分析

1. 刺参幼参生长性能

如表 3-10 所示，随着饲料中酶解海带粉含量的增加，终末体质量、增重率和特定生长率均呈现先升高后略有降低的趋势，D3、D4、D5 和 D6 组显著高于对照组（$P<0.05$）；各组间肠壁比、脏壁比和存活率均无显著性差异（$P>0.05$）。以增重率为评价指标，经一元二次回归分析（$Y=-0.198X^2+4.1029X+60.586$）得出，初始体质量为（11.40±0.04）g 的刺参幼参对饲料中酶解海带粉的最适需求量为 10.36%（图 3-2）。

表 3-10 酶解海带粉对刺参幼参生长性能的影响

项目	组别					
	D1	D2	D3	D4	D5	D6
初始体质量/g	11.41±0.02	11.41±0.02	11.39±0.02	11.40±0.03	11.41±0.02	11.43±0.04
终末体质量/g	18.46±0.67[a]	19.14±0.27[a]	20.45±0.11[b]	20.85±0.39[b]	20.56±0.88[b]	20.24±0.86[b]
增重率/%	61.96±5.89[a]	67.92±2.34[a]	79.39±0.94[b]	82.91±3.45[b]	80.36±7.68[b]	77.58±7.52[b]
特定生长率/(%/d)	0.72±0.06[a]	0.77±0.03[a]	0.87±0.01[b]	0.90±0.02[b]	0.88±0.06[b]	0.86±0.06[b]
肠壁比/%	5.25±0.18	5.49±0.64	6.45±0.07	5.24±0.53	5.17±0.46	4.93±0.29
脏壁比/%	9.28±0.49	9.95±0.73	10.12±0.64	10.30±0.67	10.07±0.24	9.58±0.12
存活率/%	96.21±4.56	96.45±5.74	97.38±3.51	96.52±4.67	97.35±5.43	98.44±4.37

注：同行数据无字母或上标相同字母表示差异不显著（$P>0.05$），上标不同字母表示差异显著（$P<0.05$）。

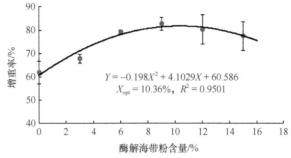

图 3-2 刺参幼参增重率与饲料中酶解海带粉含量的回归分析

2. 刺参幼参对饲料干物质及营养成分的表观消化率

如表 3-11 所示,刺参幼参对添加不同水平酶解海带粉饲料的干物质的表观消化率无显著性差异($P>0.05$);D5 和 D6 组粗蛋白的表观消化率显著低于对照组($P<0.05$);总磷的表观消化率呈现先上升后下降的趋势,在 D3 组达到最大值($P<0.05$);所有的酶解海带粉添加组均提高了总能表观消化率($P<0.05$);随着酶解海带粉含量的增加,饲料中异亮氨酸、亮氨酸、精氨酸和酪氨酸的表观消化率呈现先升高后下降的趋势,D6 组中异亮氨酸、亮氨酸、组氨酸、精氨酸和酪氨酸的表观消化率显著低于对照组($P<0.05$)。

表 3-11 酶解海带粉对刺参幼参对营养素的表观消化率的影响(%)

项目		组别					
		D1	D2	D3	D4	D5	D6
干物质		10.21±0.83	10.62±1.01	11.18±0.76	11.31±0.54	9.24±0.79	10.24±0.18
粗蛋白		53.43±0.24[b]	55.39±1.71[b]	53.34±2.03[b]	52.45±1.42[b]	48.10±1.33[a]	48.31±1.62[a]
总磷		21.78±1.06[a]	25.87±0.87[b]	29.75±2.03[c]	26.59±2.99[bc]	23.03±0.75[ab]	21.91±2.68[a]
总能		42.07±2.11[a]	46.51±2.65[ab]	51.55±2.62[cd]	52.98±2.58[d]	48.04±0.75[bc]	52.22±2.71[cd]
必需氨基酸	苏氨酸	49.92±3.19	53.59±3.92	51.16±3.74	48.94±3.86	44.88±2.56	43.18±1.68
	缬氨酸	57.74±2.07	58.32±2.87	58.41±3.16	56.22±2.45	54.30±1.44	51.04±1.70
	蛋氨酸	61.93±2.13[b]	63.24±3.12[b]	63.39±3.43[b]	63.32±2.47[b]	58.97±1.97[ab]	57.03±1.93[a]
	异亮氨酸	61.44±2.76[bc]	65.01±3.41[c]	61.86±4.88[bc]	55.76±3.68[ab]	55.30±1.59[a]	52.05±2.38[a]
	亮氨酸	61.85±3.74[c]	64.21±2.45[c]	61.77±4.30[c]	60.47±2.20[bc]	55.77±1.71[ab]	54.70±1.85[a]
	苯丙氨酸	61.79±2.93	63.91±2.43	56.19±4.29	62.31±2.04	51.02±1.55	56.28±1.41
	赖氨酸	62.49±3.78	65.94±1.89	61.56±4.21	61.51±2.70	58.06±1.73	57.47±1.42
	组氨酸	60.16±3.39[bc]	66.03±2.05[c]	59.10±4.46[b]	60.37±3.24[bc]	61.80±2.46[bc]	52.04±2.28[a]
	精氨酸	65.45±2.83[b]	68.43±1.98[b]	64.27±2.60[b]	63.63±2.48[b]	57.74±0.88[ab]	55.96±2.40[a]
非必需氨基酸	天冬氨酸	53.19±4.15	57.42±3.22	55.63±3.14	54.77±3.02	49.13±2.63	49.75±1.06
	丝氨酸	52.95±2.50	56.07±3.98	53.32±3.07	52.53±3.29	47.31±2.65	49.13±1.10
	谷氨酸	60.92±3.22	64.28±1.41	60.55±4.21	62.32±3.00	55.81±2.13	57.38±1.74
	甘氨酸	51.28±3.54	54.77±3.68	52.08±2.65	50.42±2.57	46.46±2.30	45.36±1.36

续表

项目		组别					
		D1	D2	D3	D4	D5	D6
非必需氨基酸	丙氨酸	52.01±3.39	55.40±3.51	54.45±4.04	52.85±3.92	50.18±3.15	45.07±1.51
	半胱氨酸	40.52±3.53	43.26±3.20	44.57±3.49	45.34±1.48	42.83±2.36	38.13±1.96
	酪氨酸	56.68±2.69^{bc}	60.90±2.52^{bc}	62.21±2.96^c	62.13±2.39^c	57.79±1.59^b	51.98±0.32^a
	脯氨酸	53.75±2.87	55.85±3.29	55.75±2.40	51.68±2.49	50.11±3.51	47.82±2.14
总氨基酸		52.21±2.28^b	54.72±3.41^b	53.07±3.15^b	52.87±4.12^b	48.17±2.35^a	47.17±3.17^a

注：同行数据无字母或上标相同字母表示差异不显著（$P>0.05$），上标不同字母表示差异显著（$P<0.05$）。

3. 刺参幼参体壁基本营养成分和氨基酸含量

如表 3-12 所示，饲料中添加酶解海带粉并未显著改变刺参幼参体壁的水分、粗蛋白、粗脂肪和粗灰分的含量（$P>0.05$）。随着饲料中酶解海带粉含量的增加，刺参幼参体壁蛋氨酸含量呈现先上升后下降的趋势，D2、D3、D4 和 D5 组显著高于对照组（$P<0.05$）；半胱氨酸含量呈现先上升后平稳趋势，各实验组均显著高于对照组（$P<0.05$）。

表 3-12 酶解海带粉对刺参幼参体壁营养组成及氨基酸组成的影响（干物质）（%）

项目		组别					
		D1	D2	D3	D4	D5	D6
水分		91.31±0.46	91.46±0.42	92.03±0.57	91.61±0.19	91.44±0.25	91.74±0.59
粗蛋白		48.43±1.72	48.75±0.72	48.81±0.51	47.98±1.33	48.03±1.16	47.99±1.21
粗脂肪		2.75±0.10	1.96±0.19	2.70±0.24	3.17±0.15	3.02±0.45	2.76±0.44
粗灰分		33.06±1.14	33.80±0.86	34.01±1.10	33.63±1.05	33.41±1.67	34.50±1.91
必需氨基酸	苏氨酸	2.63±0.11	2.54±0.17	2.63±0.01	2.50±0.07	2.55±0.08	2.51±0.11
	缬氨酸	2.23±0.05	2.09±0.22	2.27±0.02	2.27±0.05	2.06±0.12	2.17±0.08
	蛋氨酸	1.27±0.06^a	1.51±0.02^b	1.50±0.04^b	1.42±0.06^b	1.47±0.07^b	1.16±0.16^a
	异亮氨酸	1.32±0.10	1.38±0.05	1.35±0.05	1.29±0.10	1.47±0.11	1.28±0.05
	亮氨酸	2.57±0.13	2.62±0.11	2.59±0.02	2.50±0.09	2.60±0.06	2.48±0.08
	苯丙氨酸	1.72±0.16	1.81±0.17	1.70±0.07	1.73±0.03	1.83±0.04	1.75±0.01
	赖氨酸	2.03±0.01	2.07±0.04	2.05±0.03	2.05±0.01	2.13±0.06	2.04±0.01
	组氨酸	0.62±0.01	0.65±0.02	0.63±0.01	0.65±0.01	0.66±0.02	0.64±0.02
	精氨酸	3.07±0.01	3.10±0.07	3.02±0.01	3.06±0.14	3.26±0.14	3.15±0.01
非必需氨基酸	天冬氨酸	4.82±0.13	4.70±0.24	4.81±0.02	4.54±0.17	4.74±0.10	4.62±0.18
	丝氨酸	2.60±0.15	2.42±0.14	2.53±0.02	2.39±0.07	2.42±0.05	2.44±0.09
	谷氨酸	6.27±0.05	6.23±0.17	6.31±0.06	6.29±0.18	6.51±0.17	6.29±0.10
	甘氨酸	6.13±0.24	5.68±0.33	6.10±0.16	5.32±0.30	5.85±0.40	5.81±0.63
	丙氨酸	3.01±0.05	2.84±0.16	3.01±0.04	2.70±0.13	2.86±0.14	2.86±0.16
	半胱氨酸	1.68±0.05^a	1.80±0.06^b	1.94±0.02^c	1.93±0.03^c	1.92±0.02^c	1.89±0.01^c

续表

项目		组别					
		D1	D2	D3	D4	D5	D6
非必需氨基酸	酪氨酸	2.34±0.02	2.34±0.04	2.28±0.03	2.41±0.11	2.34±0.06	2.61±0.25
	脯氨酸	3.34±0.11	3.18±0.17	3.30±0.10	3.01±0.13	3.24±0.15	3.12±0.24
总氨基酸		47.92±0.78	46.98±0.66	48.14±0.85	46.15±0.38	47.15±0.74	46.18±0.47

注：同行数据无字母或上标相同字母表示差异不显著（$P>0.05$），上标不同字母表示差异显著（$P<0.05$）。

4. 刺参幼参肠道消化酶和代谢酶指标的影响

如表 3-13 所示，随着饲料中酶解海带粉含量的增加，刺参幼参肠道淀粉酶和蛋白酶的活性呈现先上升后下降的趋势，且均在 D5 组达到最高值；各组间肠道脂肪酶活性无显著性差异（$P>0.05$）。

表 3-13 酶解海带粉对刺参幼参肠道消化酶与代谢酶指标的影响

项目		组别					
		D1	D2	D3	D4	D5	D6
消化酶	脂肪酶/(U/g prot)	7.25±0.28	7.39±0.54	7.21±0.88	7.24±0.35	7.33±0.46	7.12±0.39
	淀粉酶/(U/mg prot)	1.43±0.03^a	1.88±0.16^{bc}	2.05±0.15^c	2.13±0.09^{cd}	2.35±0.14^d	1.69±0.25^{ab}
	蛋白酶/(U/g prot)	1.36±0.08^a	1.69±0.11^b	1.88±0.12^b	1.91±0.14^b	1.96±0.07^b	1.32±0.09^a
代谢酶	葡萄糖激酶/(ng/ml)	5.09±0.20^b	5.55±0.11^c	5.68±0.10^c	5.94±0.14^d	4.84±0.09^a	4.62±0.18^a
	丙酮酸激酶/(U/mg)	80.86±1.35^{ab}	83.53±1.62^{bc}	86.51±2.82^{bc}	89.53±1.75^c	77.03±4.95^a	74.91±2.57^a
	磷酸烯醇式丙酮酸羧激酶/(U/g prot)	167.66±11.72^a	222.63±10.36^b	281.70±17.91^c	294.48±10.67^c	344.52±18.86^d	237.49±13.23^b
	Na^+-K^+-ATPase[μmol pi/(mg prot·h)]	3.26±0.40	3.28±0.12^a	3.47±0.79^a	3.98±0.38^a	4.21±0.65^a	5.11±0.46^b
	Ca^{2+}-Mg^{2+}-ATPase[μmol pi/(mg prot·h)]	1.19±0.06^a	1.30±0.10^a	1.73±0.09^b	1.95±0.17^b	1.91±0.17^b	2.26±0.11^c
	谷丙转氨酶/(U/g prot)	1.67±0.19^a	2.05±0.13^{ab}	2.22±0.15^b	3.02±0.25^c	4.65±0.36^d	3.43±0.15^c
	谷草转氨酶/(U/g prot)	1.41±0.17^a	1.58±0.10^{ab}	1.84±0.18^b	1.94±0.08^b	2.83±0.15^c	1.96±0.34^b

注：同行数据无字母或上标相同字母表示差异不显著（$P>0.05$），上标不同字母表示差异显著（$P<0.05$）。

葡萄糖激酶、丙酮酸激酶和磷酸烯醇式丙酮酸羧激酶的活性均呈现先上升后下降的趋势，葡萄糖激酶和磷酸烯醇式丙酮酸羧激酶的活性 D2、D3 和 D4 组显著高于对照组（$P<0.05$）；Na^+-K^+-ATPase 活性 D6 组显著高于其他各组（$P<0.05$）；Ca^{2+}-Mg^{2+}-ATPase 活性随酶解海带粉含量的增加呈现上升趋势，D3、D4、D5 和 D6 组显著高于对照组（$P<0.05$）。随着饲料中酶解海带粉含量的增加，谷草转氨

酶和谷丙转氨酶的活性呈现先升高后下降的趋势,且均在 D5 组达到最高值。

5. 刺参幼参抗氧化性能

如表 3-14 所示,刺参幼参肠道总抗氧化能力和超氧化物歧化酶活性随着酶解海带粉含量的增加,均呈现先升高后下降的趋势,分别在 D4 组和 D5 组达到最高;丙二醛含量则呈现相反的趋势,D5 组和 D6 组丙二醛含量显著高于其他组（$P<0.05$）；各组间过氧化氢酶活性无明显差异（$P>0.05$）。

表 3-14　酶解海带粉对刺参幼参肠道抗氧化性能的影响

项目	组别					
	D1	D2	D3	D4	D5	D6
过氧化氢酶活性/（U/mg prot）	124.31±13.11	134.31±14.14	138.09±9.89	125.91±10.71	125.66±12.26	122.85±11.52
超氧化物歧化酶活性/（U/g prot）	10.59±0.15^b	10.76±0.41^b	11.31±0.27^c	12.27±0.62^d	12.50±0.74^d	9.71±0.39^a
总抗氧化能力/（nmol/mg prot）	67.55±1.50^a	68.45±0.90^a	73.15±2.00^b	80.38±1.43^c	78.22±1.38^c	73.92±1.42^b
丙二醛含量/（nmol/mg prot）	1.99±0.14^b	1.82±0.14^a	1.84±0.18^a	1.64±0.15^a	2.39±0.22^c	2.51±0.17^c

注：同行数据无字母或上标相同字母表示差异不显著（$P>0.05$）,上标不同字母表示差异显著（$P<0.05$）。

3.1.3.3　讨论

本研究中,饲料中添加酶解海带粉影响刺参幼参的生长性能。随着饲料中酶解海带粉含量的逐渐增加,刺参幼参生长性能呈现先升高后平稳的趋势,是因为海带经非淀粉多糖酶、纤维素酶、中性蛋白酶和风味酶复合酶制剂酶解后还原糖和酸溶蛋白的含量分别提高了 96.88% 和 1.89%（表 3-7）。饲料中还原糖和酸溶蛋白含量的显著提高,有助于刺参幼参的消化和吸收,并且其中的糖类和寡肽能调节机体代谢,进而提高刺参幼参的生长性能（江晓路等,2009；Song et al.,2016）。相关研究已经报道了藻类经酶解处理后能更好地促进海洋动物的生长。例如,秦博等（2015）发现,用纤维素酶和蛋白酶对浒苔进行酶解后干燥粉碎处理的饲料可以显著提高刺参的特定生长率；彭素晓（2017）的研究表明,海带酶解产物可显著提高凡纳滨对虾的特定生长率,并能显著降低饲料系数,提高非特异性免疫力。本研究结合表观消化率对刺参幼参的生长性能进行分析发现,饲料的粗蛋白、总磷和总能的表观消化率和增重率、特定生长率呈现一致的变化趋势,说明酶解海带粉提高了刺参幼参对饲料营养素的整体利用效率,从而促进了刺参幼参快速生长。但本研究也发现,刺参幼参的生长性能并不随着酶解海带粉的添加而进一步提高。相反,酶解海带粉添加量超过 9% 导致刺参幼参特定生长率出现下降的趋势,说明添加 9% 的酶解海带粉可能已经超过了刺参幼参机体对其代谢处理的能力范围，反而对刺参幼参生长代谢产生抑制作用。李猛等（2015）的研究表明，饲料中添加过高比例的发酵浒苔会降低刺参幼参的增重率和特定生长率，原因是过量的发酵浒苔释放的葡萄糖和氨基酸等小分子物质的含量可能超过了刺参幼参正常的吸收值，从而导致其生长性能下降。本研究的结果显示，当酶解海带粉添加量达到12%~15%时，粗蛋白的表观消化率与对照组相比显著降低，说明还原糖含量超过机体耐受极限，并影响刺参幼参肠道对营养物质的吸收，也导致刺参幼参摄入的总能量呈现降低的趋势，导致刺参幼参生长速度减缓，这也解释了过量添加酶解海带粉降低刺参幼参生长性能的原因。氨基酸的组成和表观消化率是决定饲料中的蛋白质质量的重要因素（王建学等，2021）。本研究中，蛋氨酸、异亮氨酸、亮氨酸、组氨酸、精氨酸和酪氨酸的表观消化率均有显著变化，其中，异亮氨酸、亮氨酸、精氨酸和酪氨酸的表观消化率随着酶解海带粉含量的增加呈现先升高后下降的趋势，异亮氨酸、亮氨酸、精氨酸的表观消化率均在D2组达到最高值，酪氨酸的表观消化率在D3组达到最高值，说明低酶解海带粉添加量提高了刺参幼参对以上4种氨基酸的吸收效率，从而导致氨基酸之间消化的差异性。

本研究表明，随着饲料中酶解海带粉含量的增加，各组之间刺参幼参体壁水分、粗蛋白、粗脂肪、粗灰分的含量均无显著性差异，说明添加酶解海带粉并不显著影响各营养素在刺参幼参体壁的沉积保留。已知组织中氨基酸水平受到饲料中蛋白质质量的影响（Yamamoto et al.，2000），从氨基酸角度分析，与对照组相比，饲料中酶解海带粉的适量添加提高了刺参幼参体壁中蛋氨酸和半胱氨酸的含量，原因是蛋氨酸和半胱氨酸的表观消化率发生变化，氨基酸消化吸收程度越高，对刺参幼参体壁氨基酸沉积的影响也就越大。

在饲料中添加适量的酶解海带粉能显著提高刺参幼参肠道蛋白酶和淀粉酶的活性，有助于刺参幼参对蛋白质和糖类的消化吸收，进而影响刺参幼参的生长。海带粉经酶解处理后释放了非淀粉多糖和大分子蛋白质，使还原糖和可溶性蛋白的含量显著提高，刺参幼参更容易吸收，从而诱导性地提高刺参幼参肠道中蛋白酶和淀粉酶的活性。但是，刺参幼参肠道脂肪酶活性并不受酶解海带粉添加的影响，其原因是酶解处理后的酶解海带粉脂肪的含量并没有发生明显的变化。

水产动物利用葡萄糖的方式与哺乳动物类似，都是糖酵解和糖异生协同发挥作用，以维持动物体内葡萄糖的稳态（Polakof et al.，2012）。葡萄糖激酶和丙酮酸激酶是糖酵解途径中两个关键性限制酶，由于葡萄糖激酶的底物亲和力常数K_m的值很低，只有在细胞内葡萄糖浓度足够高时才有效，因此，葡萄糖激酶活性可作为判断细胞对葡萄糖利用的指标。磷酸烯醇式丙酮酸羧激酶是糖异生途径的催化酶，可以将乳酸和氨基酸等非糖类物质通过糖异生途径转化为葡萄糖，维持机体内葡萄糖的稳态。本研究中，当酶解海带粉含量从0增加到9%时，葡萄糖激酶和丙酮酸激酶的活性显著升高，当酶解海带粉含量从12%增加到15%时，葡萄糖激酶和丙酮酸激酶的活性显著降低，这表明适量的酶解海带粉对刺参糖酵解有促进作用。磷酸烯醇式丙酮酸羧激酶活性随着酶解海带粉含量的增加呈现先升高后下降的趋势，在 D5 组达到最高值。这说明随着酶解海带粉添加量的增加，同时促进了糖酵解和糖异生代谢反应，但当添加量达到 12%时，刺参幼参肠道细胞的糖酵解减弱，而糖异生过程增强，说明 9%～12%添加水平已经满足刺参幼参对糖的需求，机体内过量的葡萄糖被转运出细胞外。ATP 酶在葡萄糖和氨基酸等营养物质的吸收过程中，起着重要的作用。Na^+-K^+-ATP 酶和 Ca^{2+}-Mg^{2+}-ATP 酶是细胞膜上重要的酶（段树丽，2021），ATP 酶活性的升高有助于细胞内能量代谢、离子转运和信息传递，对维持细胞的生理功能有重要的作用。本研究中，刺参幼参肠道中 Na^+-K^+-ATP 酶和 Ca^{2+}-Mg^{2+}-ATP 酶的活性均呈现随着酶解海带粉含量的增加逐渐升高的趋势，这表明饲料中添加酶解海带粉提高了营养物质的水解供能。另外，本研究发现，在中低添加水平下（3%～12%）谷丙转氨酶和谷草转氨酶的活性显著升高，但高添加水平则抑制二者的活性。谷丙转氨酶和谷草转氨酶是动物体内蛋白质代谢过程中两个关键的代谢酶，其活性的高低一定程度上反映了机体对蛋白质的合成和分解能力（Yan et al.，2007）。因此，这种现象表明饲料中添加适量的酶解海带粉能在一定程度上提高刺参幼参肠道对蛋白质的代谢处理能力，过量添加反而会给刺参幼参肠道增加代谢负担，不利于蛋白质的消化。

生物机体内关键性的抗氧化因子主要包括超氧化物歧化酶、总抗氧化能力和过氧化氢酶等。超氧化物歧化酶是动物体内关键的抗氧化酶之一，是超氧自由基的天然消除剂，使自由基的形成与消除处于一种动态平衡中，从而避免其对生物分子的损伤等（吴阳，2012）。总抗氧化能力的强弱与健康程度密切相关，该防御体系由酶促和非酶促 2 个体系组成，协同防护机体氧化。丙二醛是多不饱和脂肪酸受氧自由基攻击形成的脂质过氧化产物，反映细胞的受损伤程度和脂质过氧化程度（Muñoz et al.，2000）。本研究表明，超氧化物歧化酶活性随着酶解海带粉含量的增加呈现先升高后下降的趋势，可能是随着酶解海带粉含量的增加，刺参幼参肠道在代谢过程中增强了呼吸链传递通路所介导的能量代谢，从而促进了活性氧的产生，刺参幼参机体为了消除活性氧维持稳态，从而提高抗氧化酶的合成速度。当酶解海带粉含量过高时，打破了活性氧自由基的动态平衡，对细胞产生了一定的毒害作用，从而抑制了超氧化物歧化酶的产生。杜以帅（2010）将褐藻寡糖和海藻粉饲喂刺参 40d 后，刺参体腔液和体壁中超氧化物歧化酶活性有了一定程度的提高，并呈现先升高后下降的趋势，这与本研究的结果一致。Hu 等（2021）的研究表明，在草鱼饲料中添加 100～400mg/kg 褐藻寡糖，能显著提高超氧化物歧化酶和过氧化氢酶的活性，并降低丙二醛含量，从而提高草鱼的抗氧化能力。本研究中，总抗氧化能力随着酶解海带粉含量的升高呈先升高后下降的趋势，当酶解海带粉添加水平为 9%时，总抗氧化能力达到最大。与之相反的是，丙二醛含量随着酶解海带粉添加水平提高呈现下降趋势,当酶解海带粉含量达到12%时,丙二醛含量显著升高,丙二醛含量越高表明细胞毒性越大,对细胞和机体造成损伤。这表明在刺参幼参饲料中添加酶解海带粉不应超过12%,9%以下的添加量有助于提高刺参机体抗氧化能力,并降低脂质过氧化产物对细胞的伤害,从而保护机体免受自由基的损伤。

3.1.3.4 结论

综上所述,本实验条件下,以增重率为评价指标,经一元二次回归分析($Y= -0.198X^2+4.1029X+60.586$)得出,体质量为(11.40±0.04)g的刺参幼参饲料中酶解海带粉的最适需求量为 10.36%。饲料中添加 6%～12%酶解海带粉可以提高刺参幼参对营养物质的消化和代谢能力,增强刺参幼参的抗氧化性能,进而促进刺参幼参的生长。

3.2 陆生植物蛋白原料在刺参配合饲料中的应用

3.2.1 发酵豆粕对刺参幼参生长性能及体组成的影响

豆粕具有来源稳定、价格合理、消化利用率高等优点,是动物饲料中最佳的植物蛋白原料之一,但由于存在抗营养因子等因素,限制了其在高档水产饲料中的应用(李宝山等,2017)。利用现代生物工程技术,将豆粕发酵,可降低大豆凝集素、胰蛋白酶抑制因子和致甲状腺肿素的含量(张丽靖等,2008),提高粗蛋白、粗脂肪和总磷的含量(马文强等,2008),同时可将大分子蛋白质降解为小分子肽(Hong et al.,2004;欧阳亮等,2008)。此外,豆粕发酵后,矿物质及维生素的含量有所提高(Kim et al.,1999),且具有一定的芳香味和鲜味,可提高动物对其的消化利用率。研究表明,发酵豆粕可替代牙鲆(Kader et al.,2012)、黑鲷(Zhou et al.,2011b)、石斑鱼(Luo et al.,2004)等饲料中的部分鱼粉。

3.2.1.1 材料与方法

1. 实验饲料配方及配制

以鱼粉和海藻粉为蛋白源,设计粗蛋白含量为22%左右的基础饲料配方,分别以 0%、5%、10%、15%、19.35%的发酵豆粕等蛋白替代基础饲料中的海藻粉,以 7.6%、15.2%、22.8%、30.4%的发酵豆粕等蛋白替代基础饲料中的鱼粉、海藻粉混合物(2∶15),配制 9 组等氮实验饲料,记为 D1～D9 组,其中 D1 组为对照组,D2～D5 组为替代海藻粉组,D6～D9 组为替代混合物组(表 3-15)。实验中所用主要蛋白原料及等蛋白平衡后原料的氨基酸组成见表 3-16。固体原料按比例称重后粉碎过 200 目标准筛,加入鱼油及适量的蒸馏水混匀,用饲料制粒机制成直径为 0.3cm 的颗粒后 60℃烘干,用小型粉碎机粉碎,筛选 60~80 目的颗粒备用。

表 3-15 实验饲料配方及营养组成(%)

原料		组别								
		D1	D2	D3	D4	D5	D6	D7	D8	D9
原料组成 原料组成	鱼粉	10	10	10	10	10	8	6	4	2
	发酵豆粕	0	5	10	15	19.35	7.6	15.2	22.8	30.4
	海藻粉	60	44.5	29	13.5	0	45	30	15	0
	α-纤维素	5	15.5	26	36.5	45.65	14.2	23.4	32.6	41.8
	小麦粉	10	10	10	10	10	10	10	10	10
	虾粉	8.94	8.94	8.94	8.94	8.94	8.94	8.94	8.94	8.94
	鱼油	1	1	1	1	1	1.2	1.4	1.6	1.8
	磷脂	1	1	1	1	1	1	1	1	1
	磷酸二氢钙	1	1	1	1	1	1	1	1	1
	维生素预混料	1.5	1.5	1.5	1.5	1.5	1.5	1.5	1.5	1.5
	矿物质预混料	1.5	1.5	1.5	1.5	1.5	1.5	1.5	1.5	1.5
	黏合剂	0.06	0.06	0.06	0.06	0.06	0.06	0.06	0.06	0.06
营养组成	粗蛋白	22.03	22.00	22.12	22.07	21.41	22.56	22.67	22.32	22.04
	粗脂肪	2.96	3.08	2.99	3.04	3.07	2.85	2.68	2.52	2.37
	粗灰分	15.78	26.28	36.73	47.22	56.43	24.54	33.35	42.11	50.93
	蛋氨酸	0.38	0.36	0.33	0.31	0.29	0.33	0.27	0.21	0.16
	赖氨酸	0.88	0.92	0.95	0.99	1.02	0.87	0.86	0.85	0.84

注:①鱼粉为鳀鱼粉;②发酵豆粕购自北京市希普正慧生物饲料有限公司;③藻粉为鼠尾藻、大叶藻、海带 1:1:1 混合;④维生素预混料及矿物质预混料组成见文献(王际英等,2014)。

表 3-16 实验中所用主要蛋白原料及等蛋白平衡后原料的氨基酸组成(%)

项目	鱼粉	海藻粉	发酵豆粕	等蛋白海藻粉	等蛋白混合物
粗蛋白	750.38	189.96	512.43	512.43	512.43
天冬氨酸	68.16	20.18	57.53	54.44	46.62
苏氨酸	30.52	8.77	18.61	23.66	20.41
丝氨酸	31.19	8.08	24.27	21.80	19.29
谷氨酸	114.64	23.76	103.67	64.09	60.43
甘氨酸	47.39	9.43	18.69	25.44	24.29
丙氨酸	42.96	11.35	19.03	30.62	26.97
半胱氨酸	7.07	0.21	2.31	0.57	1.51
缬氨酸	36.25	10.06	21.44	27.14	23.61
蛋氨酸	21.75	2.75	3.67	7.42	8.36
异亮氨酸	31.25	7.37	20.26	19.88	18.05
亮氨酸	55.35	12.68	32.70	34.21	31.31

续表

项目	鱼粉	海藻粉	发酵豆粕	等蛋白海藻粉	等蛋白混合物
酪氨酸	24.80	3.62	10.86	9.77	10.39
苯丙氨酸	28.45	8.05	21.10	21.72	18.10
赖氨酸	55.92	5.38	23.92	14.51	18.50
组氨酸	15.66	1.27	10.16	3.43	4.76
精氨酸	47.01	6.77	27.07	18.26	19.53
脯氨酸	28.39	7.82	23.88	21.09	18.38

注：等蛋白藻粉的氨基酸含量=相应氨基酸含量×512.43÷189.96；等蛋白混合物的氨基酸含量=（鱼粉相应氨基酸含量×2+海藻粉相应氨基酸含量×15）×512.43÷（鱼粉粗蛋白含量×2+藻粉粗蛋白含量×15）。

2. 养殖实验管理

养殖实验在山东省海洋资源与环境研究院东营实验基地循环水养殖系统内进行，实验用刺参购自蓬莱安源水产有限公司。实验采用循环微流水养殖，水流流速约为100ml/min。深蓝色圆柱形养殖水桶（$h_桶$=80cm，$d_桶$=80cm）内放置刺参养殖筐1个，内嵌波纹板20张，控制水深为50cm。实验开始前，将2000头规格整齐、体质健壮的刺参幼参放养于系统中，其间投喂对照组饲料，使其适应养殖环境和实验饲料。21d后挑选体质量约为17.7g的刺参幼参810头，平均放养于27个养殖桶中，每桶30头。每种实验饲料投喂3桶，每日定时（16:00）投喂一次，投喂时将饲料与海泥（1:1）混合均匀，加水湿润后泼洒投喂，投喂量为刺参幼参体质量的4%，并根据摄食状况随时调整。投喂时停气，待饲料颗粒沉到水底或波纹板上后，恢复充气。每隔2d用虹吸法将粪便及残饵吸出，并补充1/3新水。实验期间，水温为17~22℃，溶氧量大于6mg/L，氨氮及亚硝酸氮的浓度低于0.1mg/L，pH为7.2~7.5，养殖周期为70d。

3. 样品采集与计算

养殖实验结束后，禁食48h，放掉桶内水后收集实验刺参，计数并在空气中放置30s后称重。每桶随机挑选15头刺参，置于洁净的白瓷盘中，待其恢复自然体长，测量体长、体质量，然后解剖，分离消化道，分别测量体壁重、肠道重和肠道长度。将体壁置于-20℃保存，待测。

刺参幼参增重量、增重率、特定生长率、存活率、脏壁比、肠壁比及肠长比的计算方法见附录。

4. 样品测定

饲料或刺参幼参的水分、粗蛋白、粗脂肪、粗灰分、能量及氨基酸含量的测定方法见附录；将刺参幼参体壁用浓硝酸微波消解，然后用电感耦合等离子体质谱仪（ICP-MS，Agilent 7700，美国）测定Ca、Fe、Mg、Zn、Cu、Cr的含量，用原子荧光形态分析仪（北京吉天，SA-10，中国）测定 Se、Cd、Pb、As、Hg 的含量。

5. 数据处理

实验所得数据采用 Microsoft Excel 2007 及 SPSS 11.0 软件进行单因素方差分析，结果以平均值±标准差表示，差异显著（$P<0.05$）时用邓肯多重范围检验进行多重比较分析，分别采用 SAS（statistical analysis system 9.2，Institute Inc.，美国）REG 与 NLIN 对增重率与替代比例之间的关系进行回归分析。

3.2.1.2　结果分析

1. 发酵豆粕对刺参幼参生长性能及形体指数的影响

发酵豆粕替代海藻粉后，随着替代比例增加，刺参幼参的增重率、特定生长率、脏壁比、肠壁比均呈现先升后降的趋势，在 D2 组达到最高，显著高于 D4 及 D5 组（$P<0.05$）；发酵豆粕替代鱼粉、藻粉混合物后，D8、D9 组增重率及特定生长率显著低于前三组（$P<0.05$），且前三组之间无显著性差异（$P>0.05$）。发酵豆粕替代海藻粉或鱼粉、海藻粉混合物对刺参幼参的存活率无显著影响（$P>0.05$）（表 3-17）。以增重率为评价指标，经 SAS REG 回归分析，发酵豆粕替代海藻粉的适宜比例为 29.75%（图 3-3）；经 SAS NLIN 回归分析，发酵豆粕可最高替代 46.46% 的鱼粉、海藻粉混合物而不影响其生长（图 3-4）。

表 3-17　发酵豆粕对刺参幼参生长性能及形体指标的影响

指标	D1	D2	D3	D4	D5	D6	D7	D8	D9
初始体质量/g	17.86±0.55	17.72±0.13	17.63±0.10	17.76±0.40	17.63±0.17	17.77±0.08	17.80±0.05	17.74±0.07	17.65±0.02
终末体质量/g	40.74±1.14^{cd}	41.26±0.81^{d}	40.17±0.99^{cd}	39.97±0.75^{c}	37.66±0.53^{b}	39.77±0.33^{c}	39.49±0.68^{c}	37.62±0.85^{b}	35.50±0.81^{a}
增重量/g	22.88±0.98^{cd}	23.54±0.80^{d}	22.54±0.93^{cd}	22.21±0.53^{c}	20.03±0.37^{b}	21.99±0.37^{c}	21.68±0.68^{c}	19.88±0.78^{b}	17.85±0.82^{a}
增重率/%	127.95±4.33^{cd}	132.47±4.03^{d}	126.98±4.33^{cd}	125.01±2.21^{c}	113.20±3.02^{b}	123.80±1.50^{c}	121.72±2.38^{c}	112.26±3.43^{b}	101.15±3.25^{a}
特定生长率/（%/d）	1.18±0.06^{cd}	1.21±0.05^{d}	1.18±0.05^{cd}	1.16±0.03^{c}	1.08±0.02^{b}	1.15±0.02^{c}	1.14±0.04^{c}	1.07±0.05^{b}	0.99±0.05^{a}
脏壁比/%	20.43±1.19^{d}	21.30±0.96^{d}	18.57±1.15^{c}	17.07±0.23^{abc}	16.10±0.92^{ab}	18.67±0.95^{c}	17.17±0.78^{bc}	16.33±0.55^{ab}	15.47±0.60^{a}
肠壁比/%	5.11±0.31^{cd}	5.27±0.38^{d}	4.79±0.30^{bcd}	4.62±0.030^{abc}	4.41±0.41^{ab}	4.83±0.26^{bcd}	4.74±0.19^{bcd}	4.38±0.21^{ab}	4.16±0.23^{a}
肠长比/%	3.33±0.55^{c}	3.20±0.54^{c}	2.95±0.36^{bc}	2.82±0.45^{b}	2.65±0.15^{a}	3.34±0.41^{c}	3.23±0.20^{c}	2.85±0.39^{b}	2.54±0.39^{c}
存活率/%	96.67±3.33	95.56±1.92	94.44±5.09	94.44±5.04	97.78±1.92	96.88±3.13	95.83±3.61	95.83±1.80	96.88±3.13

注：同行数据无字母或上标相同字母表示差异不显著（$P>0.05$），上标不同字母表示差异显著（$P<0.05$）。

2. 发酵豆粕对刺参幼参体壁基本成分及氨基酸含量的影响

发酵豆粕替代海藻粉后，随着替代比例增加，刺参幼参体壁中粗脂肪含量有了显著降低（$P<0.05$），水分、粗蛋白和粗灰分的含量无显著变化（$P>0.05$）；发酵豆粕替代鱼粉、藻粉混合物后，随着替代比例增加，刺参幼参体壁粗灰分的含量有了显著降低（$P<0.05$），而水分、粗蛋白及粗脂肪的含量无显著变化（$P>0.05$）（表 3-18）。

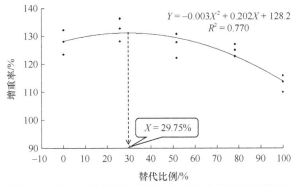

图 3-3　发酵豆粕替代海藻粉对刺参幼参增重率的影响

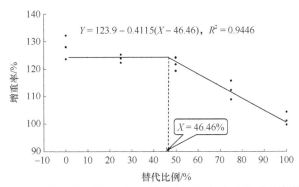

图 3-4　发酵豆粕替代鱼粉、海藻粉混合物对刺参幼参增重率的影响

表 3-18　发酵豆粕对刺参幼参体壁基本成分的影响（%）

指标	组别								
	D1	D2	D3	D4	D5	D6	D7	D8	D9
水分	92.89± 0.19	91.14± 1.12	91.83± 1.27	91.97± 0.77	92.74± 0.0.22	92.18± 1.50	92.08± 0.04	91.62± 1.13	92.39± 0.37
粗蛋白	49.25± 0.85	49.50± 0.72	49.38± 1.67	49.75± 1.25	49.38± 0.38	50.28± 1.80	48.29± 1.53	48.78± 0.42	49.82± 1.05
粗脂肪	3.88± 0.48^b	3.61± 0.29^{ab}	3.54± 0.23^{ab}	3.21± 0.53^a	3.03± 0.37^a	3.99± 0.37^b	3.68± 0.68^{ab}	3.87± 0.78^{ab}	3.64± 0.82^{ab}
粗灰分	32.18± 2.92^b	31.40± 1.49^b	31.11± 2.03^b	31.77± 1.96^b	31.87± 1.56^b	31.51± 2.26^b	30.67± 2.91^{ab}	29.69± 2.20^{ab}	29.04± 1.69^a

注：粗蛋白、粗脂肪、粗灰分的含量均为干基含量；同行数据无字母或上标相同字母表示差异不显著（$P>0.05$），上标不同字母表示差异显著（$P<0.05$）。

发酵豆粕替代海藻粉后，随着替代比例增加，刺参幼参体壁中甘氨酸和半胱氨酸的含量有了显著降低（$P<0.05$），苯丙氨酸含量有了显著升高（$P<0.05$）；发酵豆粕替代鱼粉、藻粉混合物后，随着替代比例增加，刺参幼参体壁中丝氨酸、苏氨酸、甘氨酸、蛋氨酸、赖氨酸及组氨酸的含量有了显著降低（$P<0.05$），精氨酸、脯氨酸、苯丙氨酸及半胱氨酸的含量有了显著升高（$P<0.05$）；发酵豆粕替代海藻粉或鱼粉、藻粉混合物，对刺参幼参体壁中天冬氨酸、谷氨酸、丙氨酸、缬氨酸、异亮氨酸、亮氨酸、酪氨酸、必需氨基酸及总氨基酸的含量无显著影响（$P>0.05$）（表 3-19）。

表 3-19 发酵豆粕对刺参幼参体壁氨基酸含量的影响　　（单位：mg/g）

指标	组别								
	D1	D2	D3	D4	D5	D6	D7	D8	D9
天冬氨酸	40.89±1.24	41.02±0.96	41.56±1.30	41.16±1.14	40.74±0.75	40.60±0.39	40.05±0.87	40.87±0.58	40.27±1.41
谷氨酸	67.58±2.31	65.85±3.26	66.78±2.47	68.24±2.55	67.48±1.64	68.20±1.99	68.94±1.52	68.01±2.11	68.40±2.40
丝氨酸	25.68±0.99^c	26.29±0.70^c	25.76±1.19^c	25.78±0.84^c	25.54±1.07^c	23.04±0.54^b	22.42±0.68^{ab}	21.90±0.39^a	22.33±0.41^{ab}
苏氨酸	26.08±1.91^b	26.49±1.24^b	26.43±1.08^b	26.13±1.34^b	25.92±0.77^b	24.65±0.69^a	23.91±0.93^a	23.93±0.72^a	23.16±0.81^a
精氨酸	31.96±0.85^a	31.25±1.22^a	31.42±0.73^a	30.60±1.01^a	30.52±1.84^a	33.59±0.69^b	34.86±0.52^{bc}	35.17±1.35^c	35.97±0.58^c
甘氨酸	47.63±0.77^c	46.49±1.43^{bc}	44.96±1.06^{ab}	45.12±1.28^b	45.02±0.95^b	46.35±0.72^{bc}	47.77±0.65^c	45.64±0.14^b	43.86±0.57^a
丙氨酸	21.80±0.41	21.30±0.85	21.58±0.29	21.12±0.70	21.55±0.37	21.21±0.66	20.47±1.02	19.84±0.19	21.01±0.45
脯氨酸	26.01±1.09^a	27.57±0.86^a	26.87±0.78^a	26.21±1.35^a	27.35±1.26^b	26.93±0.95^a	26.49±1.44^a	27.39±1.65^b	29.42±0.73^c
缬氨酸	18.33±0.42	18.60±1.00	19.25±1.75	18.87±1.28	18.15±0.59	18.73±1.17	18.58±0.67	18.97±1.34	17.23±1.63
蛋氨酸	5.80±0.07^{de}	5.91±0.18^e	5.69±0.25^d	5.61±0.49^d	5.53±0.19^d	5.21±0.06^c	4.67±0.31^b	4.64±0.22^b	4.03±0.16^a
异亮氨酸	12.17±0.94	12.96±0.67	12.28±0.88	12.84±1.04	12.25±0.70	12.43±0.81	13.02±1.09	13.43±1.16	13.44±0.99
亮氨酸	23.09±0.18	22.88±0.44	22.09±0.39	21.09±0.21	21.55±0.54	22.36±0.27	21.38±0.10	21.70±0.25	22.27±0.79
苯丙氨酸	18.14±1.17^a	17.80±0.93^a	18.55±0.89^a	19.77±1.20^{ab}	19.73±1.34^b	19.84±0.41^b	19.76±0.70^b	20.03±0.58^c	21.64±0.26^d
半胱氨酸	5.67±0.09^b	5.67±0.14^b	5.79±0.25^b	5.77±0.38^b	5.20±0.42^a	5.84±0.07^c	5.93±0.21^c	6.65±0.13^d	7.22±0.18^e
赖氨酸	20.15±0.12^b	20.44±0.85^b	19.46±0.34^b	20.28±0.62^b	19.82±1.20^b	19.46±0.71^b	18.18±0.88^{ab}	18.03±0.23^a	17.81±0.16^a
组氨酸	5.09±0.44^{cd}	5.14±0.29^d	5.03±0.07^{cd}	5.19±0.31^d	5.29±0.18^d	4.90±0.08^c	4.62±0.16^b	4.44±0.20^b	4.10±0.11^a

续表

指标	组别								
	D1	D2	D3	D4	D5	D6	D7	D8	D9
酪氨酸	18.90±0.99	17.64±1.23	17.68±1.07	17.80±1.34	17.96±0.85	18.93±0.69	19.00±0.49	19.22±0.52	19.54±0.93
必需氨基酸	164.81±8.32	161.46±6.75	165.20±7.39	159.48±5.88	158.76±6.24	163.17±9.69	158.98±7.22	166.34±9.78	159.65±5.63
总氨基酸	414.97±14.36	418.09±10.94	411.18±12.48	413.68±11.57	409.61±14.82	412.24±10.51	410.05±13.46	417.86±13.52	411.70±9.50

注：同行数据无字母或上标相同字母表示差异不显著（$P>0.05$），上标不同字母表示差异显著（$P<0.05$）。

3. 发酵豆粕对刺参幼参体壁矿物元素含量的影响

发酵豆粕替代海藻粉后，随着替代比例增加，刺参幼参体壁中钙、镁、铁、锌、铜、铬、锰及铅的含量有了显著降低（$P<0.05$）；发酵豆粕替代鱼粉、藻粉混合物后，随着替代比例增加，刺参幼参体壁中钙、镁、锌、铜、铬、镉及铅含量有了显著降低（$P<0.05$），铁及锰的含量有了显著升高（$P<0.05$）。发酵豆粕替代海藻粉或鱼粉、海藻粉混合物对硒、镉、砷及汞含量的影响不显著（$P>0.05$）（表3-20）。

表3-20 发酵豆粕对刺参幼参体壁矿物元素含量的影响　　（单位：mg/kg）

指标	组别								
	D1	D2	D3	D4	D5	D6	D7	D8	D9
钙	12 331.15±108.44^g	11 894.72±143.58^f	10 197.64±209.76^d	9 614.48±98.57^c	7 789.23±121.21^a	11 639.71±166.67^f	10 842.58±138.30^e	9 637.34±184.32^c	9 340.47±135.22^b
镁	23 741.30±211.76^i	20 186.39±184.72^e	18 069.10±258.41^c	15 423.88±109.56^b	13 877.44±123.93^a	22 294.57±172.58^h	21 396.85±237.49^g	20 665.72±197.25^f	19 649.81±253.31^d
铁	125.69±6.57^e	112.90±3.61^d	98.74±2.28^c	92.07±4.73^b	78.14±5.84^a	136.42±9.77^f	141.41±5.67^f	149.27±8.21^f	161.31±6.99^g
锌	51.49±1.07^g	45.34±1.32^e	42.88±0.96^d	37.46±0.67^c	29.05±1.18^a	48.80±1.42^f	42.98±0.89^d	38.24±1.21^c	35.14±0.90^b
铜	15.63±0.85^e	14.90±0.69^{de}	12.17±0.74^{bc}	10.34±0.92^b	9.45±0.57^a	15.19±0.81^e	13.26±1.01^d	12.55±0.34^c	11.70±0.17^b
硒	0.34±0.02	0.34±0.01	0.36±0.03	0.35±0.02	0.37±0.03	0.36±0.02	0.32±0.01	0.34±0.03	0.33±0.03
铬	3.33±0.09^e	3.07±0.06^d	2.86±0.11^c	2.59±0.09^b	2.30±0.14^a	3.42±0.18^f	3.27±0.09^e	3.18±0.13^d	3.09±0.06^d
锰	5.70±0.25^d	5.64±0.17^d	5.39±0.08^c	5.01±0.12^b	4.88±0.20^a	5.74±0.15^d	5.99±0.07^d	6.36±0.11^e	6.88±0.14^f
镉	0.76±0.09	0.55±0.04^{ab}	0.62±0.07^c	0.59±0.05^{bc}	0.44±0.09^a	0.71±0.03^d	0.64±0.02^c	0.56±0.04^{ab}	0.49±0.04^a
铅	1.83±0.07^d	1.74±0.11^d	1.56±0.06^c	1.28±0.09^b	1.09±0.04^a	1.69±0.06^d	1.50±0.08^c	1.46±0.11^c	1.42±0.09^c
砷	0.31±0.01	0.29±0.01	0.32±0.02	0.29±0.03	0.27±0.01	0.30±0.04	0.29±0.02	0.33±0.01	0.27±0.01

指标	组别								
	D1	D2	D3	D4	D5	D6	D7	D8	D9
汞	0.03± 0.00	0.02± 0.00	0.01± 0.00	0.03± 0.00	0.02± 0.00	0.02± 0.00	0.03± 0.00	0.01± 0.00	0.02± 0.00

注：同行数据无字母或上标相同字母表示差异不显著（$P>0.05$），上标不同字母表示差异显著（$P<0.05$）。

3.2.1.3 讨论

本实验结果显示，发酵豆粕可以部分替代刺参配合饲料中的海藻粉或鱼粉、海藻粉混合物。发酵豆粕替代藻粉后，刺参幼参的增重率呈现先升后降的趋势，这与几种蛋白原料中氨基酸模式互补有关（Kader et al.，2012），随着替代比例增加，饲料中某些氨基酸缺乏，破坏了氨基酸平衡模式（Deng et al.，2006），且可能是发酵豆粕中抗营养因子去除得不彻底（Uyan et al.，2006），导致刺参幼参生长性能的下降。此外，随着替代比例增加，刺参幼参的肠壁比及肠长比有了显著降低，说明相较于发酵豆粕，海藻粉更适合刺参幼参肠道的发育。发酵豆粕最高可替代46.46%的鱼粉、海藻粉混合物不会影响刺参幼参的生长，过高则抑制其生长，这与对其他水生动物的研究一致（Yuan et al.，2013；Azarm and Lee，2014），可能与发酵豆粕中蛋氨酸缺乏有关（杨耐德和符广才，2008）。此外，随着发酵豆粕含量的升高，饲料中粗灰分含量有了显著升高，这可能是造成刺参幼参生长性能下降的另一个原因，但需要进一步研究。

随着饲料中发酵豆粕含量的升高，刺参幼参体壁中水分及粗蛋白的含量变化不显著，粗脂肪含量有了显著降低，这与对凡纳滨对虾的研究一致（杨耐德和符广才，2008）。饲料中添加过量的豆粕，可导致牙鲆机体脂肪代谢紊乱，从而造成其脂肪蓄积降低（Ye et al.，2011）。鱼类生长性能下降会导致机体中粗灰分含量上升（段培昌等，2009；黄云等，2012）。本实验中，随着饲料中发酵豆粕含量的增加，刺参幼参体壁中粗灰分呈现下降的趋势，这可能是由于矿物元素蓄积的下降引起的。发酵豆粕替代海藻粉后，随着替代比例增加，刺参幼参体壁中8种矿物元素含量有了显著降低；发酵豆粕替代鱼粉、海藻粉混合物后，随着替代比例增加，刺参幼参体壁中7种矿物元素含量有了显著降低，仅2种有了显著升高，且降低的含量远大于升高的含量。豆粕发酵后呈酸性，可以促进铁的吸收（Pandey and Satoh，2008）。对鸡的研究表明，Fe^{2+}、Mn^{2+}、Cu^{2+}、Zn^{2+}等在肠道中通过铁诱导型二价阳离子转运蛋白1（DMT1）进行吸收（李晓丽等，2013），因而Fe^{2+}会抑制其他二价阳离子的吸收利用（Rodriguez-Matas et al.，1998）。两个实验中，发酵豆粕最高添加量分别为19.35%和30.4%，这可能是导致矿物元素沉积差异的原因。与乳山刺参相比，本实验中的刺参幼参体壁矿物元素中钙、镁的含量有了显著上升，而铁、铜、锰、锌、铬 4 种重金属元素的含量有了显著降低，由于二者之间不存在显著的地域差异性（刘小芳，2014），因此导致这种差异的原因更偏重环境及饵料，具体原因有待进一步研究。本实验中所测的镉、铬和铅 3 种重金属元素含量显著降低，这与饲料中海藻粉含量的降低有关（张永亮等，2009）。本实验所测得的刺参幼参体壁中重金属含量均低于国家相关食品卫生标准限量。

饲料中发酵豆粕含量虽然没有显著影响刺参幼参体壁中粗蛋白含量，但是影响了其氨基酸含量。本实验所测得的 17 种氨基酸中有 10 种含量发生了显著变化，其中 6 种为必需氨基酸，且体壁中氨基酸含量与饲料中氨基酸含量不存在相关性，这与对鱼（彭士明等，2012）和虾（严晶等，2012）的研究一致，表明动物是按照一定的氨基酸模式利用饲料中的蛋白质（Mai et al.，2006），氨基酸不平衡会造成饲料蛋白质的浪费。本实验中所测得的 9 种必需氨基酸含量与其他研究类似（Deng et al.，2006；王哲平等，2012），其中精氨酸含量最高，而真鲷（唐宏刚，2008）和牙鲆（Kim and Lall，2000）等海水鱼类肌肉中含量最高的必需氨基酸为赖氨酸，这既体现了种属上的差异，也表明了需求的不同（王际英等，2015）。

3.2.1.4 结论

本实验条件下，发酵豆粕替代海藻粉的最佳比例为 29.75%，而发酵豆粕最高可替代 46.46%的鱼粉、海藻粉混合物而不影响刺参幼参的生长。本实验中所测得的刺参幼参体壁中镉、铬、铅、汞、锰和砷 6 种重金属含量均低于国家相关食品卫生标准限量。

3.2.2 配合饲料中添加玉米干酒精糟及其可溶物对刺参生长、体组成及免疫指标的影响

玉米干酒精糟及其可溶物（DDGS）是玉米用于乙醇工业和饮料工业中产生的副产品，是干酒精糟和可溶干酒糟的混合物（Jie et al.，2014），它保留并浓缩了玉米中除淀粉以外的其他营养成分，蛋白质、脂肪的含量较高，氨基酸、维生素和矿物质元素丰富。Ingledew（1999）认为，玉米干酒精糟及其可溶物中酵母的生物量至少占总重的 3.9%，酵母蛋白的含量至少占玉米干酒精糟及其可溶物总蛋白的 5.3%。酿酒酵母被认为是一种良好的水生动物蛋白源，富含核苷酸、甘露寡糖和葡聚糖，在水产饲料中可作为免疫刺激剂（Refstie et al.，2010）。同时，由于生产过程中的发酵及水热处理，玉米干酒精糟及其可溶物几乎不含抗营养因子，可作为一种优质的植物蛋白饲料，其应用对于扩大蛋白饲料的来源，缓解鱼粉、发酵豆粕、海藻粉等常规蛋白原料的紧张局面有着十分重要的意义（Refstie et al.，

2010)。目前，国外有关玉米干酒精糟及其可溶物的研究已经对美国鲶鱼（Webster et al.，1992）、虹鳟（Øverland et al.，2013）、罗非鱼（Welker et al.，2014）进行了大量实验，国内相关研究才刚刚开始，在刺参配合饲料中的应用尚未见报道。因此，本实验旨在研究玉米干酒精糟及其可溶物对刺参生长性能、形体指标和免疫功能的影响，探求玉米干酒精糟及其可溶物在刺参配合饲料中的适宜添加量，对玉米干酒精糟及其可溶物在刺参配合饲料中的应用进行科学评价。

3.2.2.1 材料与方法

1. 实验饲料

本实验以鱼粉、发酵豆粕和海藻粉为主要蛋白源配制基础饲料并作为对照组，采用等蛋白替代方式，在饲料中分别添加 10%、20%、30%、40%的玉米干酒精糟及其可溶物等蛋白替代基础饲料中的海藻粉和发酵豆粕，用玉米油调节脂肪平衡，共配制 5 种等氮等能饲料，分别命名为 DDGS0（对照组）、DDGS10、DDGS20、DDGS30 和 DDGS40。实验饲料配方及营养组成见表 3-21。所有原料经超微粉碎后过 200 目标准筛，按配比称重，加入玉米油及适量蒸馏水混匀，用小型颗粒饲料挤压机制成颗粒，60℃烘干后用小型粉碎机破碎，筛选粒度为在 80～100 目的颗粒备用，于冰箱（−20℃）中保存。

表 3-21 实验饲料配方及营养组成（%）

项目		组别				
		DDGS0	DDGS10	DDGS20	DDGS30	DDGS40
原料组成	鱼粉	5	5	5	5	5
	发酵豆粕	11.2	8.4	5.6	2.8	0
	海藻粉	40	30	20	10	0
	干酒精糟及其可溶物	0	10	20	30	40
	海泥	22	22	22	22	22
	微晶纤维素	0	3.7	7.4	11.1	14.8
	虾粉	5	5	5	5	5
	α-淀粉	10	10	10	10	10
	玉米油	3.6	2.7	1.8	0.9	0
	多维	2	2	2	2	2
	多矿	1	1	1	1	1
	抗氧化剂	0.1	0.1	0.1	0.1	0.1
	三氧化二铬	0.1	0.1	0.1	0.1	0.1
营养组成	粗蛋白	21.63	21.46	21.74	21.10	21.57
	粗脂肪	5.34	5.34	5.49	5.24	5.33

续表

项目		组别				
		DDGS0	DDGS10	DDGS20	DDGS30	DDGS40
营养组成	粗灰分	35.22	32.81	29.97	25.84	22.82
	总能	12.24	12.35	12.64	12.51	12.78

注：①鱼粉和发酵豆粕购自山东日照东维饲料有限公司，粗蛋白含量为 50.56%（干重）；②海藻粉由山东升索渔用饲料研究中心提供，脱胶海带粉与马尾藻粉等量混合，粗蛋白含量为 17.1%（干重）；③干酒精糟及其可溶物购自阿德金斯能源有限公司（Adkins Energy LLC-Lena, Illinois, 美国），粗蛋白含量为 31.27%，粗脂肪含量为 12.37%（干重）；④多维和维生素预混料及多矿的矿物质预混料组成见文献（王际英等，2014）。

2. 实验动物与饲养管理

实验用刺参幼参取自山东省海洋资源与环境研究院东营实验基地当年繁育的同一批参苗，饲养实验在该基地循环水养殖系统中进行。实验开始前，将实验用刺参幼参在循环水养殖系统中暂养 14d，其间饲喂基础饲料。暂养结束后，挑选体质健康、初始体质量为（9.69±0.28）g 的刺参幼参 450 头，平均放养于 15 个深蓝色圆柱形养殖水桶（直径为 75cm，水深为 80cm）中，内放置海参养殖筐 1 个，内嵌波纹板 20 张，控制水深为 50cm，随机分为 5 组，每组 3 个重复。实验在微流水环境中进行，采用充气增氧，保证溶氧量大于 7mg/L，水温控制在 18~20℃，pH 为 7.8~8.2，盐度为 24~26，亚硝酸氮、氨氮的浓度均低于 0.05mg/L。养殖实验持续 56d，每天定时投喂 2 次（08:00 和 16:30），日投喂量为刺参幼参体质量的 2%，根据刺参幼参摄食情况及时调整投喂量。每隔 2d 换水 1 次，换水时用虹吸法将残饵及粪便吸出。

3. 样品采集与计算

养殖实验结束后，禁食 48h，统计各桶刺参幼参数量并称重。每桶随机取 10 头刺参幼参置于托盘中，待其自然舒展后测量体长，滤纸轻轻吸干体表水分后分别称重。采集体腔液后，置于冰盘上分离体壁与肠道，分别称重，并测量肠道长度。将体腔液离心（3000r/min，4℃，15min）后取上清液分装于 2ml 的离心管中，–80℃保存。

刺参幼参存活率、增重率、特定生长率、体壁指数、肠道指数及肠长比计算方法见附录。

4. 样品测定

饲料及刺参幼参体壁水分、粗蛋白、粗脂肪、粗灰分的含量及能量测定方法见附录。

溶菌酶活性采用空白对照比浊法测定，总超氧化物歧化酶活性采用羟胺法测定，碱性磷酸酶活性采用磷酸苯二钠法测定，酸性磷酸酶活性采用化学比色法测定,以上指标均采用南京建成生物工程研究所生产的试剂盒进行测定。

酚氧化酶活性的测定参考 Hernández-López 等(1996)的方法,略有改动。在 96 孔酶标板中加入 100μl 体腔液上清液,再加入 50μl 左旋多巴溶液(3mg/ml),立刻放入酶标仪中,每隔 5min 测定波长为 492nm 处的光密度值(Optical Density,OD)值,共测定 10 个点,实验设 3 个平行。以实验条件下每分钟每毫升样品吸光度值每增加 0.001 定义为 1 个酶活性单位(U/ml)。

5. 数据统计

实验所得数据采用 SPSS 13.0 软件进行单因素方差分析,若差异显著($P<0.05$),则用邓肯多重范围检验进行多重比较分析,结果以平均值±标准差表示。

3.2.2.2 结果分析

1. 配合饲料中添加玉米干酒精糟及其可溶物对刺参幼参生长性能的影响

配合饲料中添加玉米干酒精糟及其可溶物对刺参幼参生长性能的影响见表 3-22。饲料中玉米干酒精糟及其可溶物的添加量对各组刺参幼参的存活率无显著影响,且各组存活率均高于 90%。随着玉米干酒精糟及其可溶物添加量的升高,刺参幼参增重率与对照组相比分别降低了 10.53%、12.56%、11.06% 和 6.23%,但各组之间无显著性差异($P>0.05$)。与增重率的变化趋势相似,各实验组终末体质量、特定生长率、体壁指数、肠道指数和肠长比均与对照组无显著性差异($P>0.05$)。

表 3-22 配合饲料中添加玉米干酒精糟及其可溶物对刺参幼参生长性能的影响

项目	组别				
	DDGS0	DDGS10	DDGS20	DDGS30	DDGS40
初始体质量/g	9.67±0.05	9.69±0.10	9.78±0.03	9.66±0.07	9.65±0.12
终末体质量/g	15.30±0.51	14.73±0.94	14.75±0.57	14.64±0.73	14.90±0.67
增重率%	58.11±5.02	51.99±8.13	50.81±5.79	51.68±8.36	54.49±8.50
特定生长率/(%/d)	0.82±0.06	0.75±0.10	0.73±0.10	0.74±0.10	0.78±0.10
体壁指数/%	67.16±2.06	66.97±1.34	65.75±1.75	66.61±1.11	67.22±2.29
肠道指数/%	4.28±0.30	4.52±0.17	4.61±0.40	4.26±0.14	4.22±0.30
肠长比/%	3.17±0.13	3.30±0.10	3.18±0.30	3.18±0.16	3.06±0.14
存活率/%	96.00±4.00	96.00±4.00	94.00±2.00	95.33±1.15	93.33±7.02

2. 配合饲料中添加玉米干酒精糟及其可溶物对刺参幼参体壁营养成分的影响

配合饲料中添加玉米干酒精糟及其可溶物对刺参幼参体壁营养成分的影响见表 3-23。配合饲料中添加玉米干酒精糟及其可溶物后,刺参幼参体壁中水分、粗蛋白、粗脂肪和粗灰分的含量各组之间均无显著性差异（$P>0.05$）。

表 3-23　配合饲料中添加玉米干酒精糟及其可溶物对刺参幼参体壁营养成分的影响（%）

项目	组别				
	DDGS0	DDGS10	DDGS20	DDGS30	DDGS40
水分	90.99±0.29	91.31±0.36	91.06±0.45	91.19±0.39	90.85±0.25
粗蛋白	4.42±0.23	4.21±0.20	4.26±0.30	4.23±0.01	4.46±0.16
粗脂肪	0.22±0.02	0.21±0.03	0.22±0.01	0.20±0.01	0.24±0.04
粗灰分	3.00±0.21	2.91±0.08	2.93±0.12	2.96±0.11	3.01±0.07

3. 配合饲料中添加玉米干酒精糟及其可溶物对刺参幼参免疫指标的影响

配合饲料中添加玉米干酒精糟及其可溶物对刺参幼参免疫指标的影响见表 3-24。随着饲料中干酒精糟及其可溶物替代水平的升高，刺参幼参体腔液中溶菌酶活性呈现先升高后平稳的趋势，其中 DDGS20 和 DDGS40 组显著高于 DDGS0 和 DDGS10 组（$P<0.05$），DDGS30 组溶菌酶活性与其他各组无显著性差异（$P>0.05$）；酸性磷酸酶活性呈现先上升后下降的趋势，在 DDGS20 组达到最高值，其中 DDGS20 组显著高于其他各组（$P<0.05$），DDGS40 组显著高于 DDGS0 组（$P<0.05$），其他各组间无显著性差异（$P>0.05$）；酚氧化酶活性呈现上升趋势，DDGS0 组显著低于其他各组（$P<0.05$），DDGS40 组显著高于 DDGS10 组（$P<0.05$），其他各组间无显著性差异（$P>0.05$）。饲料中添加玉米干酒精糟及其可溶物对刺参幼参体腔液中碱性磷酸酶和超氧化物歧化酶的活性无显著影响（$P>0.05$）。

表 3-24　配合饲料中添加玉米干酒精糟及其可溶物对刺参幼参免疫指标的影响

项目	组别				
	DDGS0	DDGS10	DDGS20	DDGS30	DDGS40
溶菌酶/（U/ml）	41.23±3.69[a]	44.98±3.18[a]	54.09±4.84[b]	48.95±5.11[ab]	57.00±4.15[b]
碱性磷酸酶/（King unit/dl）	1.79±0.10	1.89±0.17	2.10±0.18	2.07±0.29	2.18±0.22
超氧化物歧化酶/（U/ml）	68.09±2.31	73.17±5.58	68.23±1.73	72.81±2.89	72.63±6.01
酸性磷酸酶/（U/dl）	4.42±0.74[a]	4.79±0.74[ab]	10.31±0.52[c]	5.65±0.77[ab]	6.02±0.56[b]
酚氧化酶/（U/ml）	62.5±4.01[a]	79.26±7.12[b]	89.86±7.09[bc]	91.90±6.22[bc]	98.07±10.72[c]

注：表中数据以平均值±标准差表示（$n=3$）；同行数据无字母或上标相同字母表示差异不显著（$P>0.05$），上标不同字母表示差异显著（$P<0.05$）。

3.2.2.3　讨论

1. 配合饲料中添加玉米干酒精糟及其可溶物对刺参幼参生长性能的影响

玉米干酒精糟及其可溶物是在生产燃料乙醇的过程中，微生物对玉米进行发酵产生乙醇，蒸馏后大量微生物连同玉米中的剩余成分经干燥形成的一种新型的饲料蛋白质原料。因此，玉米干酒精糟及其可溶物并不仅仅是玉米制酒精后剩余物的浓缩,还有发酵过程中产生的未知因子以及糖化曲、酵母成分(赵红霞,2009)。研究表明,水产养殖动物饲料中可以添加适量的干酒精糟及其可溶物作为蛋白源。在虹鳟饲料中使用25%玉米干酒精糟及其可溶物等蛋白替代由葵花籽粕、菜粕和全豌豆粉组成的混合蛋白源,可以显著提高其增重率,同时,50%添加组与对照组在增重率上无显著性差异(Øverland et al.,2013)。斑点叉尾鲴饲料中使用豆粕和干酒精糟及其可溶物组合替代鱼粉,对实验各组的增重率、特定生长率、饲料转化率及存活率均无显著影响(Webster et al.,1992)。与上述研究结果相似,在本研究中,玉米干酒精糟及其可溶物部分替代海藻粉后对各实验组刺参幼参的增重率、特定生长率等生长指标均无显著影响,表明刺参幼参能够较好地利用发酵副产品玉米干酒精糟及其可溶物作为其蛋白源。研究表明,发酵后植物原料可以作为水产养殖动物饲料中良好的蛋白源(Mostafizur et al.,2015;Molina-Poveda and Morales,2004)。Seo等(2011)发现,刺参能够较好地利用发酵豆粕和干酒精糟。姜燕等(2012)在以海带粉和豆粕为主要原料的刺参饲料中加入水产诱食酵母进行发酵,发酵饲料组刺参的增重率显著高于普通饲料组。玉米干酒精糟及其可溶物中氨基酸组成与发酵豆粕相近,但赖氨酸相对缺乏(王晶等,2009),这可能影响其在刺参配合饲料中的应用效果。生物发酵过程中会产生大量的微生物,包括有益的细菌和真菌,这些微生物及其产生的酶能够降解原料中的部分营养物质,使其更易被动物吸收和利用,提高了发酵基底的营养质量(Jones,1975),进而提高水产饲料利用效率,促进实验动物的生长。生物发酵过程中微生物的代谢产物可以部分或全部消除原料中的抗营养因子,进而提高原料在动物饲料中的应用比例(Refstie et al.,2005)。海参在自然条件下通常以摄取植物或动物的碎屑以及有机物的沉积物为生,谷物原料经发酵后的风味与自然食物相近,刺激了刺参的食欲和摄食活动,进而提高了刺参的生长性能(Seo et al.,2011)。Seo等(2011)研究发现,在饲料中使用10%大米干酒精糟替代鼠尾藻粉可以显著提高刺参的生长性能,但40%替代组刺参的生长性能显著降低。在本研究中,添加40%的玉米干酒精糟及其可溶物并未显著影响刺参的生长性能,这可能是因为干酒精糟及其可溶物中含有可溶干酒糟,而可溶干酒糟中包含玉米中一些可溶性营养物质及发酵过程中产生的未知生长因子、糖化物、酵母等,可以促进刺参对玉米干酒精糟及其可溶物的利用,进而提高其生长性能。

2. 配合饲料中添加玉米干酒精糟及其可溶物对刺参幼参体成分的影响

养殖动物的体成分随饲喂配合饲料组成的不同而发生变化,包括粗蛋白、粗脂肪、水分、粗灰分等,因而其被作为评定配合饲料优劣的一个指标。一般来说,优质蛋白源中的蛋白质易被饲养动物消化吸收,并及时用于机体的生长和组织的更新,对于机体成分尤其是粗蛋白和水分含量的影响不大(李二超等,2009)。已有研究表明，饲料中添加玉米干酒精糟及其可溶物对斑点叉尾鮰（Webster et al., 1993）和草鱼（钟广贤，2013）体成分均无显著影响，对草鱼幼鱼蛋白、脂肪和糖类代谢的影响也不显著（黄文庆等，2012）；在刺参饲料中添加不同水平（Seo et al., 2011）及不同来源（Jin et al., 2013）的大米干酒精糟对刺参的体组成均无显著影响。与上述研究结果一致，在本研究中刺参体壁的基本营养成分并未随饲料中玉米干酒精糟及其可溶物的增加出现显著性差异，说明刺参能够较好地利用玉米干酒精糟及其可溶物进行营养物质的代谢与累积，这可能是因为发酵过程中产生的大量微生物及产生的酶类能降解原料中的部分营养物质，使其更易被刺参吸收和利用，提高了发酵基底的营养质量（Wee，1991）。

3. 配合饲料中添加玉米干酒精糟及其可溶物对刺参幼参免疫指标的影响

刺参属于棘皮动物，缺少获得性免疫体系，因此其防御体系更多地依赖天然免疫反应。刺参的非特异性免疫防御系统包括体壁防御和体内免疫，在体内免疫中各种酶类包括溶菌酶、超氧化物歧化酶、酸性磷酸酶、碱性磷酸酶和酚氧化酶等，均为重要的免疫因子，在机体中担负着防御的重要功能，对刺参的抗病力和抗应激能力有不同程度的促进作用，因此其被作为评价刺参免疫功能的主要指标（孙永欣，2008）。刺参吞噬细胞中具有溶酶体酶，由溶菌酶、酸性磷酸酶和碱性磷酸酶等组成，其重要功能是在吞噬完成后对外源性物质进行降解（Canicattí，1990）。溶菌酶能够破坏溶解细菌细胞壁中的肽聚糖成分，从而使细菌的细胞壁破损，细胞崩解（刘晨光等，2000）。酸性磷酸酶在刺参免疫系统中起着调理素作用，能诱导阿米巴细胞对外来物质进行吞噬和包囊（Bertheussen，1982）。在本研究中，饲料中添加适量的玉米干酒精糟及其可溶物可以显著提高刺参幼参体腔液中溶菌酶和酸性磷酸酶的活性，推测是由于玉米干酒精糟及其可溶物中富含 β-葡聚糖，刺激了刺参体内的免疫系统。研究表明，饲料中添加 β-葡聚糖可以提高刺参（Chang et al., 2010）、大黄鱼（Ai et al., 2007）和凡纳滨对虾（Zhao et al., 2012）体内溶菌酶的活性，同时 β-葡聚糖还可提高奥尼罗非鱼（迟淑艳等，2006）、栉孔扇贝（孙虎山和李光友，2002）和中华绒螯蟹（王雪良，2008）体内酸性磷酸酶的活性。

超氧化物歧化酶是生物体内一种以自由基为底物的抗氧化酶，其活性可间接反映机体清除氧自由基的能力（孙虎山和李光友，2000）。研究发现，饲料中添加玉米干酒精糟及其可溶物对草鱼幼鱼（黄文庆等，2012）和奥尼罗非鱼（何晓庆，2010）血清中超氧化物歧化酶的活性均无显著影响。与上述研究结果一致，在本研究中，各组刺参幼参体腔液中超氧化物歧化酶的活性无显著性差异，表明刺参饲料中添加玉米干酒精糟及其可溶物对其抗氧化性能无不利影响。

酚氧化酶和酚氧化酶原激活系统在无脊椎动物先天免疫的识别和防御中发挥

重要作用（Haug et al.，2002）。研究表明，饲料中添加葡聚糖能够显著提高刺参（张琴，2010）体内酚氧化酶的活性。与上述研究结果一致，在本研究中，饲料中添加玉米干酒精糟及其可溶物显著提高了刺参幼参体腔液中酚氧化酶的活性，这可能归因于干酒精糟及其可溶物富含 β-葡聚糖等免疫调节剂。酚氧化酶一般以无活性的酚氧化酶原形式存在，外界的刺激和信号，如 β-葡聚糖、脂多糖和微生物多糖（Cárdenas and Dankert，1997）等，可程序性地激活酚氧化酶原，使其成为有活性的酚氧化酶（Söderhäll and Cerenius，1998），激活后产生的黑色素及其中间产物醌可抑制病原体胞外蛋白酶和几丁质酶的活性，从而杀死微生物和寄生虫（Vargas-Albores and Yepiz-Plascencia，2000）。

在本研究中，饲料中添加玉米干酒精糟及其可溶物显著提高了刺参幼参体腔液中溶菌酶、酸性磷酸酶和酚氧化酶的活性，这可能是由于玉米干酒精糟及其可溶物不仅为刺参提供了蛋白质、脂肪等营养物质，还在发酵过程中融入了大量酵母细胞和酵母细胞成分（李华磊，2014）。酵母细胞富含的 β-葡聚糖、甘露醇二酸和角质素有刺激巨噬细胞吞噬的作用；酵母细胞壁的甘露寡糖和 1,3/1,6-β-葡聚糖等成分可调节免疫性能，促进肠道微生物菌群的发育和生长（Mehdi and Hasan，2012）。

3.2.2.3 结论

在本实验条件下，配合饲料中添加 0%～40% 的玉米干酒精糟及其可溶物未对刺参幼参的生长性能和体壁组成产生显著影响，同时，添加 20%～40% 的玉米干酒精糟及其可溶物能显著提高刺参幼参体腔液中相关免疫酶的活性。在本研究中，虽然刺参幼参对玉米干酒精糟及其可溶物作为其主要蛋白源表现出良好的适应性，但不同生长阶段刺参对玉米干酒精糟及其可溶物的利用效果及长时间使用玉米干酒精糟及其可溶物是否会对刺参的生长性能产生影响，仍需进一步研究。

第 4 章 刺参配合饲料中功能性饲料添加剂的筛选

4.1 刺参配合饲料中酶制剂的筛选

4.1.1 酶制剂对刺参生长、体成分、免疫能力及氨氮胁迫下免疫酶活性和热休克蛋白 70 含量的影响

饲用酶制剂是一种以酶为主要功能性因子,通过特定生产工艺加工而成的饲料添加剂(武明欣等,2015)。饲用酶制剂可刺激内源消化酶分泌,降低肠道食糜黏度,提高饲料转化率,减少或消除饲料中的抗营养因子,提高机体免疫抵抗力,促进水产动物健康生长(明建华,2007)。目前,酶制剂在饲料中的应用主要集中于畜禽方面,在水产饲料中的应用仅见于草鱼(黄峰等,2008b;高春生等,2006a)、异育银鲫(黄峰等,2008d;黄峰等,2008c)、尼罗罗非鱼(钟国防和周洪琪,2005a)、大黄鱼(张璐等,2006;张春晓等,2008)等少数种类,在刺参上的应用鲜有报道。

4.1.1.1 材料与方法

1. 实验设计

实验设 5 个组,分别为对照组、木聚糖酶组、纤维素酶组、淀粉酶组和复合酶组,每个处理 3 个重复,每个重复放养体质量为(4.02±0.04)g 的幼参 40 头。实验所用木聚糖酶、纤维素酶和淀粉酶的活性分别为 10 000U/g、1000U/g、2000U/g。共配制粗蛋白含量为 20%的 5 种实验饲料,分别命名为 D1、D2、D3、D4、D5。首先配制不添加酶制剂的基础饲料作为对照,然后分别以 0.04%木聚糖酶、0.10%纤维素酶、0.01%淀粉酶、0.15%复合酶(由 0.04%木聚糖酶、0.10%纤维素酶和 0.01%淀粉酶组成)等量替代基础饲料中的 α-淀粉。实验饲料配方及营养组成见表 4-1。固体原料经超微粉碎后过 200 目标准筛,按配比称重,加入新鲜鱼油及适量蒸馏水混匀,用小型颗粒饲料挤压机制成直径为 0.3cm 的颗粒,60℃烘干,用小型粉碎机破碎,筛选粒度为 80~100 目的颗粒备用。

表 4-1 实验饲料配方及营养组成(干物质)(%)

项目		组别				
		D1	D2	D3	D4	D5
原料组成	鱼粉	10.00	10.00	10.00	10.00	10.00
	豆粕	15.00	15.00	15.00	15.00	15.00

续表

项目		组别				
		D1	D2	D3	D4	D5
	藻粉	20.00	20.00	20.00	20.00	20.00
	小麦粉	15.00	15.00	15.00	15.00	15.00
	海泥	25.00	25.00	25.00	25.00	25.00
	α-淀粉	9.50	9.46	9.40	9.49	9.35
	木聚糖酶	0.00	0.04	0.00	0.00	0.04
	纤维素酶	0.00	0.00	0.10	0.00	0.10
	淀粉酶	0.00	0.00	0.00	0.01	0.01
	鱼油	1.00	1.00	1.00	1.00	1.00
	磷酸二氢钙	1.00	1.00	1.00	1.00	1.00
	黏合剂	0.50	0.50	0.50	0.50	0.50
	维生素预混料	2.00	2.00	2.00	2.00	2.00
	矿物质预混料	1.00	1.00	1.00	1.00	1.00
营养组成	粗脂肪	1.66	1.61	1.72	1.69	1.83
	粗灰分	29.23	29.00	28.68	28.92	28.97
	木聚糖	12.26	12.26	12.25	12.26	12.25
	粗纤维	5.69	5.69	5.69	5.69	5.69
	淀粉	13.84	13.81	13.77	13.83	13.73
	能量	10.69	10.65	10.67	11.01	10.94

注：①藻粉脱胶海带粉与马尾藻粉等量混合，粗蛋白含量为 16.7%；②海泥经高温煅烧处理；③木聚糖酶、纤维素酶和淀粉酶的添加量为推荐添加量；④每千克维生素预混料含维生素 A 38.0mg、维生素 B_{12} 1.3mg、维生素 D_3 13.2mg、维生素 B_2 380.0mg、维生素 B_1 115.0mg、α-生育酚 210.0mg、烟酸 1030.0mg、泛酸 368.0mg、生物素 10.0mg、叶酸 20.0mg、抗坏血酸 500.0mg、盐酸吡哆醇 88.0mg、肌醇 4000.0mg；⑤每千克矿物质预混料含 KCl 3020.5mg、KAl$(SO_4)_2$ 11.3mg、NaCl 100.0mg、KI 7.5mg、NaF 4.0mg、$ZnSO_4·7H_2O$ 363.0mg、$CuSO_4·5H_2O$ 8.0mg、$MgSO_4·7H_2O$ 3568mg、$MnSO_4·4H_2O$ 65.1mg、Na_2SeO_3 2.3mg、$NaH_2PO_4·2H_2O$ 25558.0mg、$CoCl_2$ 28.0mg、$C_6H_5O_7Fe·5H_2O$ 1523.0mg、Ca-lactate 15 978.0mg。

2. 饲养管理

饲养实验在山东省海洋资源与环境研究院东营实验基地循环水养殖系统中进行，实验用刺参为该基地当年繁育的同一批参苗。实验开始前，将 1000 头刺参幼参放养于养殖系统中，暂养 15d，其间投喂基础饲料，待其完全适应饲养条件后，选择体质健壮、体质量为（4.02±0.04）g 的刺参幼参 600 头，平均放养于 15 个圆柱形玻璃钢养殖桶中。每种饲料投喂 3 个养殖桶，实验期为 56d。实验在微流水环境中进行，采用充气增氧，保证溶氧量大于 7mg/L，控制水温在 18～20℃，pH 为 7.8～8.2，盐度为 24～26，亚硝酸氮浓度低于 0.05mg/L、氨氮浓度低于 0.05mg/L。养殖实验持续 56d，每天投喂 2 次（8:00，16:30），日投喂量为刺参幼参体质量的 2%，根据刺参幼参摄食情况及时调整投喂量。每隔 2d 吸底 1 次，用虹吸法将残饵及粪便吸出。

3. 样品收集

实验结束后，禁食48h，记录每桶刺参总数并称量总重，用于存活率、增重率及特定生长率的计算。每桶取10头刺参，用滤纸轻轻吸干其体表水分，称量体质量，待其自然舒展后测体长，用于肥满度的计算。用2ml无菌注射器抽取体腔液后，置于冰盘上解剖，分别称量体壁与肠道的质量，并测量肠道长度。在4℃条件下，将体腔液以3000r/min离心15min，取上清液分装于2ml的离心管中，同体壁、肠道样品一同保存于冰箱（–80℃）中，待测。

4. 氨氮胁迫实验

收集样品后，封闭循环水系统，用氯化铵调节养殖水氨氮浓度至约5mg/L，继续饲养刺参5d，并分别在实验的0h、24h、48h、72h、96h、120h采集刺参的体腔液，用于免疫酶活性和热休克蛋白70含量的测定。

5. 测定指标和方法

刺参幼参存活率、增重率、特定生长率、肥满度、脏体比、肠体比及肠长比计算方法见附录。

饲料及刺参幼参体壁干物质、粗蛋白、粗脂肪、粗灰分的含量及能量测定方法见附录；木聚糖含量采用分光光度法测定；粗纤维含量采用过滤法［《饲料中粗纤维的含量测定》（GB/T 6434—2022）］测定；淀粉含量采用旋光法［《动物饲料中淀粉含量的测定 旋光法》（GB/T 20194—2018）］测定。

溶菌酶活性采用空白对照比浊法测定；总超氧化物歧化酶活性采用羟胺法测定，定义每毫升反应液中超氧化物歧化酶抑制率达50%时所对应的超氧化物歧化酶量为1个超氧化物歧化酶活性单位（U）；过氧化氢酶活性采用可见光分光光度计法测定，定义每毫升体腔液每秒钟分解1μmol的过氧化氢的量为一个活性单位（U）。以上指标均采用南京建成生物工程研究所生产的试剂盒测定。

热休克蛋白70的含量采用上海拜沃生物科技有限公司生产的ELISA试剂盒测定。将标准品、待测样品加入到纯化的海参热休克蛋白70单克隆抗体包被的微孔板中，加入酶联亲和物，37℃孵育60min，洗涤除去未结合的成分，依次加入底物A和B，底物在辣根过氧化物酶催化下先转化为蓝色后变为黄色。在450nm波长下测定光密度值，根据标准品和样品的光密度值计算样品中热休克蛋白70的含量。

6. 数据统计

实验所得数据采用Microsoft Excel 2007及SPSS 11.0软件进行单因素方差分析，结果以平均值±标准差表示，差异显著（$P<0.05$）时用邓肯多重范围检验进行多重比较分析。

4.1.1.2 结果分析

1. 酶制剂对刺参幼参生长性能的影响

由表 4-2 可见，饲料中添加木聚糖酶、纤维素酶和复合酶均能显著提高刺参幼参的增重率和特定生长率（$P<0.05$），且添加木聚糖酶和复合酶的促生长效果尤其明显，其增重率分别比对照组提高了 17.38%和 20.75%。木聚糖酶组的脏体比均显著大于对照组（$P<0.05$），纤维素酶组的脏体比显著大于对照组（$P<0.05$），淀粉酶组的肥满度显著大于对照组（$P<0.05$），复合酶组的肥满度、脏体比和肠体比均显著大于对照组（$P<0.05$）。饲料中添加单体酶和复合酶对刺参幼参存活率并无显著性影响（$P>0.05$）。

表 4-2 酶制剂对刺参幼参生长性能的影响

项目	组别				
	D1	D2	D3	D4	D5
初始体质量/g	4.02±0.08	4.03±0.02	4.01±0.04	4.04±0.05	4.02±0.03
终末体质量/g	4.49±0.04^a	5.19±0.07^c	4.80±0.02^b	4.56±0.09^a	5.32±0.14^c
增重率/%	12.25±1.06^a	29.63±1.94^c	19.88±0.53^b	14.13±2.30^a	33.00±3.54^c
特定生长率/(%/d)	0.21±0.02^a	0.46±0.03^c	0.32±0.01^b	0.24±0.04^a	0.51±0.05^c
肥满度/%	5.61±0.17^a	5.53±0.12^a	5.57±0.10^a	5.97±0.11^b	5.89±0.11^b
脏体比/%	7.04±0.59^a	7.53±0.47^b	8.14±0.38^b	7.12±0.60^{ab}	7.81±0.50^b
肠体比/%	4.45±0.58^a	4.90±0.64^{ab}	5.00±0.36^{ab}	4.85±0.51^a	5.78±0.35^b
肠长比/%	2.55±0.24^{abc}	2.77±0.21^{bc}	2.53±0.08^{abc}	2.40±0.22^{ab}	2.88±0.22^c
存活率/%	96.67±5.77	96.67±1.44	96.67±2.89	93.33±2.89	98.33±2.89

注：同行数据无字母或上标相同字母表示差异不显著（$P>0.05$），上标不同字母表示差异显著（$P<0.05$）。

2. 酶制剂对刺参幼参体壁基本成分的影响

由表 4-3 可见，饲料中添加不同酶制剂对刺参幼参体壁水分含量无显著影响（$P>0.05$），但添加木聚糖酶、纤维素酶、淀粉酶和复合酶均显著提高了刺参幼参体壁粗蛋白和粗灰分的含量（$P<0.05$），添加淀粉酶和复合酶均显著提高了刺参幼参体壁粗脂肪含量（$P<0.05$）。

表 4-3 酶制剂对刺参幼参体壁基本成分的影响

项目	组别				
	D1	D2	D3	D4	D5
水分	90.44±0.18	90.58±0.29	90.56±0.19	90.80±0.23	90.79±0.29
粗蛋白	47.82±0.60^a	51.13±0.74^b	52.94±0.27^b	52.59±0.48^b	52.71±0.94^b
粗脂肪	2.57±0.18^a	2.63±0.19^a	2.65±0.15^a	3.80±0.21^c	3.06±0.12^b
粗灰分	28.70±0.57^a	30.45±1.27^b	32.07±0.98^b	30.58±0.69^b	31.68±0.08^b

注：水分为鲜重基础，其余为干重基础；同行数据无字母或上标相同字母表示差异不显著（$P>0.05$），上标不同字母表示差异显著（$P<0.05$）。

3. 酶制剂对刺参幼参免疫能力的影响

由表 4-4 可见，饲料中添加木聚糖酶、纤维素酶和复合酶均显著提高了刺参幼参体腔液中过氧化氢酶的活性（$P<0.05$）；添加淀粉酶显著提高了刺参幼参体腔液中溶菌酶、超氧化物歧化酶和过氧化氢酶的活性（$P<0.05$）。刺参幼参体腔液中溶菌酶和超氧化物歧化酶的活性均以淀粉酶组最高，过氧化氢酶活性以复合酶组最高。

表 4-4 酶制剂对刺参幼参免疫酶活性的影响

项目	组别				
	D1	D2	D3	D4	D5
溶菌酶/（U/ml）	67.59±6.05[a]	69.22±6.28[a]	63.45±4.33[a]	83.25±8.94[b]	72.17±7.38[ab]
超氧化物歧化酶/（U/ml）	113.04±6.35[a]	109.65±8.24[a]	111.94±4.32[a]	121.72±3.84[b]	113.89±4.04[a]
过氧化氢酶/（U/ml）	1.50±0.02[a]	1.69±0.02[b]	1.88±0.04[c]	2.21±0.10[d]	2.32±0.03[d]

注：同行数据无字母或上标相同字母表示差异不显著（$P>0.05$），上标不同字母表示差异显著（$P<0.05$）。

4. 酶制剂对氨氮胁迫下刺参幼参体腔液中免疫酶活性和热休克蛋白 70 含量的影响

由图 4-1 可见，氨氮胁迫过程中各组刺参幼参体腔液中溶菌酶活性的总体趋势是先升高后降低，对照组、木聚糖酶组、纤维素酶组、淀粉酶组、复合酶组最大值分别出现在 72h、72h、96h、96h 和 96h。其中，淀粉酶组、复合酶组的最大值要大于其余各组。氨氮胁迫过程中各组刺参幼参体腔液中超氧化物歧化酶活性均先升高，并均在 24h 达到最大值，之后降低并于 48h 后趋于稳定，且其稳定值略低于初始值。氨氮胁迫过程中各组刺参幼参体腔液中过氧化氢酶活性均呈现先升高后降低的趋势，并均在 96h 达到最大值，其中淀粉酶组的最大值要大于其他各组。

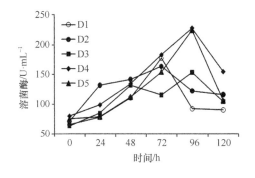
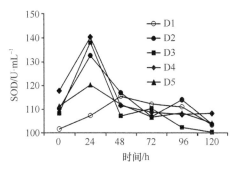

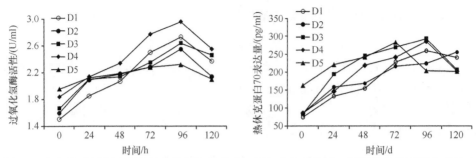

图 4-1 氨氮胁迫过程中刺参幼参体腔液溶菌酶、超氧化物歧化酶、过氧化氢酶活性及热休克蛋白 70 表达量的变化

氨氮胁迫下，实验初期复合酶组刺参幼参体腔液中热休克蛋白 70 含量明显高于其他各组。木聚糖酶组刺参幼参体腔液中热休克蛋白 70 的含量呈现一直升高的趋势，其余各组均呈现先升高后降低的趋势。复合酶组刺参幼参体腔液中热休克蛋白 70 的含量比其他组率先达到最高点，且在 96h 后趋于稳定。对照组、纤维素酶组、淀粉酶组刺参幼参体腔液中热休克蛋白 70 的含量于 96h 达到最大值。除木聚糖酶组外，其余添加酶制剂组刺参幼参体腔液中热休克蛋白 70 含量的最大值均大于对照组。

4.1.1.3 讨论

1. 酶制剂对刺参幼参生长性能和体成分的影响

本研究表明，饲料中添加酶制剂显著提高了刺参幼参的增重率和特定生长率，这与对鲫鱼（张玲等，2006）、尼罗罗非鱼（王俊丽等，2007；聂国兴等，2007b；钟国防和周洪琪，2005b）、黄河鲤（李红岗等，2010）和大西洋鲑（Carter et al.，1994）的研究结果一致。饲料中添加酶制剂促进刺参生长可能有以下几方面原因：①饲料中添加木聚糖酶能有效降低肠道食糜黏度，降解非淀粉质抗营养因子，促进养殖动物对营养物质的消化吸收，从而改善养殖动物的生产性能（何云等，2008；林少珍等，2009）；②纤维素酶能破坏细胞壁、降低饲料密集黏度，同时刺激内源酶的分泌，从而促进水产动物对营养成分的消化吸收（高春生等，2006a）；③饲料中添加淀粉酶起作用的主要为 α-淀粉酶和糖化酶，两种酶均是内切酶，其中 α-淀粉酶将淀粉大分子水解成易溶解的中等和低分子物质，有利于糖化酶的水解，而糖化酶活性的高低直接影响动物的生长（王玲鞾和瞿明仁，2006）；④复合酶综合了几种单体酶的作用，但又不仅仅是单个酶作用的复合。邓岳松（2005）在研究中发现，由纤维素酶、淀粉酶、蛋白酶等组成的复合酶对饲料中的一些抗营养因子具有灭活作用，可消除草鱼对营养物质的消化和吸收困难，从而提高鱼体的生长性能。黄峰等（2008a）的研究表明，添加复合酶促进异育银鲫生长的效果稍好于单一添加木聚糖酶，这可能得益于复合酶的叠加效应和协同效应。本实验结果也表明，添加复合酶的促生长效果优于添加单体酶。

本研究表明，饲料中添加酶制剂对刺参幼参体壁粗蛋白、粗脂肪和粗灰分的含量均有显著影响。这与王俊丽等（2007）、黄燕华等（2009）的研究结果一致。饲料中添加木聚糖酶能提高肠道蛋白酶活性，增加肠道内游离氨基酸和小肽数量，从而促进蛋白质的合成和积累。同时，添加木聚糖酶降低了肠道食糜黏度，肠道活动增强，从而促进了对脂肪的吸收，增加了体壁中脂肪的沉积。添加木聚糖酶使肠道中低聚木糖增加，低聚木糖的代谢降低了肠道 pH，增强了钙盐的溶解性，促进了钙的吸收，从而提高了粗灰分的含量（王俊丽等，2007）。而钟国防和周洪琪（2005a）、张璐等（2006）的研究证明，饲料中添加酶制剂对养殖动物的基本营养成分均无显著影响。这可能与添加酶的种类、活性及养殖动物种类有关。

2. 酶制剂对刺参幼参免疫能力的影响

刺参没有完善的特异性免疫系统，溶菌酶、超氧化物歧化酶及过氧化氢酶在其免疫及抗氧化过程中起着重要作用（Fridovich，1989）。溶菌酶活性的高低直接反映溶菌能力的强弱（Bayne and Gerwick，2001），超氧化物歧化酶活性与生物体免疫水平密切相关（Fridovich，1989），过氧化氢酶保护机体组织免受损害（臧元奇等，2012）。本实验结果表明，饲料中添加酶制剂对刺参幼参各免疫指标均有促进作用。其中，木聚糖酶组刺参幼参体腔液中过氧化氢酶的活性与对照组相比显著提高；淀粉酶组刺参幼参体腔液中溶菌酶、超氧化物歧化酶及过氧化氢酶的活性均较对照组显著提高。这与黄燕华等（2009）、钟国防和周洪琪（2005b）、黄峰等（2008d，2008c）的研究结果一致。黄燕华等（2009）的研究证实，饲料中添加淀粉酶能显著提高对虾血清中超氧化物歧化酶活性。钟国防和周洪琪（2005b）的研究表明，在尼罗罗非鱼饲料中添加木聚糖酶和复合酶能显著提高鱼体肝脏和脾脏中超氧化物歧化酶和溶菌酶的活性。目前关于刺参体腔液中溶菌酶、超氧化物歧化酶及过氧化氢酶正常活性范围的研究并不充足，尤其是酶活性升高的机制尚不清楚，因此不能单纯依据酶活性来评定其免疫及抗氧化能力的强弱，但是对大菱鲆等鱼类的研究表明，鱼体抗病能力的提高往往伴随着溶菌酶、超氧化物歧化酶及过氧化氢酶活性的升高（郝甜甜等，2014），由此可以推测，在一定范围内上述 3 种酶活性的升高提高了机体的抗逆性。

3. 酶制剂对氨氮胁迫下刺参幼参体腔液中免疫酶活性和热休克蛋白 70 含量的影响

热休克蛋白是一种应激反应的指示器，它能使细胞处于保护状态，它的产生能使细胞抵抗致死性损伤，增强细胞对外界损害的耐受力（臧元奇等，2012）。本实验过程中，在 5mg/L 的氨氮浓度胁迫下，各组刺参幼参体腔液中免疫酶活性均表现出一定的上升趋势。这主要是由于氨氮胁迫增强了刺参幼参体内水解系统的活性，表现出明显的毒物兴奋作用（Stebbing，1982）。本实验研究发现，饲料中添加酶制剂能提高刺参幼参的免疫性能，因此在短时间内各添加酶制剂组的刺参幼参表现出比对照组更迅速和强烈的免疫抵抗反应。随着胁迫时间的延长，刺参自身的水解系统活性降低，抵抗力下降（臧元奇等，2012），因此各组免疫酶活性随后表现出下降趋势。氨氮胁迫 72h 内，各组刺参幼参体腔液中热休克蛋白 70 含量均呈现上升趋势。这说明刺参幼参通过提高热休克蛋白 70 的表达量增强抗逆性，以适应氨氮胁迫的环境（臧元奇等，2012）。氨氮胁迫开始前（即 0h），复合酶组刺参幼参体腔液中热休克蛋白 70 的含量明显高于其他各组，其次为木聚糖酶组。氨氮胁迫 48h 内，纤维素酶组、淀粉酶组和复合酶组刺参幼参体腔液中热休克蛋白 70 的含量与对照组相比表现出较快的升高，且复合酶组更早地达到最高点，这可能与酶制剂促进刺参自身免疫和胁迫抵抗有关。

4.1.1.4 结论

（1）饲料中适量添加木聚糖酶、纤维素酶、淀粉酶或其复合酶均能提高刺参幼参的生长性能，提高其免疫能力，并对刺参幼参体壁粗蛋白、粗脂肪和粗灰分的含量也有一定影响，且木聚糖酶的促生长作用更明显，而淀粉酶的促免疫作用更明显。

（2）氨氮胁迫下，饲料中适量添加木聚糖酶、纤维素酶、淀粉酶或其复合酶均能使刺参幼参更迅速和强烈地表现出免疫抵抗反应。

4.1.2 木聚糖酶对刺参幼参生长、消化和体腔液酶活性的影响

木聚糖是自然界中除纤维素外多糖含量最高的物质，在植物性饲料原料中含量也较高（石军和陈安国，2002；阮同琦等，2008）。木聚糖具有抗营养作用，一般水产动物较难将其直接消化利用。木聚糖也具有较强的结合水的能力，从而增大动物肠道内的食糜黏度，阻碍对营养物质特别是蛋白质、脂肪的吸收和利用（阮同琦等，2008），并能与肠道蛋白酶和脂肪酶络合，降低内源酶活性（白雪峰，2004）。木聚糖酶是催化木聚糖水解的一类酶的统称，根据功能不同分为 β-1,4-D-内切木聚糖酶、β-1,4-D-外切木聚糖酶和 β-木糖苷酶三类（王琤韡和瞿明仁，2006）。三类酶协同作用把大分子木聚糖分解为木二糖、木三糖等低聚糖和少量阿拉伯糖（高春生等，2006b；方微等，2011），有利于动物的吸收和利用。在水产动物饲料中添加木聚糖酶能有效消除木聚糖的抗营养作用，同时破坏细胞壁，使养分充分流出，降低肠道食糜黏度（Choct et al.，1996；Preston et al.，2000），增加饲料中营

养成分与酶的反应面积,提高对营养物质的消化吸收,从而改善养殖动物的生产性能。目前,对水产动物饲料中木聚糖酶的研究主要集中在鲫(张玲等,2006)、黄河鲤(李红岗等,2010;程会昌等,2006)、尼罗罗非鱼(聂国兴等,2007a;聂国兴等,2007c)、条石鲷(刘伟成等,2010)和南美白对虾(黄燕华等,2009)等少数品种,对刺参的相关研究较少。本实验中,在基础饲料中添加 6 个水平的木聚糖酶,研究了其对刺参幼参生长性能、体成分、免疫和消化的影响,以期为确定其在刺参配合饲料中的最佳添加量提供依据(武明欣等,2018)。

4.1.2.1 材料与方法

1. 实验饲料的制备

以鱼粉、豆粕、藻粉为主要蛋白源,设计粗蛋白含量约为 24%、粗脂肪含量约为 1.5% 的基础饲料配方,分别以 0、0.03%、0.06%、0.09%、0.12% 和 0.24% 木聚糖酶替代基础饲料中的 α-淀粉,制成 6 组等氮等能的实验饲料,分别记为 D0（对照组）、D1、D2、D3、D4 和 D5 组。实验饲料配方及营养组成见表 4-5。将固体原料超微粉碎后过 200 目筛,按配比称重,加入新鲜鱼油及适量蒸馏水混匀,用小型颗粒饲料挤压机制成直径为 0.3cm 的颗粒,60℃下烘干,用小型粉碎机破碎,筛选粒度为 80～100 目的颗粒备用。

表 4-5 实验饲料配方及营养组成（风干基础）

原料		组别					
		D0	D1	D2	D3	D4	D5
原料组成	鱼粉	10	10	10	10	10	10
	豆粕	15	15	15	15	15	15
	藻粉	20	20	20	20	20	20
	小麦粉	15	15	15	15	15	15
	海泥	25	25	25	25	25	25
	α-淀粉	9.5	9.47	9.44	9.41	9.38	9.26
	木聚糖酶	0.00	0.03	0.06	0.09	0.12	0.24
	鱼油	1	1	1	1	1	1
	磷酸二氢钙	1	1	1	1	1	1
	黏合剂	0.5	0.5	0.5	0.5	0.5	0.5
	维生素预混料	2	2	2	2	2	2
	矿物质预混料	1	1	1	1	1	1
营养组成	木聚糖酶	0	0.03	0.06	0.09	0.12	0.24
	粗蛋白	23.87	23.89	23.87	23.86	23.97	24.03
	粗脂肪	1.53	1.52	1.52	1.57	1.6	1.56

原料		组别					
		D0	D1	D2	D3	D4	D5
营养组成	粗灰分	31.46	31.15	31.4	31.44	31.26	31.41
	能量/(kJ/g)	2708.38	2684.29	2687.3	2688.55	2625.67	2676.6

注：①藻粉为脱胶海带粉与马尾藻粉等量混合，粗蛋白含量为 16.7%；②海泥经高温煅烧处理；③每千克维生素预混料含维生素 A 38.0mg、维生素 B_{12} 1.3mg、维生素 D_3 13.2mg、维生素 B_2 380.0mg、维生素 B_1 115.0mg、α-生育酚 210.0mg、烟酸 1030.0mg、泛酸 368.0mg、生物素 10.0mg、叶酸 20.0mg、抗坏血酸 500.0mg、盐酸吡哆醇 88.0mg、肌醇 4000.0mg；④每千克矿物质预混料含 KCl 3020.5mg、$KAl(SO_4)_2$ 11.3mg、NaCl 100.0mg、KI 7.5mg、NaF 4.0mg、$ZnSO_4·7H_2O$ 363.0mg、$CuSO_4·5H_2O$ 8.0mg、$MgSO_4·7H_2O$ 3568mg、$MnSO_4·4H_2O$ 65.1mg、Na_2SeO_3 2.3mg、$NaH_2PO_4·2H_2O$ 25 558.0mg、$CoCl_2$ 28.0mg、$C_6H_5O_7Fe·5H_2O$ 1523.0mg、Ca-lactate 15 978.0mg。

2. 实验设计及饲养管理

实验所用木聚糖酶由武汉新华扬酶制剂有限公司提供，内附有酶活性检测报告，活性为 10 000U/g。实验用刺参参采自山东省海洋资源与环境研究院东营实验基地，为该基地当年繁育的同一批参苗。

饲养实验在山东省海洋资源与环境研究院东营实验基地循环水养殖系统中进行。实验开始前，将 1000 头刺参幼参放养于养殖系统中，暂养 15d，其间投喂基础饲料，待其完全适应饲养条件后，选择体质健壮、体质量为（7.73±0.09）g 的刺参 900 头，随机放养于室内海水循环系统的 18 个圆柱形玻璃钢桶（高 70cm，直径为 80cm）中，控制水深为 60cm。每种饲料随机投喂 3 个养殖桶的幼参。养殖车间利用遮阳布保持无光状态，每天 8:00、16:30 投喂，日投喂量根据刺参幼参摄食情况及时调整。控制水温为（19±1）℃，pH 为 7.8～8.2，盐度为 24～26，亚硝酸氮、氨氮的浓度均不高于 0.05mg/L。养殖实验持续 56d，每隔 2d 换水一次，换水量为 1/3，换水时用虹吸法将残饵及粪便吸出。

3. 样品收集

实验结束后，将刺参幼参饥饿 48h，记录每桶刺参幼参总数及总质量，用于存活率、增重率及特定生长率的计算。从每桶取 10 头刺参幼参，用滤纸轻轻吸干体表水分，称量体质量，待其自然舒展后用刻度尺测量体长。用 2ml 无菌注射器抽取体腔液后，置于冰盘上解剖，分离体壁与肠道。将体腔液在 4℃下以 3000r/min 离心 15min 后，取上清液分装于 2ml 的离心管中。样品均保存于冰箱（–80℃）中，待测。

4. 指标的测定与计算

存活率、增重率和特定生长率计算方法见附录，饲料及刺参幼参体壁营养成分含量测定方法见附录。将刺参幼参肠道解冻后用 4℃生理盐水小心冲洗，用滤纸吸干表面水分，按需要称取肠道质量，加入 9 倍生理盐水于冰水浴下匀浆，将组织悬液低温下以 7000r/min 离心 30min，上清液即粗酶液，分装后置于冰箱（4℃）中待测。采用福林-酚法、碘-淀粉比色法、滴定法和考马斯亮蓝染色法分别测定刺参幼参蛋白酶、淀粉酶、脂肪酶的活性和酶液总蛋白含量。以上指标均采用南京建成生物工程研究所生产的试剂盒测定。溶菌酶、总超氧化物歧化酶、过氧化氢酶的活性均采用南京建成生物工程研究所生产的试剂盒测定，相关酶活定义参见其说明书。谷丙转氨酶和谷草转氨酶的活性采用全自动生化分析仪测定。

5. 数据统计

实验数据用 Microsoft Excel 2007 进行统计处理，并以平均值±标准差表示。采用 SPSS 11.0 软件进行单因素方差分析，用邓肯多重范围检验进行多重比较分析，显著性水平设为 0.05。

4.1.2.2　结果分析

1. 木聚糖酶对刺参幼参生长性能的影响

从表 4-6 可见，饲料中添加木聚糖酶能显著提高刺参幼参的增重率和特定生长率（$P<0.05$），随着木聚糖酶添加量的增加呈现先升高后降低的趋势，其中 0.09% 添加组增重率和特定生长率最高；饲料中添加木聚糖酶对刺参幼参的存活率无显著影响（$P>0.05$）。

表 4-6　木聚糖酶对刺参幼参生长性能的影响

项目	组别					
	D0	D1	D2	D3	D4	D5
初始体质量/g	7.74±0.04	7.77±0.04	7.74±0.06	7.76±0.03	7.70±0.03	7.70±0.03
终末体质量/g	8.24±0.10^a	9.17±0.58^b	9.94±0.41^c	11.33±0.40^d	10.45±0.08^c	9.91±0.16^c
增重率/%	6.41±1.66^a	18.04±6.87^b	28.43±5.00^c	46.04±5.68^d	35.72±0.63^c	28.84±1.87^c
特定生长率/（%/d）	0.11±0.03^a	0.29±0.10^b	0.45±0.07^c	0.68±0.07^d	0.55±0.01^c	0.45±0.03^c
存活率/%	100	100	100	100	100	100

注：同行数据无字母或上标相同字母表示差异不显著（$P>0.05$），上标不同字母表示差异显著（$P<0.05$）。

如图 4-2 所示，以饲料中木聚糖酶含量为自变量（X）、增重率为因变量（Y）的回归分析得出，$Y=42.641+0.041（X–900）+0.01（900–X）$，当 $X<900$mg/kg 时，$0.01（900–X）=0$，此时 $R^2=0.98$，当 $X>900$mg/kg 时，$0.041（X–900）=0$，此时 $R^2=0.94$，当 $X=900$mg/kg 时，刺参幼参获得最大增重率。因此，在本实验条件下，以增重率为评价指标，初始体质量为（7.73±0.09）g 的刺参饲料中木聚糖酶最适添加量为 900mg/kg。

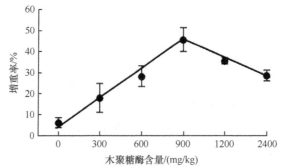

图 4-2 刺参幼参增重率与饲料中木聚糖酶含量的回归分析

2. 木聚糖酶对刺参幼参体壁基本成分的影响

从表 4-7 可见，饲料中添加木聚糖酶后不同程度地提高了刺参幼参体壁粗蛋白和粗脂肪的含量，随着木聚糖酶添加量的升高，刺参幼参体壁粗蛋白和粗脂肪的含量均呈现先升高后降低的趋势；各添加组粗蛋白含量均显著高于对照组（$P<0.05$），但各添加组间无显著性差异（$P>0.05$）；0.06%～0.24%添加组粗脂肪含量均显著高于对照组（$P<0.05$），且以 0.09% 和 0.12% 添加组受影响较为明显。

表 4-7 木聚糖酶对刺参幼参体壁基本营养成分的影响（%）

项目	组别					
	D0	D1	D2	D3	D4	D5
水分	89.90±0.59	89.80±0.14	89.39±0.27	90.31±0.41	90.29±0.61	90.17±0.77
粗蛋白	53.14±0.63^a	55.39±0.44^b	56.34±0.64^b	56.04±0.75^b	55.52±1.27^b	55.52±1.18^b
粗脂肪	2.08±0.12^a	2.15±0.24^a	2.60±0.20^b	3.75±0.16^d	3.53±0.34^d	3.05±0.25^c
粗灰分	29.11±0.71^b	26.96±1.08^{ab}	25.63±0.64^a	28.76±1.30^b	27.42±1.47^{ab}	27.36±1.65^{ab}

注：同行数据无字母或上标相同字母表示差异不显著（$P>0.05$），上标不同字母表示差异显著（$P<0.05$）。

3. 木聚糖酶对刺参幼参肠道消化酶活性的影响

从表 4-8 可见，饲料中添加木聚糖酶能明显提高刺参幼参肠道消化酶活性，随着木聚糖酶添加量的提高，蛋白酶和脂肪酶的活性呈现先升高后降低的趋势；0.06%、0.09%、0.12% 添加组刺参幼参肠道蛋白酶和脂肪酶的活性均显著高于对照组和 0.24% 添加组（$P<0.05$），0.24% 添加组刺参幼参肠道淀粉酶活性显著高于对照组和其他添加组（$P<0.05$）。

表 4-8 木聚糖酶对刺参幼参肠道消化酶活性的影响（单位：U/mg prog）

项目	组别					
	D0	D1	D2	D3	D4	D5
蛋白酶	58.83±6.76^a	60.73±8.40^a	74.19±3.78^b	78.34±2.03^b	79.82±9.96^b	60.53±1.76^a
淀粉酶	1.28±0.04^a	1.33±0.20^a	1.40±0.22^a	1.27±0.04^a	1.35±0.14^a	1.88±0.19^b
脂肪酶	0.63±0.15^a	1.32±0.17^c	1.94±0.25^d	1.22±0.21^{bc}	1.05±0.06^b	0.77±0.11^a

注：同行数据无字母或上标相同字母表示差异不显著（$P>0.05$），上标不同字母表示差异显著（$P<0.05$）。

4. 木聚糖酶对刺参幼参体腔液免疫酶活性的影响

从表 4-9 可见，饲料中添加木聚糖酶能明显提高刺参幼参体腔液免疫酶活性，随着木聚糖酶添加量的升高，体腔液中溶菌酶、过氧化氢酶、谷草转氨酶和谷丙转氨酶的活性均呈现先升高后下降的趋势；各添加组（0.03%添加组除外）溶菌酶活性显著高于对照组（$P<0.05$）；0.06%~0.12%添加组过氧化氢酶活性显著高于对照组（$P<0.05$）；0.03%~0.12%添加组谷草转氨酶活性显著高于对照组（$P<0.05$）；0.06%~0.24%添加组谷丙转氨酶活性显著高于对照组（$P<0.05$）；木聚糖酶添加组与对照组相比，超氧化物歧化酶活性有所升高，但无显著性差异（$P>0.05$）。

表 4-9 木聚糖酶对刺参幼参体腔液免疫酶活性的影响（单位：U/ml）

项目/	组别					
	D0	D1	D2	D3	D4	D5
溶菌酶	64.45 ± 1.15^a	67.48 ± 0.93^a	87.13 ± 4.89^{bc}	95.81 ± 3.97^c	93.17 ± 9.94^{bc}	84.35 ± 2.09^b
超氧化物歧化酶	107.29 ± 2.98	110.95 ± 5.90	112.21 ± 0.77	111.79 ± 2.36	111.04 ± 4.04	110.31 ± 5.84
过氧化氢酶	2.09 ± 0.07^a	2.39 ± 0.06^{ab}	2.42 ± 0.21^b	2.80 ± 0.16^c	2.86 ± 0.09^c	2.21 ± 0.29^{ab}
谷草转氨酶	3.77 ± 0.59^a	6.4 ± 0.20^c	8.5 ± 0.28^d	6.2 ± 0.14^c	5.45 ± 0.21^b	4.2 ± 0.01^a
谷丙转氨酶	2.85 ± 0.21^a	3.2 ± 0.14^{ab}	3.65 ± 0.07^c	3.50 ± 0.14^{bc}	3.45 ± 0.21^{bc}	3.45 ± 0.07^{bc}

注：同行数据无字母或上标相同字母表示差异不显著（$P>0.05$），上标不同字母表示差异显著（$P<0.05$）。

4.1.2.3 讨论

1. 木聚糖酶对刺参幼参生长性能的影响

本实验结果表明，饲料中添加木聚糖酶能显著提高刺参幼参的增重率和特定生长率，随着木聚糖酶添加量的提高，刺参幼参增重率和特定生长率呈现先升高后下降的趋势，且 0.09%添加组增重率和特定生长率显著高于其他各组。这与对凡纳滨对虾（杨志刚等，2009）、异育银鲫（黄峰等，2008b）、鲫（方微等，2011）、草鱼（高春生等，2006b）、鲤（刘凯等，2008；姜婷婷，2013）和尼罗罗非鱼（钟国防和周洪琪，2005a；明红等，2006）等的研究结果一致。杨志刚等（2009）对凡纳滨对虾的研究表明，经木聚糖酶酶解后各添加组增重率和特定生长率均显著高于对照组。黄峰等（2008b）研究发现，在饲料中添加 100mg/kg 的木聚糖酶时，异育银鲫的生长率和增重率均显著提高。高春生等（2006b）研究发现，在草鱼饲料中添加木聚糖酶能显著提高鱼体增重率和饲料利用率，降低饵料系数，且草鱼饲料中木聚糖酶的适宜添加量为 0.1%。明红等（2006）的研究显示，尼罗罗非鱼饲料中木聚糖酶的最适添加量为 0.1%。饲料中木聚糖酶添加过多反而不利于其增重率和特定生长率的提高，这是因为木聚糖酶添加过量时，过度降低了肠道食糜黏度，致使营养成分通过肠道速度过快,降低了肠道对养分的消化吸收(聂国兴等,2006)。

刺参饲料中含有较多的植物性成分,添加木聚糖酶能破坏植物细胞壁,使多糖等营养物质流出,同时降解木聚糖等抗营养因子,促进肠道对营养物质的吸收和利用。此外,木聚糖酶还能有效降低肠道食糜黏度,扩大酶与营养成分的接触面积,提高饲料利用率,并能补充外源酶,刺激内源酶的分泌,增强养殖动物的消化吸收能力(梁永等,2012),这可能是其促生长的主要原因。

2. 木聚糖酶对刺参幼参体壁基本成分的影响

本实验结果表明,饲料中添加木聚糖酶对刺参幼参体壁基本成分有一定的改善作用,这与王俊丽等(2007)对尼罗罗非鱼和刘伟成等(2010)对鲤的研究结果一致。王俊丽等(2007)的研究表明,0.05%和0.10%木聚糖酶添加组的尼罗罗非鱼肌肉粗蛋白、粗脂肪和粗灰分的含量与对照组相比均有显著提高。刘凯等(2012)对鲤的研究表明,0.05%、0.10%和0.15%木聚糖酶添加组与对照组相比,鲤肌肉粗脂肪含量有显著提高。

饲料中添加木聚糖酶能提高肠道蛋白酶活性,增加肠道内游离氨基酸和小肽数量,从而促进蛋白质的合成和积累。同时,添加木聚糖酶降低了肠道食糜黏度,肠道活动增强,从而促进了对脂肪的吸收,增加了体壁中脂肪的沉积。此外,添加木聚糖酶增加了肠道中低聚木糖含量,从而降低了肠道pH,促进了钙等矿物质沉积,进而使粗灰分含量提高(王俊丽等,2007)。钟国防和周洪琪(2005b)的研究表明,添加木聚糖酶对养殖动物体成分无显著影响。这可能是受所用木聚糖酶的活性及养殖动物种类影响所致。

3. 木聚糖酶对刺参幼参消化酶活性的影响

本实验结果表明,饲料中添加木聚糖酶对刺参幼参肠道蛋白酶、淀粉酶和脂肪酶的活性有不同程度的提高作用。刘凯等(2012)对鲤的研究结果显示,饲料中添加0.10%和0.15%的木聚糖酶显著提高了肝胰脏蛋白酶和淀粉酶的活性,并对肝胰脏中木聚糖酶、脂肪酶的活性,以及肠道中木聚糖酶、淀粉酶和脂肪酶的活性也有一定提高作用。聂国兴等(2006,2007c)在尼罗罗非鱼饲料中添加木聚糖酶,显著提高了鱼体胃蛋白酶和纤维素酶的活性,提高了肠道内多种消化酶的活性,饲料中添加木聚糖酶后肠道食糜黏度显著低于对照组,肠道绒毛表面的黏附性颗粒明显减少。

刺参肠道内消化酶主要有蛋白酶、脂肪酶、淀粉酶和纤维素酶,消化酶活性是考察其消化生理的一个重要指标(白燕和王维新,2012),其活性高低能影响刺参对营养的吸收及生长生活状况(任庆印和潘鲁青,2013)。吴韬等(2015)对中华绒螯蟹的研究表明,木聚糖阻碍消化酶与底物接触反应。木聚糖能与肠道中蛋白酶和脂肪酶结合（何云等，2008），从而降低肠道消化酶活性，影响养殖动物对食物中营养的消化吸收。饲料中添加木聚糖酶，不仅能破坏植物细胞壁，使细胞中的养分充分溶出，促进机体对其的吸收利用并有效降解木聚糖，还能减少其与内源消化酶结合，使内源消化酶活性得以提高，外源添加木聚糖酶能有效降低肠道食糜黏度，增大食物营养与酶的接触面积，刺激肠道化学感受器诱导促胰酶素的分泌，进而刺激内源酶分泌（周金敏等，2010）。

4. 木聚糖酶对刺参幼参体腔液免疫酶活性的影响

水产动物普遍没有完善的特异性免疫系统，而刺参通常被认为缺乏特异性免疫应答（李旭等，2014），非特异性免疫在其抵御外界病原体和对细菌的防御中起着尤为重要的作用（聂品，1997）。溶菌酶、超氧化物歧化酶及过氧化氢酶均是刺参免疫及抗氧化过程中重要的功能因子。本研究表明，添加木聚糖酶显著提高了刺参幼参体腔液中溶菌酶和过氧化氢酶的活性，对超氧化物歧化酶活性的提高尽管不显著，但也表现出一定的提高作用。这与黄峰等（2008d）、钟国防和周洪琪（2005b）的研究结果一致。黄峰等（2008d）在异育银鲫饲料中添加50mg/kg和100mg/kg的木聚糖酶，显著提高了异育银鲫血清、头肾、脾脏中溶菌酶和超氧化物歧化酶的活性。钟国防和周洪琪（2005b）对尼罗罗非鱼的研究表明，木聚糖酶添加组肝胰脏和脾脏中溶菌酶及超氧化物歧化酶的活性均显著高于对照组。添加酶制剂后，提高了刺参的消化吸收能力，特别是提高了鱼体对蛋白质及其他营养成分的利用率，从而也促进了相关免疫因子的合成，加强了鱼体的代谢水平，产生了更多的阳离子等自由基，相应地也提高了超氧化物歧化酶的活性和抗氧化能力。张玲等（2006）对尼罗罗非鱼的研究表明，饲料中添加木聚糖酶显著提高了谷丙转氨酶活性，但对谷草转氨酶活性影响不显著。转氨酶是催化氨基酸与酮酸间氨基转移的一种酶，谷草转氨酶和谷丙转氨酶是其中重要的两种酶，通常被用来评价动物肝脏细胞受损情况，也可以反映体内蛋白质代谢状况（孙红梅，2004；李清等，2005）。张玲等（2006）指出，谷草转氨酶和谷丙转氨酶活性的提高能促进体内蛋白质代谢，转氨酶活性高低能反映氨基酸代谢水平。本实验研究显示，添加木聚糖酶能显著提高刺参幼参体腔液中谷草转氨酶和谷丙转氨酶的活性。这说明添加木聚糖酶提高了氨基酸代谢活性，参与了营养代谢过程，进而对于机体免疫性能也产生了一定的积极作用，但其具体影响机制还需进一步研究。

综上，饲料中添加适量木聚糖酶对刺参免疫酶、抗氧化酶和转氨酶活性的提高以及刺参免疫和抗氧化性能的增强具有重要意义。

4.1.2.4 结论

饲料中添加木聚糖酶能够显著提高刺参幼参增重率和特定生长率，增强免疫和抗氧化能力，促进消化机能，对改善体成分也有一定影响。但木聚糖酶添加过少可能影响不明显，过量添加则可能产生一些副作用，导致促进作用不显著甚至抑制刺参幼参生长。在本实验条件下，以增重率为评价指标，通过回归分析得知，初始体质量为（7.73±0.09）g 的刺参幼参饲料中木聚糖酶最适添加量为 900mg/kg。

4.2 刺参配合饲料中糖制剂的筛选

4.2.1 棉子糖对刺参幼参生长性能、生理指标及糖代谢的影响

棉子糖是一种功能性低聚糖，是由 1 分子半乳糖、1 分子葡萄糖和 1 分子果糖通过 α-1,6 糖苷键、β-1,2 糖苷键连接而成的三糖，分子式为 $C_{18}H_{32}O_{16}$（邱燕，2010）。棉子糖能促进肠道双歧杆菌及乳酸杆菌等有益菌的增殖，并能够增加肠道短链脂肪酸，降低肠道 pH（Tortuero et al., 1997）。研究表明，棉子糖能提高杂交鲟（Xu et al., 2018）、尼罗罗非鱼（Abdel-Latif et al., 2020）、草鱼（邱燕，2010）等水产动物的生长性能、抗氧化和免疫能力，并调整肠道微生物分布，改善肠道结构（郭鹏等，2022）。

4.2.1.1 材料与方法

1. 实验饲料

以鱼粉、藻粉和小麦粉为蛋白源，以鱼油为脂肪源，设计粗蛋白含量为 14.00%、粗脂肪含量为 1.80%的基础饲料配方（Liao et al., 2017）。基础饲料中分别添加 0（D1）、0.04%（D2）、0.08%（D3）、0.12%（D4）、0.16%（D5）和 0.20%（D6）的包膜棉子糖（济南中棉生物科技有限公司，纯度≥99%），制成 6 组实验饲料，饲料中棉子糖实际含量分别为 0、0.02%、0.03%、0.06%、0.08%和 0.11%。包膜棉子糖的制作方法为：将棉子糖与卡拉胶 1∶1 混合，加入适量的水调至稠状，90℃加热搅拌 15min，冷却后用冻干机干燥成胶状，75℃烘干后，磨成粉状，制成包膜棉子糖待用。

实验饲料配方及营养组成见表 4-10。将固体原料超微粉碎后过 200 目筛，按照饲料配方配比进行称重，加入鱼油及适量的蒸馏水充分混匀，用小型颗粒饲料挤压机制成直径为 0.3cm 的条状饲料，60℃烘干，密封保存。

表 4-10 实验饲料配方及营养组成（干物质）

项目		组别					
		D1	D2	D3	D4	D5	D6
原料组成	鱼粉/%	10.00	10.00	10.00	10.00	10.00	10.00
	小麦粉/%	10.00	10.00	10.00	10.00	10.00	10.00
	藻粉/%	30.00	30.00	30.00	30.00	30.00	30.00

续表

项目		组别					
		D1	D2	D3	D4	D5	D6
原料组成	鱼油/%	1.00	1.00	1.00	1.00	1.00	1.00
	抗氧化剂/%	0.20	0.20	0.20	0.20	0.20	0.20
	维生素预混料/%	1.00	1.00	1.00	1.00	1.00	1.00
	矿物质预混料/%	1.00	1.00	1.00	1.00	1.00	1.00
	卡拉胶/%	0.20	0.20	0.20	0.20	0.20	0.20
	包膜棉子糖/%	0.00	0.04	0.08	0.12	0.16	0.20
	海泥/%	46.60	46.56	46.52	46.48	46.44	46.40
营养组成（干物质）	粗蛋白/%	15.73	16.11	16.10	16.18	15.79	16.01
	粗脂肪/%	1.89	1.93	1.86	1.91	1.96	1.84
	粗灰分/%	56.25	56.16	56.08	56.19	56.11	56.07
	能量/（kJ/g）	7.72	8.02	7.93	7.96	8.02	7.96
	棉子糖/%	0.00	0.02	0.03	0.06	0.08	0.11

注：①每千克维生素预混料含维生素 A 7500.00IU、维生素 D 1500.00IU、维生素 E 60.00mg、维生素 K_3 18.00mg、维生素 B_1 12.00mg、维生素 B_2 12.00mg、维生素 B_{12} 0.10mg、泛酸 48.00mg、烟酰胺 90.00mg、叶酸 3.70mg、D-生物素 0.20mg、盐酸吡哆醇 60.00mg、维生素 C 310.00mg；②每千克矿物质预混料含锌 35.00mg、锰 21.00mg、铜 8.30mg、铁 23.00mg、钴 1.20mg、碘 1.00mg、硒 0.30mg。

2. 饲养管理

养殖实验在山东省海洋资源与环境研究院东营实验基地的循环水养殖系统中进行，共计 67d。实验用刺参幼参购自山东安源种业科技有限公司。实验开始前，刺参幼参在养殖系统中暂养 2 周，其间投喂基础饲料。暂养结束后控食 36h，挑选个体健壮、体质量为（11.46±0.06）g 的刺参幼参 540 头，随机分配到 18 个养殖桶（直径为 60cm，高 80cm）中，每桶 30 只刺参。每组饲料随机投喂 3 桶刺参，每天投喂 1 次（16:00），初始投喂量为刺参幼参体质量的 3%，根据每日摄食情况调整次日投喂量，每 3 天清理 1 次残饵和粪便并换水，换水量为 1/3，养殖 1 个月时更换海参养殖筐。实验在弱光环境中进行，控制水温为 16~18℃，pH 为 7.6~8.3，溶氧量大于 6mg/L，氨氮与亚硝酸盐浓度低于 0.05mg/L。

3. 样品采集与分析

养殖实验结束后，控食 48h，统计各桶刺参幼参数量并称重，计算存活率、增重率和特定生长率。每桶随机选 8 头刺参幼参置于干净托盘中，轻轻擦干表面水分，称量体质量。之后进行解剖，清理体腔液和肠道内容物，收集体壁及肠道并称重，计算肠壁比，样品保存于–20℃。生长指标计算方法见附录。

选取刺参幼参中前肠部位 0.4cm，置于伯恩（Bouin）氏液中固定 24h 后，转入 70%的乙醇中长期保存。将固定后的肠道组织经脱水、透明、透蜡包埋后，切成 6μm 厚的切片并用苏木精-伊红染色，用中性树脂封片。每个样品选取 10 个非连续的切片，在 40×物镜下观察，采用图像采集系统测量肌层厚度（tM）及皱襞高度（hF）。

常规营养成分指标测定方法见附录。棉子糖含量采用高效液相色谱法[《水苏糖》（QB/T 4260—2018）]测定。刺参幼参肠道蛋白酶、脂肪酶、淀粉酶、超氧化物歧化酶、碱性磷酸酶、酸性磷酸酶、葡萄糖激酶、磷酸果糖激酶（PFK）、丙酮酸激酶、磷酸烯醇式丙酮酸羧激酶的活性和丙二醛含量使用南京建成生物工程研究所生产的试剂盒测定，方法参照试剂盒说明书。

使用 Trizol 法提取刺参幼参肠道总 RNA，并用核酸蛋白仪（Nanoprop，2000，中国）检测 RNA 浓度，将 RNA 浓度稀释至 500～1000ng/nl，并用琼脂糖凝胶电泳检测 RNA 完整性。使用 Evo M-MLV 反转录试剂盒去除 gDNA，并反转录，于 −80℃保存待用。基于本实验室已有的刺参转录组数据，设计葡萄糖激酶、磷酸果糖激酶和丙酮酸激酶的基因序列与相近物种进行 BLAST 同源性比对，选用肌动蛋白基因作为内参基因，基因表达的引物序列如表 4-11 所示。实时荧光定量 PCR 试剂盒为 TAKARA 的 TB GreenTM Premix ExTaqTM II。通过荧光定量 PCR 仪得到各基因 Ct 值，按照 $2^{-\Delta\Delta Ct}$ 计算目的基因的相对表达量。

表 4-11　基因表达的引物序列

引物名称	序列（5'-3'）
肌动蛋白-F	TTATGCTCTTCCTCACGCTATCC
肌动蛋白-R	TTGTGGTAAAGGTGTAGCCTCTCTC
葡萄糖激酶-F	TGTCGGTCAAGTCCACTCCTTAGG
葡萄糖激酶-R	GATCGTCGGCCAATCCTGTAACC
磷酸果糖激酶-F	TCTACCACAGCACAAAGTCACCAAAG
磷酸果糖激酶-R	GAAGAGCCAATCAGCACCACAGG
丙酮酸激酶-F	TGATGTTGACCTTCCAGCGTTATCC
丙酮酸激酶-R	TTTGCCTTGTTCTCCCAACTCCTTC

4. 数据分析

实验数据采用 SPSS 17.0 软件进行单因素方差分析，若各组间差异显著（$P<0.05$），则用邓肯多重范围检验进行多重比较分析。统计结果以平均值±标准差表示。采用一元二次回归分析，确定刺参幼参饲料中棉子糖的最适添加量。

4.2.1.2　结果分析

1. 棉子糖对刺参幼参生长性能及体成分的影响

饲料中添加不同含量的棉子糖对刺参幼参存活率无显著影响（$P>0.05$）。随着棉子糖含量的增加,刺参幼参的增重率和特定生长率均呈现先上升后下降的趋势,D4 和 D5 组显著高于 D1 组($P<0.05$),均在 D4 组达到最大值;肠壁比无显著性差异($P>0.05$)(表 4-12)。

表 4-12 棉子糖对刺参幼参生长性能的影响

指标	组别					
	D1	D2	D3	D4	D5	D6
初始体质量/g	11.48±0.02	11.47±0.06	11.45±0.03	11.43±0.06	11.49±0.03	11.45±0.03
终末体质量/g	18.73±0.82^a	19.28±0.79^a	19.77±0.61^{ab}	21.08±0.41^c	20.70±0.83^{bc}	18.62±0.35^a
增重率/%	63.22±6.90^a	68.11±6.31^a	72.66±4.90^{ab}	84.41±3.35^c	80.20±7.13^{bc}	62.59±2.70^a
特定生长率/(%/d)	0.73±0.07^{ab}	0.77±0.06^{ab}	0.82±0.04^{bc}	0.91±0.03^d	0.88±0.06^{cd}	0.72±0.03^a
肠壁比/%	7.98±1.44	8.09±1.79	7.87±0.97	7.49±0.98	7.75±1.44	9.24±0.67
存活率/%	98.89±1.92	95.56±5.09	96.67±3.33	94.44±1.93	97.78±1.92	100.00±0.00

注:同行数据无字母或上标相同字母表示差异不显著($P>0.05$),上标不同字母表示差异显著($P<0.05$)。

以增重率为评价指标,经一元二次回归分析得出,初始体质量为(11.46±0.06)g 的刺参幼参饲料中棉子糖的最适添加量为 0.063%(图 4-3)。

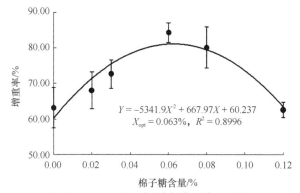

图 4-3 刺参幼参增重率与饲料中棉子糖含量的回归分析

饲料中不同含量的棉子糖对刺参幼参体壁的水分、粗蛋白、粗脂肪和粗灰分的含量无显著影响($P>0.05$)(表 4-13)。

表 4-13 棉子糖对刺参幼参体壁基本成分的影响(%)

指标	组别					
	D1	D2	D3	D4	D5	D6
水分	91.33±0.06	91.19±0.12	91.17±0.28	91.27±0.48	91.27±0.05	91.21±0.02
粗蛋白	46.86±0.13	47.11±0.25	46.86±0.09	46.48±0.16	47.20±0.17	47.24±0.25
粗脂肪	4.00±0.36	4.13±0.19	4.02±0.06	3.98±0.17	4.02±0.13	4.03±0.14
粗灰分	32.92±0.12	32.87±0.77	32.48±0.06	33.08±0.14	32.49±0.10	32.44±0.08

注:粗灰分、粗蛋白和粗脂肪的含量为干基含量。

2. 棉子糖对刺参幼参肠道消化酶及免疫酶活性的影响

随着饲料中棉子糖含量的增加，刺参幼参肠道蛋白酶、脂肪酶和淀粉酶的活性均呈现先上升后下降的趋势。D3、D4 组蛋白酶活性显著高于 D1、D5、D6 组（$P<0.05$）；D2、D3、D4 组脂肪酶活性显著高于其他组（$P<0.05$）；D3～D5 组淀粉酶活性显著高于 D1 组（$P<0.05$）（表 4-14）。

表 4-14 棉子糖对刺参幼参肠道消化酶活性的影响 （单位：U/mg prot）

指标	组别					
	D1	D2	D3	D4	D5	D6
蛋白酶	1267.90±30.61[bc]	130.53±96.76[cd]	1372.33±17.62[de]	1443.67±37.53[e]	1231.33±45.61[b]	1094.67±32.04[a]
脂肪酶	4.22±0.72[a]	5.44±0.11[b]	5.72±0.11[b]	5.81±0.41[b]	4.72±0.07[a]	4.40±0.07[a]
淀粉酶	1.28±0.41[a]	1.51±0.06[ab]	1.62±0.03[bc]	1.78±0.04[bc]	1.86±0.02[c]	1.30±0.03[a]

注：同行数据无字母或上标相同字母表示差异不显著（$P>0.05$），上标不同字母表示差异显著（$P<0.05$）。

随着饲料中棉子糖含量的增加，刺参幼参肠道超氧化物歧化酶活性呈现先上升后下降的趋势，在 D4 组达最大值；丙二醛含量呈现先下降后上升的趋势，且 D3、D4 组显著低于 D1、D2、D6 组（$P<0.05$）。D3～D6 组刺参幼参肠道酸性磷酸酶、碱性磷酸酶的活性显著低于 D1 组（$P<0.05$）（表 4-15）。

表 4-15 棉子糖对刺参幼参肠道免疫酶活性的影响

指标	组别					
	D1	D2	D3	D4	D5	D6
酸性磷酸酶活性/（U/mg prot）	147.14±6.96[c]	136.31±7.13[bc]	131.09±5.15[ab]	126.69±6.40[ab]	130.84±8.72[ab]	119.03±10.00[a]
碱性磷酸酶活性/（U/mg prot）	633.64±34.74[c]	602.18±18.25[bc]	579.35±17.21[b]	686.67±8.08[b]	504.21±11.37[a]	497.90±20.51[a]
超氧化物歧化酶活性/（U/mg prot）	7644.36±226.89[a]	8032.12±56.16[ab]	7698.55±47.00[a]	8381.33±191.68[b]	8084.76±450.60[ab]	7867.15±249.06[a]
丙二醛含量/（nmol/mg prot）	4.42±0.01[c]	3.86±0.35[b]	3.45±0.08[a]	3.27±0.19[a]	3.58±0.13[ab]	4.96±0.23[d]

注：同行数据无字母或上标相同字母表示差异不显著（$P>0.05$），上标不同字母表示差异显著（$P<0.05$）。

刺参幼参肠道皱襞高度随棉子糖含量的增加呈现先增大后减小的趋势，D3、D4 组显著大于 D1、D6 组（$P<0.05$）；肌层厚度无显著性差异（$P>0.05$）（表 4-16）。D2、D3 组细胞核排列较整齐，D5、D6 组肠道皱襞出现空泡和炎症细胞浸润（图 4-4）。

表 4-16 棉子糖对刺参幼参肠道结构的影响 （单位：μm）

指标	组别					
	D1	D2	D3	D4	D5	D6
肌层厚度	6.96±0.63	6.50±0.96	7.25±0.50	7.32±0.50	7.14±0.66	8.34±0.46
皱襞高度	176.93±20.34[a]	203.02±8.54[ab]	233.36±10.39[c]	262.59±18.70[d]	219.93±16.04[bc]	194.39±8.09[ab]

注：同行数据无字母或上标相同字母表示差异不显著（$P>0.05$），上标不同字母表示差异显著（$P<0.05$）。

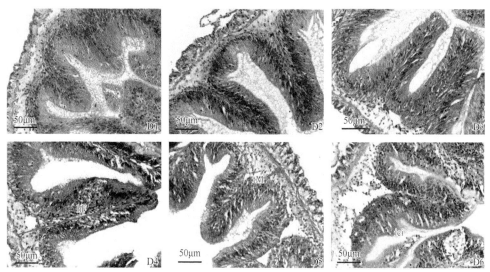

图 4-4 棉子糖对刺参肠道组织形态的影响（×40）

hF-皱襞高度；tM-肌层厚度；ici-炎症细胞浸润

3. 棉子糖对刺参幼参肠道代谢酶活性及相关基因相对表达的影响

随着饲料中棉子糖含量的增加，刺参幼参肠道葡萄糖激酶含量、磷酸果糖激酶和丙酮酸激酶的活性均呈现先上升后下降的趋势，其中葡萄糖激酶含量、磷酸果糖激酶活性在 D4 组达最高值，丙酮酸激酶活性在 D3 组达最高值；磷酸烯醇式丙酮酸羧激酶活性呈现先上升后平稳的趋势，D4~D6 组显著高于其他组（$P<0.05$）（表 4-17）。

表 4-17 棉子糖对刺参幼参肠道代谢酶相关指标的影响

指标	组别					
	D1	D2	D3	D4	D5	D6
葡萄糖激酶含量/（ng/ml）	3.22±0.09^a	3.41±0.17^a	3.96±0.05^b	4.65±0.13^c	3.23±0.48^a	3.22±0.21^a
磷酸果糖激酶活性/（U/mg prot）	24.80±1.43^a	26.84±0.53^{bc}	27.25±0.47^c	28.71±1.18^c	25.22±0.96^{ab}	25.04±1.38^{ab}
丙酮酸激酶活性/（U/g prot）	170.23±2.56^a	210.03±3.84^b	222.28±6.02^b	221.86±8.00^b	221.26±10.54^b	180.59±5.58^a
磷酸烯醇式丙酮酸羧激酶活性/（U/mg prot）	40.91±2.96^a	44.47±0.65^b	47.81±0.93^c	55.73±1.49^d	53.77±1.70^d	56.38±1.12^d

注：同行数据无字母或上标相同字母表示差异不显著（$P>0.05$），上标不同字母表示差异显著（$P<0.05$）。

刺参幼参肠道葡萄糖激酶、磷酸果糖激酶和丙酮酸激酶的基因相对表达量均随着棉子糖含量的增加呈现先上升后下降的趋势。D4 和 D5 组葡萄糖激酶基因相对表达量显著高于其他组（$P<0.05$），D1、D2 和 D6 组无显著性差异（$P>0.05$）；各组磷酸果糖激酶基因相对表达量均显著高于 D1 组（$P<0.05$），且 D3、D4 组显

著高于其他各组（$P<0.05$）；各组丙酮酸激酶基因相对表达量均显著高于 D1 组（$P<0.05$），D4 组达最高值（图 4-5～图 4-7）。

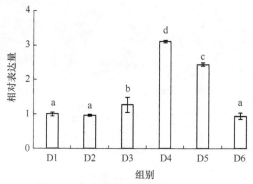

图 4-5　葡萄糖激酶基因相对表达量

上标相同字母表示差异不显著（$P>0.05$），上标不同字母表示差异显著（$P<0.05$），下同

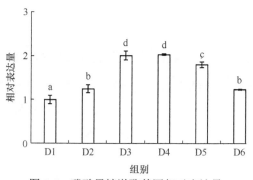

图 4-6　磷酸果糖激酶基因相对表达量

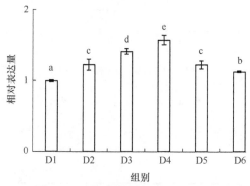

图 4-7　丙酮酸激酶基因相对表达量

4.2.1.3　讨论

1. 棉子糖对刺参幼参生长性能的影响

已有研究表明，饲料中添加低聚糖能增加畜禽动物肠道双歧杆菌和乳酸杆菌

的数量（Flickinger et al., 2003），改善黄颡鱼肠道结构，调节肠道菌群平衡（Wu et al., 2014）。本实验中，饲料中添加适量棉子糖时，刺参参肠道结构显著优于不添加组，且生长性能显著提高，这与对草鱼（邱燕，2010）和尼罗罗非鱼（Abdel-Latif et al., 2020）的研究结果一致。肠道黏膜褶皱数量众多，微绒毛丰富，增加了肠上皮细胞吸收面积，从而提高了生长性能（Sørensen et al., 2011）。刺参蛋白酶、脂肪酶和淀粉酶活性的提高，也表明棉子糖通过改变肠道形态，促进了其对食物的消化吸收，并促进了机体生长，这与对杂交鲟的研究结果一致（Xu et al., 2018）。已有的研究表明，棉子糖与肠道的双歧杆菌等有益菌作用产生短链脂肪酸，可为肠道黏膜生长提供能量（Archer et al., 1998）；肠道皱襞高度及肠上皮细胞数量显著增加，能增强对营养物质的消化吸收能力（张彩霞等，2010）；肠道皱襞高度的增加，可为更多的消化酶提供附着之地，促进糖类和蛋白质的消化分解（Refstie et al., 1998），这可能是添加适量的棉子糖促进刺参生长的主要原因。以增重率为评价指标，采用一元二次回归分析得出刺参幼参饲料中棉子糖的最适添加量为 0.063%。各组刺参幼参存活率无显著性差异，且棉子糖对刺参幼参体壁水分、粗蛋白、粗脂肪和粗灰分的含量无显著影响。

2. 棉子糖对刺参幼参消化生理及免疫的影响

本实验中，棉子糖含量在 0.03%以上时，刺参幼参肠道酸性磷酸酶和碱性磷酸酶的活性显著低于对照组，酸性磷酸酶作为溶菌酶的标志酶，反映了刺参非特异性免疫中对病原体的水解能力受到抑制（高晓莉等，2003）。而在体外培养中性粒细胞的研究却表明，适量的棉子糖能显著提高其细胞活性，并能提高大肠杆菌和金黄色葡萄球菌的吞噬能力（窦江丽等，2008）。这可能与动物的种属及结构差异有关。超氧化物歧化酶作为重要的抗氧化酶，可保护在过氧化和噬菌过程中受到损伤的组织，反映了机体抗应激水平（刘存歧等，2005），丙二醛则反映机体自由基氧化水平。在本实验中，添加适量的棉子糖显著提高了刺参幼参肠道超氧化物歧化酶活性，降低了丙二醛含量，这可能是因为棉子糖和双歧杆菌和乳酸杆菌的代谢产物氧化还原电势较低，可提供质子并与体内自由基结合，以减少自由基对机体造成的损伤（邱燕，2010）。已有的研究也证明，寡糖能提高机体阳离子多肽、补体和活性氧含量，并提高免疫球蛋白水平和溶菌酶活性，以此提高鱼类对病原体的防御能力（Meng et al., 2017；Das et al., 2009；Maqsood et al., 2010），这也可能是棉子糖提高机体抗氧化能力和免疫能力的原因之一。

3. 棉子糖对刺参幼参糖代谢的影响

葡萄糖激酶、磷酸果糖激酶和丙酮酸激酶均属于糖酵解中的关键酶（Brouwers et al., 2015；Pilkis and Claus, 1991），磷酸烯醇式丙酮酸羧激酶则是糖异生中的关键限速酶（孙永欣，2008），糖酵解和糖异生在糖代谢过程中调节葡萄糖稳态，提供能量（张傲，2020）。葡萄糖激酶具有感应葡萄糖的功能，当葡萄糖含量较高时，其活性会增强，同时能储存和利用葡萄糖，并磷酸化葡萄糖形成葡萄糖-6-磷酸（glucose-6-phosphate，G-6-P），为合成糖原或生成丙酮酸提供底物（吕永智，2018），磷酸果糖激酶和丙酮酸激酶则能控制该过程的方向与速率（张傲，2020）。本实验中，随着棉子糖含量的增加，刺参幼参肠道葡萄糖激酶含量、磷酸果糖激酶和丙酮酸激酶的活性均呈现先上升后下降的趋势，刺参幼参肠道中与之相关的葡萄糖激酶、磷酸果糖激酶和丙酮酸激酶呈现相同的变化趋势，葡萄糖激酶基因在D4、D5组相对表达量显著提高，各实验组磷酸果糖激酶和丙酮酸激酶基因相对表达量均高于不添加组；在低浓度水平下，磷酸烯醇式丙酮酸羧激酶活性呈现上升趋势，当继续增加棉子糖含量，磷酸烯醇式丙酮酸羧激酶活性趋于稳定。目前，关于棉子糖对水生动物肠道微生物影响的研究表明，棉子糖对肠道双歧杆菌和乳酸杆菌的增殖是有促进作用的。棉子糖含有α-半乳糖苷键，能被肠道双歧杆菌中的α-半乳糖苷酶分解产生短链脂肪酸和葡萄糖（Hu et al.，2015；Louis et al.，2014），葡萄糖含量升高时，葡萄糖激酶、磷酸果糖激酶及丙酮酸激酶的活性也会增强。添加棉子糖显著提高了刺参幼参的糖酵解和糖异生能力，但当棉子糖含量超过0.06%时，刺参幼参糖代谢平衡被打破，过高的糖异生反应使体内的丙酮酸和糖原分解为葡萄糖，超出了刺参幼参生理所需，从而被排出，并产生过量能量（李平凡和钟彩霞，2012），产物过量也会减缓糖酵解反应，过量的产物也会通过糖异生途径而还原成葡萄糖。由于糖代谢过程中过量的能量并未作用在生理代谢上，这也会导致肠道内细胞活性降低，从而影响生长性能。综上所述，棉子糖对调节刺参幼参糖代谢有一定作用，适量添加棉子糖能通过糖代谢为机体生理活动提供能量，并提高生长性能。

4.2.1.4 结论

本实验表明，在饲料中添加0.03%～0.06%棉子糖提高了刺参糖代谢的效率，并增强了刺参的消化和抗氧化能力，进而提高刺参的生长性能。以WGR当作评价指标，通过一元二次回归分析可知，初始质量为（11.46±0.06）g的刺参幼参配合饲料中棉子糖的最适添加量为0.063%。

4.2.2 水苏糖对刺参幼参生长、消化生理与糖代谢的影响

水苏糖（stachyose，STA）是一种天然的非还原性功能性低聚糖，是由1分子葡萄糖、1分子果糖和2分子半乳糖通过α-1,6糖苷键、β-1,2糖苷键连接而成的四糖，分子式为$C_{24}H_{42}O_{21}$，结构稳定，具有良好的热稳定性（谢瑾，2018；韩诗雯等，2019）。由于动物小肠缺少α-半乳糖苷酶，水苏糖不能被肠道直接吸收，

而是被肠道中的双歧杆菌和乳酸杆菌水解,并改善肠道菌群、促进矿物质吸收和调节糖类与脂类代谢(杨东升,2004;李治龙和孟良玉,2010)。与水苏糖相似的低聚糖有棉子糖、大豆低聚糖、低聚木糖及低聚果糖等,其被统称为功能性低聚糖(任红立等,2016)。研究表明,水苏糖对大菱鲆(Hu et al.,2015;Yang et al.,2018)、大西洋鲑(Sørensen et al.,2011)、异育银鲫(王文娟等,2010)等水产动物的生长、抗氧化和免疫有促进作用,并能改善肠道结构。

4.2.2.1 材料与方法

1. 实验饲料

以鱼粉、藻粉和小麦粉为主要蛋白源,设计粗蛋白含量为14.00%、粗脂肪含量为1.80%的基础饲料。在基础饲料中分别添加0、0.12%、0.24%、0.36%、0.48%和0.60%的包膜水苏糖(购买自陕西森悦生物科技有限公司,纯度≥98%),配制水苏糖含量分别为0、0.04%、0.11%、0.15%、0.21%和0.27%的6组等氮等脂的实验饲料,分别命名为D1(对照组)、D2、D3、D4、D5和D6组。

将等质量的水苏糖与卡拉胶混合,加入适量的水调至稠状,90℃加热搅拌15min,冷却后用冻干机干燥成胶状,经75℃烘干后磨成粉状,制成包膜水苏糖待用。在10ml蒸馏水中加入1g包膜水苏糖,静置5h后过滤,测定滤液中总糖含量[《饲料中总糖的测定 分光光度法》(DB12/T 847—2018)],计算包膜水苏糖5h的溶失率为36%。

将固体原料超微粉碎后经200目标准筛过滤,按照饲料配方配比进行称重,加入鱼油及适量的蒸馏水,充分混匀,用小型颗粒饲料挤压机制成直径为0.3cm的条状饲料,60℃烘干,密封保存。实验饲料配方及营养组成见表4-18。

表4-18 实验饲料配方及营养组成(干物质)

项目		组别					
		D1	D2	D3	D4	D5	D6
原料组成	白鱼粉/%	10.00	10.00	10.00	10.00	10.00	10.00
	小麦粉/%	10.00	10.00	10.00	10.00	10.00	10.00
	藻粉/%	30.00	30.00	30.00	30.00	30.00	30.00
	鱼油/%	1.00	1.00	1.00	1.00	1.00	1.00
	抗氧化剂/%	0.20	0.20	0.20	0.20	0.20	0.20
	维生素预混料[1]/%	1.00	1.00	1.00	1.00	1.00	1.00
	矿物质预混料[2]/%	1.00	1.00	1.00	1.00	1.00	1.00
	卡拉胶/%	0.20	0.20	0.20	0.20	0.20	0.20
	包膜水苏糖/%	0.00	0.12	0.24	0.36	0.48	0.60
	海泥/%	46.60	46.48	46.36	46.24	46.12	46.00
营养组成(干物质)	粗蛋白/%	15.73	15.70	15.67	15.68	16.16	15.77
	粗脂肪/%	1.89	1.98	1.90	1.85	1.94	1.87

续表

项目		D1	D2	D3	D4	D5	D6
营养组成（干物质）	粗灰分/%	56.25	56.11	55.89	55.94	55.78	55.62
	能量/(kJ/g)	7.72	7.83	7.96	7.74	7.98	8.02
	水苏糖/%	0.00	0.04	0.11	0.15	0.21	0.27

注：①每千克维生素预混料含维生素 A 7500.00IU、维生素 D 1500.00IU、维生素 E 60.00mg、维生素 K_3 18.00mg、维生素 B_1 12.00mg、维生素 B_2 12.00mg、维生素 B_{12} 0.10mg、泛酸 48.00mg、烟酰胺 90.00mg、叶酸 3.70mg、D-生物素 0.20mg、盐酸吡哆醇 60.00mg、维生素 C 310.00mg；②每千克矿物质预混料含锌 35.00mg、锰 21.00mg、铜 8.30mg、铁 23.00mg、钴 1.20mg、碘 1.00mg、硒 0.30mg。

2. 实验管理及样品采集

养殖实验在山东省海洋资源与环境研究院东营实验基地的循环水养殖系统中进行，所用刺参幼参购自山东安源种业科技有限公司。养殖实验前，将刺参幼参暂养于养殖系统中，以基础饲料暂养 2 周，控食 36h 后，挑选个体健壮、体质量为（11.46±0.03）g 的刺参幼参 540 头，随机分配到 18 个循环水养殖桶（直径为 70cm，高 65cm）中，每桶 30 只，各实验组随机投喂 3 桶，共 6 组，每桶放置一个海参养殖筐。养殖周期为 67d，养殖期间，每天 16:00 饱食投喂 1 次，初始投喂量为刺参幼参初始体质量的 3%，根据每日摄食情况调整次日投喂量，每 3 天清理 1 次残饵和粪便并换水，换水量为 1/3，养殖 1 个月时更换海参养殖筐。养殖室保持弱光，控制水温为 16~18℃，pH 为 7.6~8.3，溶氧量大于 6mg/L，氨氮与亚硝酸盐浓度低于 0.05mg/L。

养殖实验结束后，控食 48h，统计各桶刺参幼参数量并称重，计算存活率、增重率和特定生长率。每桶随机选取 8 头刺参幼参置于干净托盘中，轻轻擦干表面水分，称量体质量。之后进行解剖，清理体腔液和肠道内容物，收集体壁及肠道并称重，计算肠壁比，样品保存于 −20℃。

切取刺参幼参中前肠部位（位置相同）0.4cm，置于伯恩（Bouin）氏液中固定 24h，之后转入 70% 的乙醇中长期保存。

3. 检测指标与方法

生长指标计算方法及常规成分测定方法见附录。水苏糖含量采用高效液相色谱法 [《水苏糖》（QB/T 4260—2018）] 测定。

刺参幼参肠道蛋白酶、脂肪酶、淀粉酶、超氧化物歧化酶、碱性磷酸酶、酸性磷酸酶、葡萄糖激酶、磷酸果糖激酶、丙酮酸激酶、磷酸烯醇式丙酮酸羧激酶的活性和丙二醛含量使用南京建成生物工程研究所生产的试剂盒测定，方法参照各试剂盒说明书。

4. 肠道组织切片的制作与观察

将固定后的肠道组织经脱水、透明、透蜡后包埋，之后切成 6μm 厚的切片并用苏木精-伊红染色，用中性树脂封片。每个样品选取 10 个非连续的切片，在 40×物镜下观察，采用 Leica DM500 图像采集系统测量肌层厚度及皱襞高度。

5. 数据统计分析

实验数据采用 SPSS 17.0 软件进行单因素方差分析，若各组间差异显著（$P<0.05$），则用邓肯多重范围检验进行多重比较分析。统计结果以平均值±标准差表示。

4.2.2.2 结果分析

1. 水苏糖对刺参幼参生长性能及体成分的影响

饲料中不同含量的水苏糖对刺参幼参存活率无显著影响（$P>0.05$）；除 D6 组外，各组刺参幼参的增重率和特定生长率均显著高于对照组（$P<0.05$），且随着水苏糖含量的增加，呈现先上升后下降的趋势，在 D3 组达到最大值；肠壁比无显著性差异（$P>0.05$）（表 4-19）。

表 4-19 水苏糖对刺参幼参生长性能的影响

指标	组别					
	D1	D2	D3	D4	D5	D6
初始体质量/g	11.48±0.02	11.45±0.04	11.43±0.04	11.48±0.02	11.45±0.04	11.47±0.04
终末体质量/g	18.73±0.82^a	20.43±0.85^b	21.82±0.51^c	21.06±0.32^{bc}	20.52±0.45^b	18.75±0.32^a
增重率/%	63.22±6.90^a	78.48±6.80^b	90.88±4.57^c	83.48±2.50^{bc}	79.08±4.46^b	63.51±2.67^a
特定生长率/(%/d)	0.73±0.07^a	0.87±0.06^b	0.96±0.04^c	0.90±0.02^{bc}	0.86±0.04^b	0.73±0.02^a
肠壁比/%	7.98±1.44	7.29±0.43	7.04±0.13	7.97±1.92	7.71±0.39	7.76±0.34
存活率/%	98.89±1.92	97.78±1.92	98.89±1.92	97.78±1.92	100.00±0.00	97.78±1.92

注：同行数据无字母或上标相同字母表示差异不显著（$P>0.05$），上标不同字母表示差异显著（$P<0.05$）。

以增重率为评价指标，经一元二次回归分析得出，初始体质量为（11.46±0.03）g 刺参幼参饲料中水苏糖的最适添加量为 0.129%（图 4-8）。

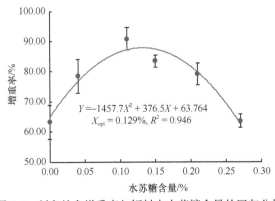

图 4-8 刺参幼参增重率与饲料中水苏糖含量的回归分析

饲料中不同含量的水苏糖对刺参幼参体壁水分、粗蛋白、粗脂肪和粗灰分的含量无显著影响（$P>0.05$）（表 4-20）。

表 4-20　水苏糖对刺参幼参体壁基本成分的影响（%）

指标	组别					
	D1	D2	D3	D4	D5	D6
水分	91.33±0.06	91.34±0.12	91.34±0.48	91.37±0.05	91.24±0.03	91.3±0.26
粗蛋白	46.86±0.13	46.85±0.23	46.82±0.18	46.79±0.07	46.68±0.1	46.68±0.11
粗脂肪	4.00±0.36	4.18±0.16	4.09±0.06	4.08±0.06	3.93±0.13	4.12±0.11
粗灰分	32.92±12	32.87±0.09	33.1±1.78	32.94±0.32	32.62±0.22	33.24±0.97

注：粗灰分、粗蛋白和粗脂肪的含量为干基含量。

2. 水苏糖对刺参幼参肠道消化酶活性的影响

随着饲料中水苏糖含量的增加，刺参幼参肠道蛋白酶和脂肪酶的活性均呈现先上升后下降的趋势。D3、D4 组蛋白酶活性显著高于 D1、D5、D6 组（$P<0.05$）；D3 组脂肪酶活性显著高于除 D2 组外的其他组（$P<0.05$）。水苏糖对刺参幼参肠道淀粉酶活性的影响不显著（$P>0.05$）（表 4-21）。

表 4-21　水苏糖对刺参幼参肠道消化酶活性的影响

指标	组别					
	D1	D2	D3	D4	D5	D6
蛋白酶/(U/mg prot)	1267.90±30.61^a	1388.46±34.07^{bc}	1456.08±36.86^c	1445.00±43.27^c	1351.26±67.53^b	1326.96±25.55^{ab}
脂肪酶/(U/g prot)	4.22±0.72^a	4.87±0.17^{bc}	5.27±0.08^c	4.65±0.09^{ab}	4.55±0.03^{ab}	4.52±0.14^{ab}
淀粉酶/(U/mg prot)	1.28±0.41	1.26±0.11	1.33±0.05	1.31±0.04	1.33±0.27	1.27±0.01

注：同行数据无字母或上标相同字母表示差异不显著（$P>0.05$），上标不同字母表示差异显著（$P<0.05$）。

刺参幼参肠道的皱襞高度随水苏糖含量的增加而呈现先增大后减小的趋势，其中 D3、D4 组显著大于其他组（$P<0.05$）；肌层厚度无显著性差异（$P>0.05$）（表 4-22）。

表 4-22　水苏糖对刺参幼参肠道结构的影响　　　　（单位：μm）

指标	组别					
	D1	D2	D3	D4	D5	D6
肌层厚度	6.96±0.63	6.86±1.73	6.10±0.62	6.37±1.55	6.61±0.27	6.46±0.35
皱襞高度	176.93±20.34^{ab}	204.08±20.8^b	274.91±25.02^c	306.16±25.95^c	187.43±15.63^{ab}	150.15±10.82^a

注：同行数据无字母或上标相同字母表示差异不显著（$P>0.05$），上标不同字母表示差异显著（$P<0.05$）。

D3、D4 组细胞核排列整齐、紧密，D6 组肠道皱襞出现异常，有炎症细胞浸润的现象（图 4-9）。

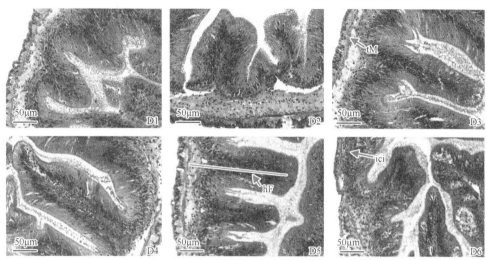

图 4-9 水苏糖对刺参幼参肠道组织形态的影响（×40）

tM-肌层厚度；hF-皱襞高度；ici-炎症细胞浸润

3. 水苏糖对刺参幼参肠道免疫酶相关指标的影响

随着饲料中水苏糖含量的增加，刺参幼参肠道超氧化物歧化酶活性呈现先上升后下降的趋势，D3、D4、D5 组超氧化物歧化酶活性显著高于对照组（$P<0.05$）；丙二醛含量呈现先下降后上升的趋势，且 D3 组显著低于其他组（$P<0.05$）。酸性磷酸酶、碱性磷酸酶的活性不受饲料中水苏糖含量的影响（$P>0.05$）（表 4-23）。

表 4-23 水苏糖对刺参幼参肠道免疫酶相关指标的影响

指标	组别					
	D1	D2	D3	D4	D5	D6
酸性磷酸酶活性/（g prot）	147.14±6.96	157.80±10.17	154.92±7.64	143.46±2.16	141.92±8.97	153.85±3.10
碱性磷酸酶活性/（g prot）	633.64±34.74	606.33±31.94	631.68±28.02	617.87±35.00	630.50±24.25	668.88±48.59
超氧化物歧化酶活性/（U/mg prot）	7644.36±226.89[a]	7784.66±26.77[ab]	8269.63±310.50[bc]	8472.09±359.90[c]	8044.62±376.50[b]	7776.57±330.50[ab]
丙二醛含量/（nmol/mg prot）	4.42±0.01[d]	3.48±0.13[b]	2.77±0.21[a]	3.49±0.15[b]	3.81±0.11[bc]	4.06±0.38[c]

注：同行数据无字母或上标相同字母表示差异不显著（$P>0.05$），上标不同字母表示差异显著（$P<0.05$）。

4. 水苏糖对刺参幼参肠道代谢酶相关指标的影响

随着饲料中水苏糖含量的增加，刺参幼参肠道葡萄糖激酶含量呈现先上升后平稳的趋势，且 D3~D6 显著高于对照组，D5、D6 组显著高于 D3、D4 组（$P<0.05$）；磷酸果糖激酶和丙酮酸激酶的活性呈现先上升后下降的趋势，分别在 D4 和 D3 组达最高值；磷酸烯醇式丙酮酸羧激酶活性呈现上升趋势，D2~D5 组显著

高于对照组，D6 组显著高于其他各组（$P<0.05$）（表 4-24）。

表 4-24 水苏糖对刺参幼参肠道代谢酶相关指标的影响

指标	组别					
	D1	D2	D3	D4	D5	D6
葡萄糖激酶含量/（ng/ml）	3.22±0.09[a]	3.42±0.10[a]	3.81±0.10[b]	4.00±0.14[b]	4.35±0.15[c]	4.25±0.18[c]
磷酸果糖激酶活性/（U/mg prot）	24.80±1.43[a]	27.23±1.35[b]	30.98±1.37[c]	31.31±0.84[c]	29.52±1.06[c]	25.70±1.35[ab]
丙酮酸激酶活性/（U/g prot）	170.23±2.56[a]	207.32±15.37[bc]	222.86±33.30[c]	207.49±12.99[bc]	177.26±1.90[ab]	162.98±12.52[a]
磷酸烯醇式丙酮酸羧激酶活性/（U/mg prot）	40.91±2.96[a]	54.29±2.81[b]	55.30±3.04[b]	58.24±0.66[b]	58.68±1.93[b]	64.37±1.85[c]

注：同行数据无字母或上标相同字母表示差异不显著（$P>0.05$），上标不同字母表示差异显著（$P<0.05$）。

4.2.2.3 讨论

1. 水苏糖对刺参幼参生长性能及体成分的影响

本实验中，刺参幼参的存活率未受饲料中水苏糖含量的影响，且饲料中添加适量水苏糖，刺参幼参的增重率和特定生长率显著提高，但随着水苏糖含量的进一步增加，生长性能呈下降趋势，这与对大菱鲆的研究类似（Hu et al., 2015; Yang et al., 2018）。水苏糖和壳寡糖等低聚糖作为双歧因子，可促进肠道中双歧杆菌及乳酸杆菌增殖（司滨，2017），水苏糖的促生长作用可能与之相关。首先，水苏糖被双歧杆菌和乳酸杆菌转运至其细胞内，被其所含有的 α-半乳糖苷酶水解为单糖（邬佳颖等，2021），影响了机体糖代谢，为生命活动提供能量，提高了刺参的生长性能；其次，水苏糖的代谢产物短链脂肪酸（乙酸、丙酸等）可以降低肠道 pH，促进钙、镁等矿物质元素的吸收，促进有益菌增殖，抑制内源和外源有害菌的生长（Louis et al., 2014；崔洪斌，2000；蔡琨等，2012）；然后，水苏糖能诱导双歧杆菌和乳酸杆菌大量增殖，这些有益菌能自身合成或促进合成多种维生素，如维生素 B 和叶酸等（蔡琨等，2012）；此外，水苏糖还可促进肠道蠕动，提高肠道对营养物质的吸收能力（Sørensen et al., 2011）。本实验中，随着水苏糖含量进一步增加，刺参幼参生长性能呈现下降趋势，与对牙鲆（蔡英华，2006）的研究结果一致，由于水苏糖无法被动物肠道消化吸收，需被双歧杆菌和乳酸杆菌水解，当水苏糖含量超过机体耐受极限时，会破坏肠道菌群平衡，改变肠道上皮细胞的渗透压，导致肠道胀气或产生炎症，并影响肠道对营养物质的吸收，最终抑制机体生长（Wiggins, 1984）。以增重率为评价指标，采用一元二次回归分析得出，刺参幼参饲料中水苏糖的最适添加量为0.129%，远低于大菱鲆（1.25%）（Hu et al., 2015），这可能与刺参的肠道结构和糖代谢能力有关。水苏糖对刺参幼参体壁水分、粗蛋白、粗脂肪和粗灰分含量的影响不显著，该结果与对大菱鲆（Hu et al.，2015）、异育银鲫（王文娟等，2010）和大西洋鲑（Sørensen et al.，2011）的研究结果一致。

2. 水苏糖对刺参幼参消化生理及抗氧化酶活性的影响

研究表明，功能性低聚糖（含水苏糖）作为营养物质被肠道的双歧杆菌和乳酸杆菌等消化利用，这些有益菌可以促进肠道上皮细胞增殖，从而提高消化酶活性（Balcázar et al.，2006）。本实验中，刺参幼参肠道的淀粉酶活性未受水苏糖的影响，这与对牙鲆的研究结果（蔡英华，2006）一致，其原因可能和淀粉酶无法水解水苏糖的半乳糖苷键有关（杨东升，2004）；而蛋白酶和脂肪酶的活性随水苏糖含量增加，呈现先上升后下降的趋势，这与对大菱鲆的研究结果（Hu et al.，2015）一致，同为功能性低聚糖的壳寡糖和低聚木糖在对刺参的研究中也有相同的结果（司滨，2017；赵丽丽等，2019）。进一步增加水苏糖含量会导致消化酶活性降低，这可能是由于水苏糖与双歧杆菌、乳酸杆菌等产生了过量的代谢产物，刺激了肠道蠕动（Kihara and Sakata，2002），营养物质和代谢产物在肠道停留的时间缩短，消化酶被抑制或排出体外（李君华等，2016）。

刺参不具备专门的消化酶分泌器官，且消化酶多位于肠道黏膜的柱状细胞内，所以肠道黏膜能为消化酶分泌提供特定位点，是决定营养物质利用率的基础（李宝山等，2019）。添加 0.11%～0.15%的水苏糖，刺参肠道的皱襞高度变长且细胞核排列更整齐，表明水苏糖可通过改善刺参肠道结构，促进消化酶的分泌，从而提高营养物质的消化利用率。当饲料中水苏糖含量过高时，刺参肠道的皱襞高度变短，且出现炎症细胞浸润的现象，此时刺参肠道渗透压增加，这会有利于病原菌在肠道上皮的定植，导致肠道过度蠕动，营养物质消化率降低，从而抑制刺参生长（谷珉，2010）。

刺参是无脊椎动物，主要依靠非特异性免疫，对外来物质的识别与清除多依靠吞噬细胞进行（张琴，2010）。碱性磷酸酶和酸性磷酸酶参与磷酸单酯的水解和磷酸基团的转移，可代谢机体磷化物及参与机体解毒，可作为巨噬细胞溶菌体的标志酶，可反映吞噬细胞的活性，是非特异性免疫的重要指标（孙静秋等，2007；Wang and He，2009）。在本实验中，水苏糖含量对刺参幼参肠道碱性磷酸酶和酸性磷酸酶的活性无显著影响，这表明水苏糖对（刺参吞噬作用）刺参吞噬细胞的影响较弱。超氧化物歧化酶是重要的抗氧化酶，可以清除机体的超氧阴离子自由基，丙二醛则是被自由基氧化后的产物，超氧化物歧化酶和丙二醛均能直接反映机体自由基的代谢状况和组织损伤情况（Bagnyukova et al.，2003；Martinez-Álvarez et al.，2002）。当水苏糖含量为 0.11%～0.15%时，刺参肠道超氧化物歧化酶活性显著提高，丙二醛含量显著降低，这表明水苏糖的代谢产物能提高刺参抗氧化能力，这与其他功能性低聚糖[低聚木糖（李君华等，2016）、甘露寡糖（谷珉，2010）]的作用一致。适量的水苏糖可促进肠道双歧杆菌等有益菌的增殖，提高刺参对营养物质的代谢利用能力，这有利于免疫因子的合成，以提高机体抗氧化能力（姚朋波，2018）；过量的水苏糖则会破坏刺参肠道菌群平衡，降低营养物质的利用率，体内代谢平衡被打破，从而导致刺参抗氧化能力降低（强俊等，2009；Chesson，1994）。

3. 对刺参幼参肠道水苏糖代谢的影响

糖代谢相关途径包括糖酵解、三羧酸循环、磷酸戊糖途径、糖原的生成及分解和糖异生等，通过这些复杂系统的正常运行可以控制动物体内葡萄糖稳态，并为机体提供能量（张傲，2020）。葡萄糖激酶、磷酸果糖激酶和丙酮酸激酶是糖酵解途径中3个关键的限速酶，反映细胞中葡萄糖的分解代谢能力（Brouwers et al.，2015；Pilkis and Claus，1991）。葡萄糖激酶能将葡萄糖磷酸化为葡萄糖-6-磷酸，葡萄糖-6-磷酸无法转运出细胞外，但可作为底物在各催化酶催化下一步步合成糖原或转化为丙酮酸，磷酸果糖激酶和丙酮酸激酶则是控制该过程方向与速率的关键酶（张傲，2020）。在本实验中，随着水苏糖含量的增加，D1~D4组刺参幼参肠道葡萄糖激酶、磷酸果糖激酶和丙酮酸激酶的活性呈现上升趋势，这表明适量的水苏糖可促进肠道糖酵解，增强葡萄糖分解代谢能力，并为机体提供能量。水苏糖作为双歧因子，能诱导双歧杆菌产生更多的α-半乳糖苷酶水解水苏糖（邬佳颖等，2021）。相关研究表明，水苏糖在双歧杆菌中的代谢产物短链脂肪酸（如乙酸、丙酸等）能增强细胞活性，并为机体提供能量，促进肠道双歧杆菌的增殖（孙纪录等，2003）。在稳定的糖代谢条件下，机体会通过三羧酸循环、磷酸戊糖循环分别产生短链脂肪酸和免疫因子，短链脂肪酸可调节肠道微生物系统和改善肠道结构（Louis et al.，2014），而免疫因子作为抗氧化系统中重要的组成部分，对提高机体抗氧化能力起重要作用（姚朋波，2018）。当水苏糖含量超过刺参的耐受能力时，葡萄糖激酶活性趋于平稳，而磷酸果糖激酶和丙酮酸激酶的活性显著降低，这表明丙酮酸生成途径出现异常（张琴，2010）。丙酮酸作为糖代谢及多种物质相互转化的重要中间体，影响了刺参体内的糖、脂肪与氨基酸之间的相互转化和免疫因子的产生（干懿洁和丁树哲，2006）。磷酸烯醇式丙酮酸羧激酶是糖异生中重要的催化酶，可反映机体血糖水平的稳定性（Mcleod et al.，2019；Oh et al.，2013），也可作为糖异生过程的催化酶，将丙酮酸、乳酸等非糖类物质转化为糖，维持体内葡萄糖稳定（Oh et al.，2013）。本实验中，刺参幼参肠道磷酸烯醇式丙酮酸羧激酶的活性呈现显著上升的趋势，这表明随着水苏糖含量的增加，其代谢产生的能量超过了刺参生理活动所需的能量，机体糖异生水平升高，丙酮酸等非糖类物质会转变为葡萄糖或糖原，过量的葡萄糖会转运至细胞外（张琴，2010），此时刺参对营养物质的利用能力和抗氧化能力均显著下降，从而降低刺参的生长性能。综上所述，饲料中添加水苏糖可影响刺参肠道糖酵解与糖异生途径，并以此调节机体能量代谢，影响刺参的生长性能。这表明适量的水苏糖对刺参糖代谢有促进作用。结合相关研究，水苏糖促进刺参糖代谢途径的机制如下：刺参肠道中的双歧杆菌和乳酸杆菌可通过转运蛋白质和透性酶将水苏糖转运至细胞内，并利用其细胞质中的 α-半乳糖苷酶水解水苏糖为葡萄糖，以提高细胞内葡萄糖水平（邬佳颖等，2021）。

4.2.2.4 结论

本实验表明，在饲料中添加 0.11%～0.15%的水苏糖提高了刺参幼参的糖代谢效率，并增强了刺参的消化和抗氧化能力，进而提高刺参的生长性能。以增重率为评价指标，经一元二次回归分析得出初始体质量为（11.46±0.03）g 的刺参幼参配合饲料中水苏糖的最适添加量为 0.129%。

4.2.3 半乳甘露寡糖对刺参幼参生长性能、体壁营养组成及免疫力的影响

寡糖，又称低聚糖，是指 2～10 个单糖通过糖苷键连接而成的直链或支链的一类糖。近年来，越来越多的研究表明，寡糖作为新型绿色水产饲料添加剂，能够提高水产动物免疫相关酶的活性，增强机体的免疫功能，是一类稳定、安全和环保的抗生素替代物。目前在水产动物中研究和应用的功能性寡糖主要有果寡糖、甘露寡糖、葡萄糖寡糖、木聚糖以及低聚乳糖（张琴，2010）。半乳甘露寡糖，又称为半乳甘露低聚糖，由 D-半乳糖和 D-甘露寡糖组成，其来源主要有田菁胶、葫芦巴胶、长角豆胶、瓜尔豆胶和卡拉胶，是一种无污染、无残留的新型添加剂（胡静和林英庭，2012），在畜禽养殖中其被认为具有激活免疫系统、优化肠道菌群和黏附、排除有害微生物的作用（王吉潭等，2003；王彬等，2006）。半乳甘露寡糖在水产养殖中的应用研究较少，仅有的研究表明，基础饲料中添加 0.2%的半乳甘露寡糖具有显著增强异育银鲫幼鱼非特异性免疫功能，提高增重率和特定生长率的作用（王锐等，2008）。目前半乳甘露寡糖在刺参养殖中的应用还未见报道。本研究首次将不同浓度半乳甘露寡糖添加到基础饲料中饲喂刺参，以刺参体壁、体腔液及体腔细胞中超氧化物歧化酶、碱性磷酸酶和溶菌酶 3 种免疫酶作为指标反映刺参的免疫水平，探讨半乳甘露寡糖对刺参机体免疫特性的影响，通过生长性能以及免疫相关酶活性等的变化来反映半乳甘露寡糖对刺参机体生长性能、体组成以及非特异性免疫的影响，旨在为将半乳甘露寡糖开发为刺参免疫增强剂提供理论依据，同时为刺参的健康养殖和病害防治工作提供一些参考资料。

4.2.3.1 材料和方法

1. 实验动物驯化

刺参幼参选自山东省海洋资源与环境研究院东营实验基地当年同批健康苗种，随机选取 720 头体质健康、活力旺盛的刺参幼参用于实验，体质量为（1.79±0.06）g，随机置于 18 个养殖桶（直径 80cm×高度 70cm）中驯化 7d。驯化期间水质条件为：水温（18.0±0.5）℃，溶氧量大于 7.0mg/L，氨氮浓度低于 0.05mg/L。实验期间每天投饵一次，2d 吸底换水一次，换水量为 1/2，在塑料水槽底部放置波纹板以供刺参幼参附着。

2. 实验饲料的配制

以大豆脱脂蛋白、花生粕和海藻粉为主要蛋白源，添加脱脂鱼粉、贝壳粉、虾粉、小麦粉等原料，以木薯淀粉调节饲料平衡，以基础饲料为对照组（E0），添加组中半乳甘露寡糖的添加量分别为 0.2%（E1）、0.4%（E2）、0.8%（E3）、1.2%（E4）和 1.6%（E5），将所有原料经超微粉碎后，过 300 目筛绢，并将过滤后的原料充分混匀，喷水使各成分黏合，然后晾干，再经超微粉碎，过 100 筛绢，未通过的滤渣再经粉碎直到全部通过，使饲料成为粉末状混合物，备用。实验饲料配方及营养组成见表 4-25。

表 4-25 实验饲料配方及营养组成

	原料	组别					
		E0	E1	E2	E3	E4	E5
原料组成	脱脂鱼粉/%	5	5	5	5	5	5
	大豆脱脂蛋白/%	10	10	10	10	10	10
	贝壳粉/%	16	16	16	16	16	16
	花生粕/%	14	14	14	14	14	14
	海藻粉/%	17	17	17	17	17	17
	木薯淀粉/%	14	13.8	13.6	13.2	12.8	12.4
	虾粉/%	8	8	8	8	8	8
	小麦粉/%	13	13	13	13	13	13
	复合维生素/%	2	2	2	2	2	2
	复合矿物质/%	1	1	1	1	1	1
	半乳甘露寡糖/%	0.0	0.2	0.4	0.8	1.2	1.6
营养组成	粗蛋白/%	39.15	38.92	37.99	38.94	38.97	38.50
	粗脂肪/%	1.59	1.59	1.56	1.58	1.62	1.59
	灰分/%	36.26	34.11	35.12	34.88	35.52	34.50
	总能（cal/g）	3318.53	3352.15	3333.12	3315.09	3298.67	3340.06

注：①每千克复合维生素含维生素 A 38.0mg、维生素 D_3 13.2mg、α-生育酚 210.0mg、维生素 B_1 115.0mg、维生素 B_2 380.0mg、盐酸吡哆醇 88.0mg、泛酸 368.0mg、烟酸 1030.0mg、生物素 10.0mg、叶酸 20.0mg、维生素 B_{12} 1.3mg、肌醇 4000.0mg、抗坏血酸 500.0mg；②每千克复合矿物质含 $MgSO_4 \cdot 7H_2O$ 3568.0mg、$NaH_2PO_4 \cdot 2H_2O$ 25 568.0mg、KCl 3020.5mg、$KAl(SO_4)_2$ 8.3mg、$CoCl_2$ 28.0mg、$ZnSO_4 \cdot 7H_2O$ 353.0mg、Ca-lactate 15 968.0mg、$CuSO_4 \cdot 5H_2O$ 9.0mg、KI 7.0mg、$MnSO_4 \cdot 4H_2O$ 63.1mg、Na_2SeO_3 1.5mg、$C_6H_5O_7Fe \cdot 5H_2O$ 1533.0mg、NaCl 100.0mg、NaF 4.0mg。

3. 实验管理

养殖实验在山东省海洋资源与环境研究院东营实验基地进行。将驯化后的刺参幼参随机分配至 18 个圆形养殖桶中，每桶 40 头，实验设置对照组（E0）及添加组（E1～E5），分别投喂配合饲料，每组 3 个重复，进行养殖实验，其间水质条件与驯化期保持一致。实验期间每天 18:00 投喂饲料一次，日投喂量占刺参幼参体质量的 5%左右，并根据摄食情况适当调整以确保刺参幼参饱食。实验周期为 56d。

4. 样品采集与处理

采样前禁食 24h，将刺参从养殖桶中取出后，先用纱布吸干表面水分，然后将其置于灭菌玻璃培养皿上称重。每组随机取 10 头刺参，用解剖剪沿刺参腹面剪开以去除体腔液，在食道与泄殖腔开口以去除肠道与呼吸树。剩余体壁冷冻保存，以备常规成分及氨基酸分析测定使用。

每组随机取 8 头刺参置于灭菌玻璃培养皿上，用解剖剪沿腹面剪开，在相同部位取一块体壁（1g 左右），置于 1.5ml 离心管中；在食道与泄殖腔开口以取出肠道，将肠道纵向剪开，然后用生理盐水冲洗以去除内容物，用定性滤纸擦净后置于 1.5ml 离心管中。刺参体腔液以灭菌枪头吸取，然后置于 5ml 离心管中。样品放入液氮速冻后，于-80℃条件下保存待测。取 0.2～0.5g 体壁或肠道样品剪碎，加入 4 倍体积的冰冷生理盐水（0.86%），制成 20%匀浆，在 4℃条件下 10 000r/min 离心 20min，取出上清液分装待测。体腔液样品则先置于冰箱（4℃）中解冻后，再用超声波粉碎仪进行细胞破碎，然后在 4℃条件下 376r/min 离心 10min，收集上清体腔液及底层体腔细胞分装待测。

5. 测定指标与方法

刺参幼参生长性能计算及常规营养成分测定方法见附录。刺参幼参肠道蛋白酶、淀粉酶和纤维素酶以及体壁、体腔液等的超氧化物歧化酶、溶菌酶、碱性磷酸酶均采用南京建成生物工程研究所生产的试剂盒测定，具体测定参照试剂盒说明书。酸性黏多糖以岩藻聚糖硫酸酯为标准品，采用次甲基蓝分光光度法测定（刘红英等，2002）。糖醛酸采用间羟基联苯比色法测定（Meseguer et al., 1998）。

6. 数据统计分析

采用 SPSS 18.0 软件进行单因素方差分析，差异显著（$P<0.05$）时用邓肯多重范围检验进行多重比较分析。统计数据以平均值±标准差表示。

4.2.3.2 结果分析

1. 半乳甘露寡糖对刺参幼参生长性能的影响

饲料中不同添加量的半乳甘露寡糖对刺参幼参增重率及特定生长率具有显著

影响，各添加组均显著高于对照组（$P<0.05$），其中 1.2%（E4）添加组，刺参幼参增重率及特定生长率最高。添加半乳甘露寡糖对刺参幼参脏壁比及肠壁比无显著影响（$P>0.05$）（表 4-26）。

表 4-26 半乳甘露寡糖对刺参幼参生长性能的影响

生长性能	组别					
	E0	E1	E2	E3	E4	E5
初始体质量/g	2.75±0.01	2.75±0.02	2.75±0.03	2.75±0.03	2.75±0.01	2.76±0.01
终末体质量/g	5.81±0.54	6.92±0.24	6.61±1.32	6.69±0.73	7.36±0.68	7.26±0.88
增重率/%	111.27±5.93^a	151.64±6.78^b	140.36±5.62^b	143.27±4.97^b	167.64±6.36^b	163.04±6.64^b
特定生长率/（%/d）	1.33±0.06^a	1.64±0.14^b	1.57±0.15^b	1.59±0.12^b	1.76±0.19^b	1.73±0.18^b
脏壁比/%	13.83±5.70	17.04±5.43	15.23±6.33	14.82±3.12	11.94±5.78	14.39±5.53
肠壁比/%	5.95±1.15	5.58±1.44	6.21±1.17	7.59±1.32	4.91±1.46	5.65±0.76

注：同行数据无字母或上标相同字母表示差异不显著（$P>0.05$），上标不同字母表示差异显著（$P<0.05$）。

2. 半乳甘露寡糖对刺参幼参肠道消化酶活性的影响

饲料中不同添加量的半乳甘露寡糖对刺参幼参肠道蛋白酶、淀粉酶、纤维素酶的活性均无显著影响（$P>0.05$）（表 4-27）。

表 4-27 半乳甘露寡糖对刺参幼参肠道消化酶活性的影响

消化酶	组别					
	E0	E1	E2	E3	E4	E5
蛋白酶/（U/mg prot）	19.90±2.93	16.10±1.14	16.75±0.88	18.14±1.73	18.30±1.84	19.79±0.37
淀粉酶/（U/mg prot）	0.88±0.07	0.77±0.05	0.76±0.05	0.77±0.14	0.77±0.13	0.80±0.11
纤维素酶/（U/g prot）	14.77±2.13	12.33±1.06	13.40±0.22	14.78±1.72	12.97±2.07	14.71±1.39

注：同行数据无字母表示差异不显著（$P>0.05$）。

3. 半乳甘露寡糖对刺参幼参体壁糖类含量的影响

随着饲料中半乳甘露寡糖添加量的增加，刺参幼参体壁总糖含量呈现先上升后降低趋势，0.8%（E3）、1.2%（E4）和 1.6%（E5）添加组中总糖含量均显著低于对照组及 0.2%（E1）、0.4%（E2）添加组（$P<0.05$）。饲料中不同添加量的半乳甘露寡糖对刺参幼参体壁酸性黏多糖的含量无显著影响（$P>0.05$）。1.2%（E4）和 1.6%（E5）添加组糖醛酸含量显著低于对照组及其他添加组（$P<0.05$）（表 4-28）。

表 4-28 半乳甘露寡糖对刺参幼参体壁糖类含量的影响　　（单位：mg/g）

糖类	组别					
	E0	E1	E2	E3	E4	E5
总糖	3.54±0.30^c	3.66±0.33^c	3.83±0.24^c	3.07±0.27^b	2.75±0.21^{ab}	2.53±0.04^a
酸性黏多糖	1.09±0.08	1.03±0.09	1.05±0.08	1.04±0.11	1.07±0.03	1.05±0.10
糖醛酸	0.87±0.03^b	0.81±0.05^b	0.82±0.04^b	0.89±0.06^b	0.76±0.09^a	0.74±0.05^a

注：同行数据无字母或上标相同字母表示差异不显著（$P>0.05$），上标不同字母表示差异显著（$P<0.05$）。

4. 半乳甘露寡糖对刺参幼参体壁氨基酸含量的影响

由表 4-29 可知，半乳甘露寡糖添加量为 0.2%（E1）、0.4%（E2）、0.8%（E3）、1.2%（E4）的添加组，刺参幼参体壁天冬氨酸、谷氨酸、精氨酸、胱氨酸的含量及氨基酸总量显著高于对照组（$P<0.05$）和 1.6%（E5）添加组，而对照组（$P<0.05$）和 1.6%（E5）添加组之间差异不显著（$P>0.05$）。

表 4-29 半乳甘露寡糖对刺参幼参体壁氨基酸组成的影响（%）

氨基酸种类	E0	E1	E2	E3	E4	E5
天冬氨酸	3.24±0.13^a	4.21±0.03^b	4.36±0.21^b	4.19±0.33^b	4.50±0.05^b	3.42±0.17^a
谷氨酸	5.88±0.24^a	7.72±0.16^b	7.73±0.38^b	7.055±0.65^b	7.49±0.18^b	6.07±0.15^a
丝氨酸	1.55±0.18^a	1.80±0.09^{ab}	1.98±0.27^b	1.99±0.15^b	2.12±0.055^b	1.61±0.02^a
组氨酸	0.17±0.09^a	0.15±0.03^a	0.20±0.04^a	0.36±0.06^b	0.35±0.03^b	0.50±0.12^c
甘氨酸	3.14±0.12^a	4.35±0.30^b	4.70±0.01^b	4.79±0.39^b	4.83±0.29^b	4.31±0.16^b
苏氨酸	3.47±0.91^{bc}	3.56±0.08^{bc}	3.73±0.33^c	2.78±0.87^{ab}	3.62±0.08^c	2.58±0.20^a
精氨酸	2.58±0.37^a	3.57±0.10^b	3.80±0.20^b	3.28±0.36^b	3.81±0.18^b	2.56±0.05^a
丙氨酸	1.76±0.09^a	2.66±0.09^b	2.88±0.22^b	2.37±0.58^{ab}	2.95±0.21^b	1.67±0.06^a
酪氨酸	1.01±0.02^a	1.22±0.02^{ab}	1.58±0.05^{ab}	1.41±0.31^{ab}	1.78±0.20^b	1.03±0.02^a
胱氨酸	0.09±0.05^a	0.36±0.03^b	0.32±0.03^b	0.30±0.02^b	0.35±0.03^b	0.11±0.04^a
缬氨酸	1.33±0.11^a	1.63±0.01^{bc}	1.63±0.01^{bc}	1.60±0.17^{bc}	1.73±0.03^c	1.43±0.09^{ab}
蛋氨酸	0.55±0.05^a	1.12±0.09^b	0.99±0.14^b	0.90±0.17^b	0.96±0.05^b	0.85±0.21^b
苯丙氨酸	1.29±0.05	1.44±0.06	1.42±0.06	1.43±0.10	1.51±0.03	1.30±0.30
异亮氨酸	1.09±0.00^a	1.56±0.05^b	1.48±0.19^b	1.37±0.23^{ab}	1.52±0.01^b	1.10±0.11^a
亮氨酸	1.81±0.17^a	1.88±0.03^{ab}	2.18±0.27^{bc}	2.10±0.16^{abc}	2.25±0.07^c	1.92±0.09^{abc}
赖氨酸	1.47±0.03^{ab}	1.88±0.02^b	1.82±0.13^b	1.59±0.57^{ab}	1.08±0.23^a	1.58±0.02^{ab}
氨基酸总量	30.72±2.25^a	39.06±0.48^b	40.67±2.53^b	37.34±2.63^b	40.90±0.65^b	32.05±1.67^a

注：同行数据无字母或上标相同字母表示差异不显著（$P>0.05$），上标不同字母表示差异显著（$P<0.05$）。由于本测定方法未对半胱氨酸和色氨酸进行保护，因此未测得其含量。

5. 半乳甘露寡糖对刺参幼参免疫指标的影响

饲料中添加半乳甘露寡糖对刺参幼参体壁及体腔液中超氧化物歧化酶活性具有显著影响（表 4-30）。0.2%（E1）添加组刺参幼参体壁超氧化物歧化酶活性显著提高（$P<0.05$），随着饲料中半乳甘露寡糖添加量的增加，体壁中超氧化物歧化酶活性呈现降低趋势。各添加组刺参幼参体腔液超氧化物歧化酶活性显著高于对照组（$P<0.05$），但各添加组之间差异不显著（$P>0.05$）。饲料中添加半乳甘露寡糖对体腔细胞中超氧化物歧化酶活性无显著影响（$P>0.05$），随着饲料中半乳甘露寡糖添加量的增加，各添加组超氧化物歧化酶活性与对照组均无显著性差异（$P>0.05$）。

表 4-30 半乳甘露寡糖对刺参幼参免疫指标的影响

	免疫指标	组别					
		E0	E1	E2	E3	E4	E5
体壁	超氧化物歧化酶/（U/mg prot）	23.35±1.92^a	41.67±2.87^c	35.83±2.07^{bc}	35.75±1.88^{bc}	31.77±2.53^{abc}	27.01±1.78^{ab}
	溶菌酶/（U/mg prot）	6.09±0.23^a	4.33±0.28^a	12.18±0.51^b	9.43±0.37^b	8.44±0.61^b	3.43±0.12^a
	碱性磷酸酶/（U/g prot）	21.41±1.38^a	44.27±2.39^b	53.63±2.16^b	37.36±2.41^a	40.64±3.01^a	39.36±1.28^a
体腔液	超氧化物歧化酶/（U/ml）	40.39±2.89^a	48.53±3.12^b	47.95±2.62^b	47.82±3.01^b	48.77±2.79^b	47.50±3.41^b
	溶菌酶/（U/ml）	18.66±0.97^a	18.29±1.31^a	13.79±0.48^a	19.31±1.24^a	24.29±1.35^b	27.98±1.15^b
	碱性磷酸酶/（U/L）	10.55±0.49^a	14.24±0.92^b	15.93±0.78^b	8.70±0.24^a	12.10±0.79^a	10.74±0.39^a
体腔细胞	超氧化物歧化酶/（U/ml）	14.58±0.93	16.15±0.72	13.01±0.51	12.65±0.79	14.98±0.82	15.76±0.43
	溶菌酶/（U/ml）	6.94±0.21^b	3.33±0.17^a	4.34±0.36^a	2.82±0.11^a	3.04±0.15^a	3.76±0.27^a
	碱性磷酸酶/（U/L）	18.55±1.21	16.67±1.39	24.33±2.82	19.7±1.68	16.1±1.34	18.65±1.99

注：同行数据无字母或上标相同字母表示差异不显著（$P>0.05$），上标不同字母表示差异显著（$P<0.05$）。

饲料中添加半乳甘露寡糖对刺参幼参体壁、体腔液及体腔细胞中溶菌酶活性均具有显著影响（$P<0.05$）。0.4%（E2）、0.8%（E3）和1.2%（E4）添加组刺参幼参体壁中溶菌酶活性显著高于对照组及其余添加组（$P<0.05$）；1.2%（E4）和1.6%（E5）添加组刺参幼参体腔液中溶菌酶活性显著高于对照组及其余添加组（$P<0.05$）；各添加组刺参幼参体腔细胞中溶菌酶活性均显著低于对照组（$P<0.05$），但各添加组之间差异不显著（$P>0.05$）。

饲料中添加半乳甘露寡糖对刺参幼参体壁和体腔液中碱性磷酸酶活性均具有显著影响（$P<0.05$）。低添加量0.2%（E1）和0.4%（E2）添加组刺参体壁和体腔液中碱性磷酸酶活性显著提高（$P<0.05$），而添加量高于0.4%时，刺参幼参体壁和体腔液中碱性磷酸酶活性与对照组相比差异不显著（$P>0.05$）。饲料中添加半乳甘露寡糖对刺参幼参体腔细胞中碱性磷酸酶活性无显著影响（$P>0.05$），随着添加量的增加，各添加组体腔细胞中碱性磷酸酶活性与对照组均无显著性差异（$P>0.05$）。

4.2.3.3 讨论

1. 半乳甘露寡糖对刺参幼参生长和消化的影响

本研究中，饲料中添加半乳甘露寡糖能促进刺参参的生长，各添加组刺参幼参增重率及特定生长率均显著高于对照组，且在1.2%（E4）添加组达到最大值。

一些学者已经报道了糖类免疫增强剂能够促进水产动物的生长。例如，半乳甘露低聚糖能够提高异育银鲫幼鱼增重率及特定生长率，并明显降低饲料系数（王锐等，2008）；β-葡聚糖能够提高凡纳滨对虾的增重率（谭北平等，2003）；肽聚糖可以提高鲈鱼的特定生长率（张璐等，2008）；免疫多糖可以提高养殖刺参的体增重等（马跃华和胡守义，2006）。本实验首次应用半乳甘露寡糖进行刺参饲料的添加，显示其对刺参的生长具有明显的促进作用，这与上述研究结果一致。对水产动物及畜禽类的研究表明，半乳甘露寡糖等寡糖类物质基本不能被动物自身消化吸收，直接进入肠道从而被消化道微生物选择性地作为营养物，能被消化道有益生物利用而不能被有害微生物利用，从而调节微生态平衡，达到促进生长的效果（王锐等，2008；曹功明等，2010）。

蛋白酶、淀粉酶、纤维素酶是刺参重要的消化酶，其活性的高低反映了其对营养物质的利用能力。本实验研究结果显示，饲料中添加半乳甘露寡糖对刺参消化酶活性没有显著影响，表明其作为添加剂使用并不能提高刺参对营养物质的消化能力。目前关于甘露寡糖对水产动物肠道消化酶活性的研究报道很少。于艳梅（2010）研究发现，在黄颡鱼饲料中添加魔芋甘露寡糖，显著提高了黄颡鱼肝胰脏和肠道中蛋白酶、脂肪酶及淀粉酶的活性。刘爱君等（2009）研究发现，在饲料中添加 0.25%、0.5%和 0.75%的甘露寡糖，均能显著提高奥尼罗非鱼胃蛋白酶和肠蛋白酶的活性，推测这可能与甘露寡糖改善胃、肠道黏膜结构，促进消化酶的分泌有关。在本研究中，半乳甘露寡糖对刺参幼参肠道消化酶活性没有显著影响，产生这种差异的原因可能与所作用的动物种类及其摄食习性有关，刺参摄食的天然饵料和鱼类有很大不同，导致其消化酶的诱导机制存在较大的差别，其原因有待进一步研究。

2. 半乳甘露寡糖对刺参幼参体壁多糖含量及氨基酸组成的影响

海参体壁含有多种生物活性物质，如海参多糖、海参皂苷、胶原蛋白、脂肪酸、多肽、海参神经节苷脂等，并且具有药理活性（牛娟娟和宋扬，2009；赵芹等，2008）。多糖是刺参体壁的重要成分，其中酸性黏多糖为刺参的有效生理活性物质，系由 D-N-乙酰氨基半乳糖、D-葡萄糖醛酸和 L-岩藻糖组成的分支杂多糖，具有抗凝血、抗血栓、降血脂、降低血黏度、抗肿瘤、免疫调节、抗菌、抗病毒及促进细胞生长等作用（牛娟娟和宋扬，2009）。本实验研究表明，在饲料中添加半乳甘露寡糖对刺参幼参体壁中酸性黏多糖含量没有显著影响，低添加量半乳甘露寡糖对总糖和糖醛酸含量的影响不显著，但高添加量则显著降低刺参幼参体壁中总糖和糖醛酸的含量。

刺参体壁氨基酸组成及含量是刺参的食（药）性的重要指标（赵芹等，2008）。本实验首次发现，添加半乳甘露寡糖对刺参体壁氨基酸组成及含量具有显著影响，在0.2%~1.2%的添加量下，刺参幼参体壁氨基酸总量显著升高，且重要的药效氨基酸甘氨酸、谷氨酸、精氨酸含量均明显增加。目前其机制尚不明确，可能与半乳甘露寡糖调控肠道发育及生长因子分泌有关。有研究表明，摄入一定量寡糖能显著增加哺乳动物肠绒毛的长度及肠壁厚度（何亚男，2006），提高对某些必需氨基酸及非必需氨基酸的净吸收量（王彬等，2006）；同时提高某些生长因子的浓度，通过介导作用而起到调节氨基酸组成、促进生长的作用（Tang et al.，2005）。

3. 半乳甘露寡糖对刺参幼参免疫指标的影响

刺参是一种棘皮动物，防御机制主要是非特异性免疫，包括体壁防御和体内免疫，体壁和体内分布的免疫相关酶发挥着重要作用（孙永欣，2008）。超氧化物歧化酶是无脊椎动物体内重要的抗氧化酶，在消除氧自由基、防止生物分子损伤方面发挥着重要的生理作用。碱性磷酸酶是溶酶体的重要组成部分，在体内直接参与磷酸基团的转移和代谢，并与一些营养物质的消化吸收有关（Zhang et al.，2000），是刺参体内参与免疫防御等活动的重要水解酶。溶菌酶是一种专门作用于微生物细胞壁的水解酶，其活性高低反映了水生生物非特异性免疫水平的高低（常杰等，2011）。

研究结果显示，饲料中添加适量的半乳甘露寡糖能显著提高刺参幼参体壁和体腔液中超氧化物歧化酶、碱性磷酸酶和溶菌酶的活性，0.2%~0.8%添加量可以显著提高刺参幼参体壁和体腔液中超氧化物歧化酶活性；0.0%~0.4%添加量可以显著提高刺参幼参体壁和体腔液中碱性磷酸酶活性；0.4%~1.2%添加量可以显著提高体壁中溶菌酶活性。这说明半乳甘露寡糖对刺参机体具有明显的免疫刺激作用，在提高刺参抗氧化防御能力、细胞内溶酶体酶的水解能力方面具有显著的作用。而碱性磷酸酶和溶菌酶除了具有免疫防御能力，还具有体内物质代谢能力（Zhang et al.，2000），其活性被半乳甘露寡糖非特异性诱导增强，表现为特定生长率升高，亦与本研究结果相符。

寡糖具有一定的免疫原性，能够刺激机体免疫应答，而且能与某些毒素、病毒和真菌细胞的表面结合而作为这些外源抗原的佐剂，减缓抗原的吸收，增加抗原的效价，从而增强动物体的细胞和体液免疫反应（Yoshida et al.，1995；Bornet and Brouns，2002）。研究表明，在饲料中添加甘露寡糖能够显著提高虹鳟（Staykov et al.，2007）、黑鲈（Torrecillas et al.，2007）和斑点叉尾鮰（Welker et al.，2007）的抗病力。在饲料中添加褐藻低聚糖可以提高大菱鲆鱼苗血液中免疫酶的活性，对大菱鲆非特异性免疫具有促进作用（王鹏等，2006）。在饲料中添加牛蒡寡糖亦对大菱鲆表现出良好的促生长作用，且显著提高血清酸性磷酸酶、碱性磷酸酶、溶菌酶、超氧化物歧化酶的活性（郝林华等，2007）。何四旺等（2003）报道，在饲料中添加0.2%异麦芽寡糖可以显著提高罗非鱼血清中溶菌酶活性。在饲料中添加低聚果糖、甘露寡糖均可以显著提高刺参体腔液中酚氧化酶以及酸性磷酸酶的活性（张琴，2010）。在刺参基础饲料中添加0.1%的褐藻寡糖，均能提高刺参体腔液和体壁中过氧化物酶、酸性磷酸酶、碱性磷酸酶和溶菌酶的活性，显著提高刺参非特异性免疫水平（江晓路等，2009）。经体壁注射后，κ-卡拉胶寡糖能显著提高体腔液中溶菌酶、碱性磷酸酶和超氧化物歧化酶的活性（马悦欣等，2010）。

4.2.3.4 结论

综上所述，本研究结果表明，以饲料添加剂的形式在刺参配合饲料中添加适量的半乳甘露寡糖不仅可以提高其特定生长率、改善体壁营养组成，还可以提高刺参机体免疫力，0.2%添加量即可起到促进生长、提高免疫力的作用，当添加量高于1.2%时，降低了刺参的生长性能及相关免疫酶的活性。

第 5 章 国 际 交 流

PART I: Optimal dietary methionine requirement for juvenile sea cucumber *Apostichopus japonicus*

5.1 Introduction

Methionine is the sulfur-containing amino acid with the side chain $-CH_2CH_2SCH_3$, which is required for normal growth of many fishes. Among the 20 protein amino acids, only methionine and cystine have sulfur (Kensei, 2015). It is a substrate for protein synthesis, meanwhile, it also serves as the major methyl group donor for synthesis of polyamines, methylation of phospholipids, and DNA and RNA intermediates in vivo (Loy and Lundy, 2019). Usually, methionine is the most limiting amino acid in aquatic diet, especially in a feed with high proportion of legume protein (Chen et al., 2018b). Many studies have been reported on the methionine requirements of marine aquatic species, which illustrated that dietary optimal methionine varied from 0.72% (*Litopenaeus vannamei*) (Felipe et al., 2016) to 2.58% (*Pseudosciaena crocea*) (Yu et al., 2013; Li et al., 2021).

Sea cucumber, *Apostichopus japonicus* Selenka, is one kind of high value marine cultured animals in northeast China. The output of sea cucumber has been tripled in the last decade, and the annual yield got 171,700 tons in 2019. Studies have been reported on the requirements of some nutrients, such as protein (Seo and Lee, 2011), lipid (Liao et al., 2017), carbohydrate (Xia et al., 2015a), ascorbic acid (Ren et al., 2016), riboflavin (Okorie et al., 2011), and vitamin E (Wang et al., 2015b). Up to now, there have been no reports on methionine requirement of sea cucumber. In order to develop the formulate feed for sea cucumber, the methionine requirement was determined in this study.

5.2 Materials and methods

5.2.1 Diet preparation

The basic diet formulation included 209g/kg crude protein and 47g/kg crude lipid, using fermented soybean meal and kelp powder as the main protein sources (Seo and Lee, 2011; Liao et al., 2017). Six feeds were formulated by adding coated methionine to a final concentration of 0.18% (basal diet), 0.37%, 0.57%, 0.78%, 0.96%, 1.10% diet, which were named as D1, D2, D3, D4, D5, D6 group. Ingredients and proximate composition of the experimental diets, and amino acids composition of experimental diets were listed in Table 5-1 and Table 5-2 respectively.

Table 5-1 Ingredients and proximate composition of the experimental diets

	Items	\multicolumn{6}{c}{Groups}
		D1	D2	D3	D4	D5	D6
Ingredients	Wheat meal /%	5	5	5	5	5	5
	Fermented soybean meal /%	30	30	30	30	30	30
	Kelp powder /%	15	15	15	15	15	15
	Glycine /%	0.8	0.7	0.6	0.5	0.4	0.3
	Coated methionine /%	0	0.2	0.4	0.6	0.8	1
	Multi-vitamin /%	0.5	0.5	0.5	0.5	0.5	0.5
	Multi-mineral /%	0.5	0.5	0.5	0.5	0.5	0.5
	Fish oil /%	1	1	1	1	1	1
	Ethoxyquinoline /%	0.1	0.1	0.1	0.1	0.1	0.1
	Soybean lecithin /%	1	1	1	1	1	1
	Sea mud /%	45.3	45.1	44.9	44.7	44.5	44.3
Proximate composition (dry matter)	Crude protein /%	20.97	21.01	20.98	21.06	21.04	21.09
	Crude lipid /%	4.74	4.31	4.80	4.72	4.98	5.06
	Crude ash /%	48.77	48.81	49.22	49.37	49.46	49.64
	Gross energy /(kJ/g)	11.80	11.83	11.86	11.92	12.07	12.24

Table 5-2 Amino acids composition of experimental diets (%)

Amino acids	\multicolumn{6}{c}{Groups}
	D1	D2	D3	D4	D5	D6
Methionine	0.18	0.37	0.57	0.78	0.96	1.10
Arginine	1.01	0.98	0.98	1.08	0.99	1.05
Histidine	0.36	0.34	0.33	0.37	0.36	0.36
Isoleucine	0.50	0.46	0.48	0.55	0.49	0.55
Leucine	1.32	1.32	1.31	1.43	1.33	1.42
Lysine	0.81	0.76	0.76	0.84	0.80	0.83
Phenylalanine	0.79	0.78	0.77	0.85	0.79	0.82
Threonine	0.67	0.61	0.60	0.70	0.63	0.67
Valine	0.57	0.52	0.55	0.60	0.57	0.60

Continued

Amino acids	Groups					
	D1	D2	D3	D4	D5	D6
Cysteine	0.56	0.51	0.58	0.56	0.55	0.54
Glycine	1.46	1.32	1.23	1.24	1.07	1.03
Aspartic acid	1.82	1.78	1.75	1.92	1.85	1.87
Serine	0.92	0.90	0.89	0.97	0.90	0.95
Alanine	0.86	0.83	0.83	0.88	0.83	0.87
Tyrosine	0.48	0.46	0.45	0.51	0.48	0.48
Glutamic acid	3.70	3.65	3.54	3.92	3.78	3.82
Proline	0.83	0.81	0.82	0.88	0.82	0.85
TAA	16.83	16.39	16.44	18.07	17.30	17.79

Note: TAA data in the table have been rounded off and are not the simple addition of all amino acids

All solid ingredients were ground into powder (0.075mm), and mixed thoroughly. Fresh fish oil was added into the powder and mixed sufficiently, and then appropriate distilled water was added and made them into dough. Feed pellets were prepared using a moist pelletizer, and dried at 60℃ for 24h in an oven and grounded into desirable particle sizes (0.18-0.25mm). All diets were packed and stored at −20℃ (Xia et al., 2015b).

Coated methionine was prepared by the following methods: Firstly, methionine (purity 99.99%, Sinopharm Chemical Reagent Co., Ltd, Shanghai, China) and β-cyclodextrin (Huaxing Biochemstry Co. Ltd, Mengzhou, China) were mixed equally. Secondly, distilled water was added and stirred into a paste. Thirdly, heat and stir in a water bath at 90℃ for 30min. Fourthly, the paste was dried in an oven at 90℃ overnight. Finally, the mixture was crushed into particles less than 0.0177mm in diameter by a small feed mill (Chen et al., 2008). There was about 40% methionine in the final product.

Determination of the dissolution rate: Accurately weigh 50mg coated methionine, fix to 100ml with distilled water, shake well, stand for 5min, and filter for the supernatant. The content of methionine dissolved in water was determined by ninhydrin ratio method (Wang et al., 2007). The dissolution loss of coated methionine was about 21.18%.

5.2.2 Experimental design

Experimental sea cucumbers were obtained from a local commercial aquaculture farm in Penglai, China, and transported into the laboratory at 4℃. Prior to the beginning of the experiment, juvenile sea cucumbers were reared in a recirculating aquatic system for 2 weeks to acclimate to the experimental conditions. Animals were fed basal diets during this period.

At the beginning of the trial, sea cucumbers were starved for 24h and weighted. Sea cucumbers with similar body weight (12.14g±0.14g) were randomly distributed into eighteen 300L blue cylindrical fiberglass tanks (70cm × 80cm) and each tank was stocked with 30 juveniles. Each diet was randomly assigned to triplicate tanks. Water depth was kept at 60cm, and the feeding trial lasted for 8 weeks. There were two culture baskets with 20 corrugated boards in each tank, and the basket was replaced every 20 days. During this period, temperature ranged from 17.0-19.0℃, water flow rate was kept at 2L/min, salinity was held at 26-28, pH was maintained at 7.6-8.2, dissolved oxygen was above 5 mg/L, and total ammonia nitrogen was maintained at less than 0.05mg/L. Sea cucumbers were fed once at 16:00 and up to 3% of their initial total biomass per day. Uneaten diet and feces were siphoned out of the culture system every 3 days. The trial was performed in low light (∼150lx).

5.2.3 Sampling collection and chemical analysis

At the end of the experiment, sea cucumbers were starved for 24h before harvest. Total number and body weight were measured for calculating survival rate (SR), weight gain rate (WGR) and specific growth rate (SGR). Ten sea cucumbers per tank were randomly selected, and body weight was measured after absorbing the water from the body surface with filter papers. Intestinal tract and body wall were separated on ice plate after body fluid being removed and weighed. Body wall and intestine samples were stored at –80℃ for further analysis.

Proximate composition analyses on experimental diets and body wall samples were performed using standard methods (AOAC, 2003). Moisture was determined by drying in an oven at 105℃ overnight. Crude protein ($N \times 6.25$) was determined by the Kjeldahl method after acid digestion. Crude lipid was determined by the ether-extraction method. Energy was determined by an adiabatic bomb calorimeter (PARR 6100, USA). Amino acids were determined by acids hydrolysis with an automatic amino acid analyzer (L-8900, Hitachi, Japan).

Intestine was homogenate with 9 fold cold physiological saline after cutting in ice bath. Homogenate solution was collected by centrifugation at 8000 ×g for 10min at 4℃ using a high-speed refrigerated centrifuge (Hitachi Crg Series, Japan). The supernatant was separated into 2ml centrifugal tube and stored in refrigerator at –80℃ for further analysis.

Protease, lipase, amylase, total superoxide dismutase (T-SOD), catalase (CAT) activities, total antioxidant capacity (T-AOC), and malondialdehyde (MDA) contents were measured by the commercial kits provided by Nanjing Jiancheng Bioengineering Institute, Nanjing, China, in accordance with the instructions of the manufacturer. Alkaline phosphatase (AKP), acid phosphatase (ACP), aspartate aminotransferase (AST) and alanine aminotransferase (ALT) activities were analyzed by the commercial kits provided by Shanghai Enzyme-linked Industrial Co., Ltd., Shanghai, China. The protein contents in coelomic fluid were determined by coomassie brilliant blue method using commercial kit (Nanjing Jiancheng Bioengineering Institute, Nanjing, China).

5.2.4 Statistical analysis

Data from the trial were subjected to one-way analysis of variance (ANOVA), using the SPSS program version for windows (SPSS Inc., Chicago, IL, USA). When the ANOVA identified differences among groups, multiple comparisons among means were made with Duncan's multiple-range test at $P < 0.05$. The results were presented as means ± SD.

The optimal dietary methionine requirement of sea cucumbers was estimated with linear regression using WGR or methionine content in body wall as the dependent variable against dietary methionine levels using Statistical Analysis System 9.2 (SAS institute, USA) NLIN regression or quadratic regression analysis (Microsoft excel, USA).

5.3 Results

5.3.1 Growth

At the end of the trial, there were no differences in survival rate (SR) among all groups (Table 5-3). Both weight gain rate (WGR) and specific growth rate (SGR) were presented the trend of increasing firstly and then decreasing with the increase of dietary methionine levels, and these of D3 group were significantly higher than others. FCRs dramatically decreased and then increased afterwards. The ratio of intestine weight to body weight (IBR) and the ratio of intestinal length to body length (IBL) were not affected by dietary methionine levels.

Table 5-3 Effects of dietary methionine on growth performance and figure index of juvenile sea cucumbers *Apostichopus japonicus* Selenka

Items	Groups					
	D1	D2	D3	D4	D5	D6
IBW /g	12.00±0.01	12.21±0.09	12.09±0.14	12.18±0.02	12.13±0.12	12.22±0.26
FBW /g	13.10±0.30^a	14.17±0.38^b	15.69±0.53^e	15.34±0.14^d	14.67±0.33^c	14.70±0.48^c
WGR /%	9.20±1.12^a	15.90±2.29^b	29.83±3.29^e	25.93±1.15^d	21.00±2.04^c	20.33±1.48^c
FCR	18.33±0.05^d	10.72±1.69^c	5.68±0.66^a	6.48±0.29^a	8.06±0.75^b	8.31±0.63^b
SGR /(%/d)	0.15±0.00^a	0.25±0.04^b	0.45±0.05^e	0.40±0.02^d	0.33±0.03^c	0.32±0.03^c
IBR /%	4.64±0.30	4.67±0.06	4.73±0.15	4.46±0.14	4.50±0.18	4.53±0.11
IBL /%	2.38±0.12	2.33±0.20	2.34±0.06	2.39±0.07	2.36±0.09	2.35±0.13
SR /%	93.35±4.74	92.20±1.91	90.00±3.30	95.57±1.96	98.35±2.33	94.47±3.87

Notes: Values (mean ± SD) (n=3) with the different letters in the same line are significantly different at $P<0.05$.
IBW (g), initial body weight; FBW (g), final body weight; WGR (weight gain rate, %)=100× (FBW–IBW)/IBW; FCR =feeding amount / [(FBW–IBW) × 30]; SGR (specific growth rate, %/d)=100×[(lnFBW–lnIBW)/56]; IBR (the ratio of intestine weight to body weight, %)= 100× intestine weight / body weight; IBL (the ratio of intestinal length to body length, %)= intestine length / body length; SR (survival rate, %)=100×final fish number/initial fish number.

SAS NLIN analysis based on WGR and dietary methionine levels showed that the requirement of methionine for juvenile sea cucumbers was 0.58% diet (Figure 5-1).

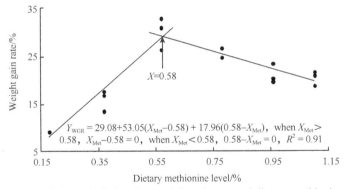

Figure 5-1 SAS NLIN analysis based on weight gain rate and dietary methionine level. The requirement of dietary methionine for juvenile sea cucumbers *Apostichopus japonicus* was 0.58% diet (2.76% dietary protein)

5.3.2 Proximate composition of body wall

The effects of dietary methionine on proximate composition of body wall of juvenile sea cucumbers are presented in Table 5-4. There were no significant differences in the moisture contents of body wall among all groups. Both crude protein and crude lipid increased firstly and then decreased with the increase of dietary methionine levels, and those of D3-D5 groups were significantly higher than others, but there were no differences among them. Crude ash decreased and then increased with dietary methionine, and the lowest value appeared in D4 group.

Table 5-4 Effects of dietary methionine on proximate composition of body wall of juvenile sea cucumbers *Apostichopus japonicus* Selenka (%)

Items	Groups					
	D1	D2	D3	D4	D5	D6
Moisture	91.47±0.30	90.65±0.39	91.26±0.67	90.78±0.62	91.03±0.04	90.69±0.13
Crude protein	44.10±0.27^a	44.42±0.26^a	45.76±0.32b^c	46.24±0.24^c	46.01±0.18b^c	45.64±0.26^b
Crude lipid	3.47±0.11^a	3.88±0.14^b	4.51±0.08^c	4.46±0.14^c	4.28±0.12^c	3.55±0.17^a
Crude ash	33.25±0.29^c	31.57±0.26^b	31.72±0.21^b	30.49±0.13^a	31.28±0.18^b	30.65±0.19^{ab}

Notes: In the above table, the crude protein, crude lipid and crude ash contents of body wall are based on dry basis. Values (mean ± SD) (n=3) with the different letters in the same line are significantly different at $P < 0.05$.

5.3.3 Amino acids profiles of body wall

The effects of dietary methionine on amino acids profiles of body wall of juvenile sea cucumbers are presented in Table 5-5. Most of amino acids were not affected by dietary methionine, except methionine, aspartic acid and cysteine. The contents of these three amino acids increased firstly and then decreased with the increase of dietary methionine contents. Based on quadratic regression analysis between methionine contents in body wall and dietary methionine levels, it was showed that optimum dietary methionine was 0.72% diet for juvenile sea cucumbers *Apostichopus japonicus* (Figure 5-2).

Table 5-5 Effects of dietary methionine on amino acids profiles of body wall of juvenile sea cucumbers *Apostichopus japonicus* Selenka (%)

Amino acids	Groups					
	D1	D2	D3	D4	D5	D6
Methionine	0.56±0.02^a	0.60±0.03^{ab}	0.66±0.02^{bc}	0.68±0.02^c	0.64±0.03^{bc}	0.61±0.03^{ab}
Arginine	3.04±0.03	3.11±0.02	3.12±0.07	3.14±0.05	3.09±0.09	3.11±0.07
Histidine	0.55±0.06	0.58±0.08	0.57±0.08	0.62±0.05	0.57±0.07	0.59±0.07
Isoleucine	1.16±0.07	1.18±0.06	1.20±0.05	1.21±0.03	1.20±0.05	1.22±0.04
Leucine	2.20±0.07	2.31±0.08	2.28±0.08	2.31±0.07	2.32±0.09	2.30±0.08
Lysine	1.74±0.05	1.80±0.08	1.77±0.08	1.82±0.04	1.80±0.07	1.85±0.08
Phenylalanine	1.24±0.07	1.26±0.05	1.26±0.07	1.38±0.06	1.26±0.07	1.25±0.09
Threonine	2.19±0.07	2.22±0.09	2.25±0.07	2.24±0.09	2.21±0.07	2.23±0.06
Valine	1.73±0.07	1.78±0.07	1.83±0.07	1.83±0.06	1.84±0.06	1.82±0.09
Cysteine	1.29±0.03^a	1.37±0.04^{ab}	1.41±0.06^b	1.44±0.06^b	1.40±0.06^b	1.36±0.04^{ab}

Continued

Amino acids	Groups					
	D1	D2	D3	D4	D5	D6
Aspartic acid	4.23±0.07^a	4.28±0.09^{ab}	4.31±0.07^b	4.68±0.09^c	4.40±0.09^b	4.36±0.07^b
Serine	2.11±0.08	2.14±0.09	2.19±0.07	2.14±0.08	2.21±0.06	2.20±0.06
Glycine	5.38±0.03	5.36±0.10	5.46±0.03	5.35±0.06	5.45±0.05	5.39±0.07
Alanine	2.79±0.07	2.78±0.06	2.79±0.04	2.85±0.08	2.85±0.08	2.80±0.08
Tyrosine	1.06±0.08	1.12±0.07	1.10±0.07	1.16±0.07	1.12±0.06	1.11±0.08
Glutamic acid	7.04±0.04	7.09 0.02	7.11±0.08	7.12±0.07	7.12±0.06	7.14±0.06
Proline	3.06±0.05	3.01±0.04	3.03±0.07	3.06±0.08	3.04±0.05	3.04±0.07
TAA	41.37±0.24^a	41.99±0.30^{ab}	42.34±0.42^{bc}	43.03±0.44^c	42.52±0.41^{bc}	42.38±0.36^{bc}

Notes: Values (mean ± SD) (n=3) with the different letters in the same line are significantly different at $P<0.05$.

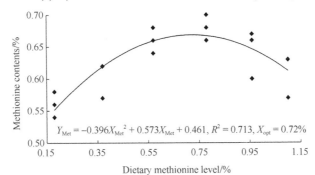

Figure 5-2 Quadratic regression analysis based on methionine contents in body wall and dietary methionine level. The requirement of dietary methionine for juvenile sea cucumbers *Apostichopus japonicus* was 0.72% diet (3.43% dietary protein)

Taking methionine as quantitative 1, essential amino acids pattern can be calculated and presented in Table 5-6. It can be concluded that there were no significant differences on ratios of essential amino acids to methionine in body wall of juvenile sea cucumbers among experimental groups.

Table 5-6 Pattern of essential amino acids of body wall of juvenile sea cucumbers *Apostichopus japonicus,* based on the methionine contents (%)

Amino acids	Groups					
	D1	D2	D3	D4	D5	D6
Methionine	1.00	1.00	1.00	1.00	1.00	1.00
Histidine	0.98±0.10	0.96±0.09	0.99±0.13	0.96±0.04	0.99±0.08	0.97±0.10
Isoleucine	2.16±0.18	2.18±0.18	2.20±0.10	2.16±0.03	2.22±0.06	2.27±0.05
Phenylalanine	2.50±0.18	2.49±0.04	2.47±0.15	2.50±0.11	2.46±0.01	2.46±0.17
Valine	3.07±0.13	2.98±0.02	3.02±0.18	2.99±0.02	2.97±0.05	3.00±0.16
Lysine	3.59±0.08	3.51±0.07	3.49±0.19	3.52±0.14	3.54±0.07	3.55±0.14
Threonine	4.49±0.11	4.43±0.11	4.38±0.19	4.31±0.02	4.45±0.06	4.39±0.13

Continued

Amino acids	Groups					
	D1	D2	D3	D4	D5	D6
Leucine	4.87±0.18	4.85±0.06	4.79±0.19	4.76±0.05	4.81±0.05	4.79±0.05
Arginine	5.68±0.16	5.62±0.20	5.51±0.20	5.66±0.22	5.54±0.09	5.59±0.13

5.3.4 Enzymes activities of intestinal homogenate

Effects of dietary methionine on digestive, antioxidant and metabolism enzymes activities in intestinal homogenate are shown in Table 5-7. Activities of protease and lipase significantly increased firstly and then decreased when dietary methionine was up to 0.78%-0.96%, and amylase activities were not affected by dietary methionine. Both total antioxidant capacity (T-AOC) and total superoxide dismutase (T-SOD) activities rose firstly and then stabilized, and D1 group was significantly lower than other groups. Catalase (CAT) activities increased firstly and then decreased afterwards, and D3 group was significantly higher than others. Malondialdehyde (MDA) contents decreased firstly and then stabilized afterwards, and D1 group was significantly higher than others, and there were no differences among other groups. Both alanine aminotransferase (ALT) and aspartate aminotransferase (AST) activities increased firstly and then decreased afterwards with the increase of dietary methionine levels, and got highest values in D3 group. Acid phosphatase (ACP) activities significantly increased firstly and then decreased with dietary methionine, and that of D1 group was significantly lower than others. Alkaline phosphatase (AKP) activities increased firstly and then stabilized afterwards with dietary methionine, and that of D3-D6 groups was significantly higher than D1-D2 group, but there were no differences among them.

Table 5-7 Effects of dietary methionine on digestive capacity and antioxidant capacity of intestine homogenate of juvenile sea cucumbers *Apostichopus japonicus* Selenka

Items	Groups					
	D1	D2	D3	D4	D5	D6
Protease /(U/mg prot)	1050.93±16.18^a	1108.13±18.00^b	1238.36±17.41^d	1265.66±15.76^d	1163.06±15.77^c	1158.47±15.94^c
Lipase /(U/g prot)	53.17±1.73^a	54.68±1.63^{ab}	58.59±1.67^c	58.16±1.82^c	57.51±1.89^{bc}	56.42±1.64^{bc}
Amylase /(U/mg prot)	0.30±0.05	0.33±0.03	0.33±0.02	0.34±0.03	0.34±0.03	0.35±0.02
T-AOC /(U/g prot)	0.28±0.02^a	0.30±0.02^b	0.31±0.01^b	0.33±0.01^b	0.33±0.01^b	0.33±0.01^b
T-SOD /(U/mg prot)	100.68±2.85^a	105.66±2.97^b	111.02±2.55^b	110.06±2.77^b	108.94±2.56^b	110.16±2.89^b
CAT /(U/mg prot)	21.65±1.27^a	24.87±0.92^{bc}	26.69±1.31^c	24.03±0.96^b	23.88±1.03^b	21.30±1.17^a

Continued

Items	Groups					
	D1	D2	D3	D4	D5	D6
MDA /(nmol/mg prot)	0.95±0.06^b	0.78±0.05^a	0.75±0.05^a	0.81±0.04^a	0.83±0.04^a	0.81±0.05^a
ALT /(U/g prot)	2.34±0.14^a	2.25±0.17^a	3.28±0.13^c	3.12±0.10^{bc}	2.92±0.16^b	2.43±0.11^a
AST /(U/g prot)	2.46±0.12^a	2.93±0.13^{bc}	3.03±0.13^c	2.76±0.14^b	2.45±0.15^a	2.34±0.13^a
ACP /(U/g prot)	1.08±0.07^a	1.24±0.02^{bc}	1.31±0.05^c	1.19±0.07^b	1.15±0.07^b	1.21±0.06^{bc}
AKP /(U/g prot)	1.59±0.08^a	1.61±0.05^a	1.79±0.07^b	1.78±0.04^b	1.74±0.06^b	1.74±0.07^b

Notes: Values (mean ± SD) (n=3) with the different letters in the same line are significantly different at $P < 0.05$. T-AOC is the abbreviation of total antioxidant capacity, T-SOD is the abbreviation of total superoxide dismutase, CAT is the abbreviation of catalase, MDA is the abbreviation of malondialdehyde, ALT is the abbreviation of alanine aminotransferase, AST is the abbreviation of aspartate aminotransferase, ACP is the abbreviation of acid phosphatase, and AKP is the abbreviation of alkaline phosphatase.

5.4　Discussions

5.4.1　Effects of dietary methionine on growth of juvenile sea cucumbers

Methionine is one of the essential amino acids for animals. Lack of methionine in feed can significantly reduce the growth of animals (Felipe et al., 2016; Zhang et al., 2019). In this experiment, the survival rates of juvenile sea cucumbers were not affected by dietary methionine, but weight gain rate and specific growth rate significantly increased with dietary methionine. It was proved that the growth of sea cucumbers could be promoted by dietary methionine. But with the increase of dietary methionine, the growth decreased significantly, which indicated that dietary appropriate methionine was necessary for sea cucumbers. Growth inhibition could be caused by dietary lower or upper concentration of methionine, which were in agreement with other species (Millamena et al., 1996; Klatt et al., 2016). In addition to acting as a substrate for protein synthesis, methionine is the most important methyl-donor in vivo (Courtney-Martin and Pencharz, 2016). Otherwise, the toxic of methionine caused by the accumulation of S-adenosylmethionine will affect the growth of animals (Baker, 2006). Meanwhile, methionine, as with lysine, is a precursor to the synthesis of carnitine, which is involved in the transport of fatty acids into the mitochondria for oxidation (Walton et al., 1982; Ala et al., 2019). A distinct oversupply of methionine + cystine can lead to toxic effects, like metabolic acidosis in weanling rates (Wamberg et al., 1987) or a hepatic coma in dogs (Merino et al., 1975). So it is one of key factors that optimum methionine intake can ensure the normal growth of animals.

Sea cucumber is one of aquatic animals whose weight gain rate is affected greatly by husbandry, especially under the experimental condition. It was reported that the weight gain rate varied from 5% to 50% (Yuan et al., 2006; Gao et al., 2011; Wang et al., 2015a; Chen et al., 2018b) during experiment period. The data obtained under the same experimental condition are reliable.

Overfeeding is a common feeding husbandry in sea cucumber culture, and part of diet was dissolved in water or lost with flow. FCR of sea cucumber was not accurate, but it can predict the efficiency of diets. There were a few reports on the FCR about sea cucumber, such as Xia et al. (2012a) (5.25-6.24), Yuan et al. (2006) (40-50), and Liao et al. (2017) (2.23-4.01). In this trial, FCR varied from 6.48 to 18.33. Dietary methionine increased the feed efficiency greatly.

5.4.2 Effects of dietary methionine on proximate composition of body wall of juvenile sea cucumbers

Moisture contents were not affected by dietary methionine levels, which were in agreement with other researches (Huang and Lin, 2002; Xie et al., 2018; Wang et al., 2019). It has been reported that both crude protein and crude lipid in carcass were affected significantly by dietary methionine levels, such as yellow tail *Seriola quinqueradiata* (Ruchimat et al., 1997), Indian catfish *Heteropneustes fossilis* (Ahmed, 2014), rainbow trout *Oncorhynchus mykiss* (Alami-Durante et al., 2018) and cobia *Rachycentron canadum* (Wang et al., 2016). Generally, dietary methionine can affect growth and lipid metabolism via GCN2 pathway (Wang et al., 2016) and increase the expression of IGF-1(Espe et al., 2016; Shan et al., 2017). But there were no effects on crude protein and crude lipid contents in carcass of softshell turtle *Pelodiscus sinensis* (Huang and Lin, 2002) and white shrimp *Litopenaeus vannamei* (Niu et al., 2018; Wang et al., 2019). This may be related to the experimental treatment, especially the dietary methionine levels, and may also be related to the experimental animal species. More studies are needed to demonstrate the physiological regulation by dietary methionine for different animals.

Regression analysis illustrated that optimal dietary methionine for juvenile sea cucumber was 0.58% diet (2.76% dietary protein) and 0.72% diet (3.43% dietary protein), respectively. It was higher than most fish species (Mai et al., 2006; Chu et al., 2014; Elmada et al., 2016) and soft turtle (Huang and Lin, 2002), but close to the tiger shrimp (Millamena et al., 1996), calculated by % dietary protein. Numerous reports have illustrated that nutrients requirement was affected by animal species, size, age (Ko et al., 2009; Wang et al., 2015a; Li et al., 2020) and nutrients form (Powell et al., 2017; Niu et al., 2018). But more importantly, the ingest rates of sea cucumber and tiger shrimp were much slower than that of fish, some of methionine were lost into water, so the data obtained from this trial may be higher than it's actually need.

5.4.3 Effects of dietary methionine on amino acids profiles and pattern of essential amino acids of juvenile sea cucumbers

Most of amino acids were not affected by dietary methionine levels, except methionine, cystine and aspartic acid in the present study. Similar results had been reported on yellow catfish *Pelteobagrus fulvidraco* (Elmada et al., 2016), Chinese sucker *Myxocyprinus asiaticus* (Chu et al., 2014) and large yellow croaker *Pseudosciaena crocea* (Mai et al., 2006). Both methionine and cystine were sulfur-containing amino acid, and the activated form of methionine, S-adenosylmethionine, can be hydrolyzed to release adenosine and homocystine which ultimately formed cystine (Brosnan et al., 2007; Grillo and Colombatto, 2008). Methionine can be spared when dietary oversupply cystine (Goff and Gatlin III, 2004). Aspartic acid plays an important role in the citric acid cycle during which other amino acids and biochemical, such as asparagine, arginine, lysine, methionine, threonine, and isoleucine, are synthesized (Bellamy, 1961). With the increasing of dietary methionine, more cystine and aspartic acid were accumulated or spared in body wall.

Cysteine is considered dispensable because it can be synthesized by the fish from the indispensable amino acid methionine. Thus, the dietary requirement for methionine is determined either in the absence of cysteine or with test diets containing very low levels of cysteine. Till now, there was no information on dietary cystine requirement of sea cucumbers, and it was hard to control dietary cystine levels in formula. It can be concluded from the variations of cystine of body wall, dietary methionine levels can be decreased if there were enough cystine in diet.

Amino acids balance in diets is necessary for growth of animals. Little information is available for the optimum dietary amino acid pattern. Until now, it is derived from the whole body protein essential amino acid (EAA) pattern (Kaushik, 1998). Alam et al. (2002) reported the influences of different dietary amino acids patterns on growth responses and the whole body protein-bound amino acid composition of the juvenile Japanese flounder *Paralichthys olivaceus*. In the present study, essential amino acids patterns were drawn based on the methionine contents and it can be concluded that dietary methionine had no effects on the pattern of essential amino acids of body wall of sea cucumbers. In other words, it is difficult to change the pattern of essential amino acids of body wall of sea cucumber. The pattern can be recommended to feed producers.

5.4.4 Effects of dietary methionine on enzymes activities of juvenile sea cucumbers

Intestinal tract is the main place where nutrients can be digested and absorbed for sea cucumbers, and digestive capacity is mainly regulated by digestive enzymes. It was reported that dietary optimal methionine can stimulate the digestive enzymes activities in many other species, such as Atlantic salmon (*Salmo salar*) (Nordrum et al., 2000), grass carp (*Ctenopharyngodon idella*) (Jiang et al., 2016), Chinese mitten crab (*Eriocheir sinensis*) (Sun et al., 2013) and red swamp crayfish (*Procambarus clarkii*) (Zhu et al., 2014). In the present study, dietary optimal methionine can stimulate the activities of protease and lipase, hence increase the growth, which was in accordance with the results obtained from growth trial. Well-balanced indispensable amino acids (IAA) profiles can reduce the dietary protein requirement for aquatic animals (Oliva-Teles, 2012), and dietary optimal methionine can stimulate the protein retention (Abidi and Khan, 2011; Abimorad et al., 2009). Besides genetic variation, the ability to utilize food resources is related to the growth rate limitations (Blier et al., 1997). With the increase of dietary methionine levels, amino acids balance was broken down, and protease activities decreased slightly. Every tissue possesses the methionine cycle. Therefore, each can synthesize AdoMet, employ it for transmethylation, hydrolyze AdoHcy, and remethylate homocysteine (Finkelstein, 1998). The methylation of homocysteine will be inhibited by the increase of methionine concentration in intestinal cells, and damage of DNA methylation of cells will be caused by the methylation of homocysteine, thus the expression of intestinal protein will be significantly reduced (Waterland, 2006). Nordrum et al. (2000) has reported that methionine supplementation increased fat digestibility, which was in agreement with the lipase variations in this study.

Methionine can act as a scavenger which protects cells from oxidative stress by oxidation of its sulphur to sulfoxide and repairing methionine sulfoxide through methionine sulfoxide reductases (Campbell et al., 2016). Therefore, additional dietary methionine supplement may contribute to the antioxidant status of Jian carp (*Cyprinus carpio*) (Xiao et al., 2012). Amino acid residues in proteins represent one of the major targets of reactive oxygen species (ROS) and cellular oxidants, while the sulfur-containing amino acids, cysteine and methionine, are especially sensitive to ROS mediated oxidation (Brot and Weissbach, 2000; Stadtman and Berlett, 1999).

Oxidations of cysteine and methionine residues in proteins are noteworthy in that the reactions can be physiologically reversible. The specific oxidation and reduction of methionine residues in proteins is expected to have profound consequences for protein function and may constitute a mechanism for protein regulation (Hoshi and Heinemann, 2001). In this study, dietary optimal methionine enhanced the scavenging oxygen free radicals and antioxidant capacity of sea cucumbers.

Both transaminase and phosphatase are important physiological metabolizing enzymes in sea cucumbers. Transamination represents one of the main pathways for synthesis and deamination of amino acids, thereby allowing interplaying between carbohydrate and protein metabolism during the fluctuating energy demands of the organism in various adaptive situations (Babalola et al., 2009). The changes of AST and ALT activities may contribute to a disturbance in the Krebs cycle (Salah El-Deen and Rogers, 1993). Phosphatase is an enzyme that accelerates the hydrolysis and synthesis of organic esters of phosphoric acid and the transfer of phosphate groups to other compounds, and is closely related to nutrient metabolism (Prescott and Topham, 2013; Ghosh et al., 2013). In the present study, dietary optimal methionine (0.57%-0.78%) can enhance the metabolism of amino acids and promote the assimilation of lipid, calcium and phosphorus.

5.5 Conclusions

Methionine can increase the growth, enhance the accumulation of protein of body wall, and increase the antioxidant capacity of juvenile sea cucumbers. Optimum dietary methionine for juvenile sea cucumbers was 0.58%-0.72% diet (2.76%-3.43% dietary protein).

PART II : Requirement of vitamin E of growing sea cucumber *Apostichopus japonicus*

5.6 Introduction

Vitamin E is one of the essential fat-soluble nutrients, which acted as inter- and intracellular antioxidant to maintain homeostasis of labile metabolites in the cell and tissue plasma (Halver, 2002). They can donate their phenolic hydrogen to lipid-free radicals, thus neutralizing the radical, terminating the autocatalytic lipid peroxidation processes and protecting cell membranes (Liebler, 1993; Kamal-Eldin and Appelqvist, 1996). Furthermore, the resulting tocopherol radicals can be reconverted to the corresponding tocopherol by reacting with other antioxidants such as ascorbate or glutathione (Fukuzawa et al., 1982). Additionally, vitamin E is very important for normal function of the immune system (Pekmezci, 2011). In aquatic animals, vitamin E is also thought to affect disease resistance and health through modulation of the immune responses (Trichet, 2010). It was reported that the vitamin E requirement measured in different fish species ranges from 25 mg to 150mg all-rac-a-TOAc kg/dry diet (Roem et al., 1990; Sau et al., 2004; Lin and Shiau, 2005; Peng et al., 2009). Wan et al. (2004) has reported that dietary 50IU/ kg vitamin E and 0.6mg/kg Se can keep the antioxidant enzymes in a balanced level for abalone *Haliotis discus hannai* INO (Li et al., 2020).

Sea cucumber, *Apostichopus japonicus* Selenka, is one of valuable commercial species cultured in Northeast China. The outcome of cultured sea cucumber was 219,907 metric ton in 2017. Many researchers have focused on its nutrition requirement, such as protein and lipid (Seo and Lee, 2011; Liao et al., 2017), carbohydrate (Xia et al., 2015a), L-ascorbyl-2- polyphosphate (Okorie et al., 2008; Ren et al., 2016), and riboflavin (Okorie et al., 2011). Besides, Ko et al. (2009) and Wang et al. (2015b) reported that the vitamin E requirements for fry (body weight 1.48g) and juvenile (body weight 7.96g) were 23.1-44mg/kg diet and 114.7mg/kg diet respectively. It seemed that the requirement of vitamin E increased with the increasing of body weight. In this study, an experiment was conducted to investigate the vitamin E requirement for growing sea cucumbers.

5.7 Materials and methods

5.7.1 Diet preparation

The composition of the basal diet that used in this trial is shown in Table 5-8. The basal diet included 23.72% protein and 4.69% lipid (Seo and Lee, 2011; Liao et al., 2017), meanwhile there were 484mg/kg vitamin C and 0.72mg/kg selenium in basal diet. Graded levels of dl-a-tocopheryl acetate (purity 500g/kg, Zhejiang NHU Company Ltd., China) were added into basal diet to formulate six isonitrogenous and isoenergetic experimental diets containing vitamin E 6.7mg/kg, 81.2mg/kg, 159.3mg/kg, 237.8mg/kg, 314.6mg/kg, 395.9mg/kg respectively, which were named as D1, D2, D3, D4, D5, D6 group.

Table 5-8　Composition and nutrient levels of experimental diets

	Items	Groups					
		D1	D2	D3	D4	D5	D6
Ingredients	Fish meal /%	5	5	5	5	5	5
	Fermented soybean meal /%	10	10	10	10	10	10
	Wheat flour /%	23	23	23	23	23	23
	Shrimp meal /%	15	15	15	15	15	15
	Kelp powder /%	6	6	6	6	6	6
	α-starch /%	2	2	2	2	2	2
	Fish oil /%	1	1	1	1	1	1
	Soybean lecithin /%	1	1	1	1	1	1
	Vitamin premix /%	0.5	0.5	0.5	0.5	0.5	0.5
	Vitamin E /(mg/kg)	0	160	320	480	640	800
	Mineral premix /%	0.5	0.5	0.5	0.5	0.5	0.5
	Sea mud /%	36	36	36	36	36	36
Proximate composition	Crude protein /%	23.72	23.5	23.5	23.37	23.18	23.57
	Crude lipid /%	4.69	4.32	4.48	4.56	4.59	4.84
	Crude ash /%	42.69	42.75	42.76	42.68	42.82	42.78
	Energy /(kJ/g)	14.64	14.70	14.67	14.87	14.66	14.63
	Vitamin E /(mg/kg)	6.7	81.2	159.3	237.8	314.6	395.9
	Vitamin C /(mg/kg)	484	491	487	489	485	490
	Selenium /(mg/kg)	0.72	0.71	0.67	0.71	0.69	0.69

Notes: ①Raw materials were supplied by Shengsuo Feed Company (Shandong, China), fish meal (crude protein, 71.69%; crude lipid, 7.36%), fermented soybean meal (crude protein, 50.17%; crude lipid, 0.81%), shrimp meal (crude protein, 60.17%; crude lipid, 10.11%), kelp powder(crude protein; 19.53%, crude lipid, 2.02%). ② Vitamin premix (mg/100g): vitamin A, 380; vitamin D_3, 132; thiamin, 1150; riboflavin, 3800; pyridoxine HCl, 880; pantothenic acid, 3680; niacin acid, 10,300; biotin, 100; folic acid, 200; vitamin B_{12}, 13; inositol, 40,000; ascorbic acid, 5000. All ingredients were diluted with α-cellulose to 100g. ③ Same contents of mineral premix as reference Wang et al. (2014).

All solid ingredients were grounded into powder (0.075mm), and mixed thoroughly. Fresh fish oil was added into the powder and mixed sufficiently, and then appropriate distilled water was added and made them into dough. Feed pellets were prepared using a moist pelletizer, and dried at 60℃ for 24h in an oven and grounded into desirable particle sizes (0.18-0.25mm). All diets were packed and stored at −20℃ (Xia et al., 2015a).

5.7.2　Experimental design

Growing sea cucumbers were obtained from a local commercial aquaculture farm in Penglai, China, and transported into the laboratory at 4℃. Prior to the beginning of the experiment, growing sea cucumbers were reared in a recirculating aquatic system for 15 days to acclimate to the experimental conditions. All sea cucumbers were fed diet with 6.7mg/kg vitamin E during this period.

At the beginning of the trial, sea cucumbers were starved for 48h and weighted. Sea cucumbers with similar body weight (15g) were randomly distributed into eighteen 300L blue cylindrical fiberglass tanks (70cm × 80cm) and each tank was stocked with 30 growing sea cucumbers. Each diet was randomly assigned to triplicate tanks. Water depth was kept at 60cm, and the feeding trial lasted for 8 weeks. There were two culture baskets with 20 corrugated boards in each tank, and the basket was replaced every 20 days. During this period, temperature ranged from 16.9-18.4℃, water flow rate was kept at 2L/min, salinity was held at 26-28, pH was maintained at 7.6-8.2, dissolved oxygen was approximately 6mg/L, and the total ammonia nitrogen was maintained at less than 0.05mg/L. The sea cucumbers were fed once at 16:00 and up to 2% of their initial total biomass per day. Uneaten diet and feces were siphoned out of the culture system every 3 days. The trial was performed in low light (～150lx).

5.7.3　Sampling collection and chemical analysis

At the end of the experiment, the sea cucumbers were starved for 48h before harvest. Total number and body weight were measured for calculating survival rate (SR), weight gain rate (WGR) and specific growth rate (SGR). Eight sea cucumbers per tank were randomly selected and put on clean foam board for 5min, and body weight was measured after absorbing the water from the body surface with filter paper. Intestinal tract and body wall were separated on ice plate after removal of body fluid. Intestine was wiped and weighed after washing with cold distilled water. All samples were stored at −80℃ for further analyses.

Proximate composition analyses on experimental diets and body wall samples were performed using standard methods (AOAC, 1990). Moisture was determined by drying in an oven at 105℃ overnight. Crude protein ($N \times 6.25$) was determined by the Kjeldahl method after acid digestion. Crude lipid was determined by the ether-extraction method. Vitamin E contents in diets were determined by high-performance liquid chromatography (Agilent 1200, USA). Vitamin C contents in diets were analyzed by reverse-phase high-performance liquid chromatography (Agilent 1100LC, USA). Selenium contents in diet and body wall were digested in concentrated nitric acid and analyzed by inductively coupled plasma mass spectrometry (ICP-MS,

Agilent, 7700). Vitamin E contents in body wall were analyzed by spectrophotometer method (Hitachi U3900H, Japan). Hydroxyproline contents in body wall were analyzed by amino acids auto analyzer (Hitachi L-8900, Japan).

Intestine was homogenated with 9 fold cold physiological saline after cutting in ice bath. Homogenate solution was collected by centrifugation at 8000 ×g for 10min at 4℃ using a high-speed refrigerated Centrifuge (Hitachi Crg Series, Japan). The supernatant was separated into 2ml centrifugal tube and stored in refrigerator at −80℃ for further analysis.

Protease, lipase, amylase, total superoxide dismutase (T-SOD), catalase (CAT) and malondialdehyde (MDA) were measured by the commercial kits provided by Nanjing Jiancheng Bioengineering Institute, Nanjing, China in accordance with the instructions of the manufacturer. Alkaline phosphatase (AKP), acid phosphatase (ACP), glutamic oxaloacetic transaminase (GOT), lactate dehydrogenase (LDH), and pyruvate kinase (PK) were analyzed by the commercial kits provided by Shanghai Enzyme-linked Industrial Co., Ltd., Shanghai, China. The protein contents in serum were determined by coomassie brilliant blue method using commercial kit (Nanjing Jiancheng Bioengineering Institute, Nanjing, China).

5.7.4　Statistical analysis

Data from the trial were subjected to one-way analysis of variance (ANOVA), using the SPSS program version for windows (SPSS Inc., Chicago, IL, USA). When the ANOVA identified differences among groups, multiple comparisons among means were made with Duncan's multiple-range test at $P < 0.05$. The results were presented as means ± SD.

The optimal dietary vitamin E requirement of sea cucumbers was estimated with a broken-line regression using weight gain rate and vitamin E contents in body wall as the dependent variable against dietary methionine level using Statistical Analysis System 9.2 (SAS institute, USA) NLIN regression.

5.8　Results

5.8.1　Growth

At the end of the trial, there were no differences in survival rate (SR) among all groups (Table 5-9). Weight gain rate (WGR) and specific growth rate (SGR) were presented the trend of increasing first and then kept stable when dietary Vitamin E was higher than 159.3mg/kg. The ratios of intestine weight to body weight (IBR) increased with the increasing of the dietary vitamin E levels.

Table 5-9 Effects of dietary vitamin E on growth of growing sea cucumbers *Apostichopus japonicus* Selenka

Growth performances	Groups						P-value
	D1	D2	D3	D4	D5	D6	
IBW /g	15.36±0.04	15.45±0.11	15.41±0.11	15.34±0.05	15.43±0.03	15.41±0.03	0.886
FBW /g	20.61±0.06^a	20.78±0.05^a	21.35±0.05^b	21.64±0.09^c	21.46±0.06^{bc}	21.26±0.06^b	0.014
WGR /%	34.11±0.12^a	34.50±0.15^a	38.46±0.20^b	40.31±0.29^c	39.38±0.56^{bc}	38.63±0.59^b	0.001
SGR /(%/d)	0.49±0.00^a	0.49±0.00^a	0.54±0.00^b	0.56±0.00^c	0.55±0.01^b	0.55±0.01^b	0.000
IBR /%	3.16±0.41^a	3.61±0.50^{ab}	3.77±0.49^{ab}	4.15±0.48^b	4.18±0.94^b	3.60±0.48^{ab}	0.040
SR /%	85.78±3.67	85.11±6.19	88.33±7.07	84.44±1.92	83.33±0.00	82.00±2.83	0.152

Notes: Values (mean ± SD) (*n*=3) with the different letters in the same row are significantly different at $P < 0.05$. IBW (g), initial body weight; FBW (g), final body weight; WGR (weight gain rate, %)=100×(FBW−IBW)/IBW; SGR (specific growth rate, %/d)=100×[(lnFBW−lnIBW)/56]; IBR (the ratio of intestine weight to body weight, %)= 100× intestine weight/body weight; SR (survival rate, %)=100×final fish number/initial fish number.

SAS NILN analysis based on weight gain rate and dietary vitamin E levels showed that the requirement of vitamin E for growing sea cucumbers was 187.2mg/kg (Figure 5-3).

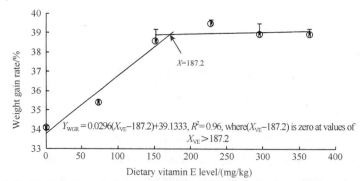

Figure 5-3 SAS NILN analysis based on weight gain rate and dietary vitamin E level. The requirement of vitamin E for growing sea cucumbers *Apostichopus japonicus* Selenka was 187.2mg/kg diet.

5.8.2 Proximate composition of body wall

The effects of dietary vitamin E on proximate composition of body wall of growing sea cucumbers are presented in Table 5-10. There were no significant differences among all groups in the moisture, crude protein and crude ash contents of body wall. Crude lipid contents significantly increased with dietary vitamin E.

Hydroxyproline contents of body wall increased firstly and decreased afterwards with the increasing of dietary vitamin E levels. Vitamin E and selenium contents of body wall were increased significantly with dietary vitamin E.

Table 5-10 Effects of dietary vitamin E on proximate composition of body wall of growing sea cucumbers *Apostichopus japonicus* Selenka (wet weight)

Body compositions	Groups						P-value
	D1	D2	D3	D4	D5	D6	
Moisture /%	93.32±0.01	92.32±0.01	92.33±0.01	92.33±0.02	92.34±0.01	92.32±0.01	0.201
Crude protein /%	3.50±0.16	3.51±0.10	3.51±0.05	3.50±0.06	3.47±0.05	3.52±0.12	0.269
Crude lipid /%	0.25±0.01^a	0.32±0.01^{bc}	0.34±0.0^c	0.34±0.00^c	0.35±0.01^c	0.30±0.01^b	0.014
Crude ash /%	2.86±0.09	2.85±0.01	2.78±0.11	2.71±0.04	2.86±0.02	2.87±0.03	0.311
Hydroxyproline /%	0.43±0.03^a	0.47±0.05^{ab}	0.53±0.06^{bc}	0.67±0.06^d	0.55±0.04^c	0.57±0.06^c	0.000
Vitamin E /(mg/kg)	21.77±0.73^a	27.07±0.73^b	33.97±0.20^c	34.90±0.57^c	32.99±1.15^c	34.80±0.44^c	0.000
Selenium /(mg/kg)	0.27±0.00^a	0.32±0.01^b	0.34±0.00^c	0.35±0.01^{cd}	0.36±0.00^d	0.39±0.02^e	0.010

Notes: Values (mean ± SD) (n=3) with the different letters in the same row are significantly different at $P < 0.05$.

SAS NILN analysis based on vitamin E contents of body wall and dietary vitamin E levels showed that the requirement of vitamin E for growing sea cucumbers was 165.2mg/kg (Figure 5-4).

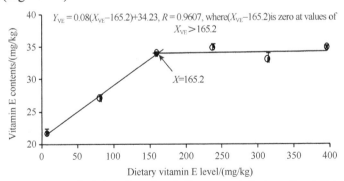

Figure 5-4 SAS NILN analysis based on vitamin E contents in body wall and dietary vitamin E level. The requirement of vitamin E for growing sea cucumbers *Apostichopus japonicus* Selenka was 165.2mg/kg diet

5.8.3 Digestive, antioxidant and metabolic enzymes

Effects of dietary vitamin E on digestive enzymes activities of intestine of growing sea cucumbers are listed in Table 5-11. Protease activities were not affected by dietary vitamin E, but lipase activity of the highest dosage group was significantly lower than that of 81.2mg/kg-237.8mg/kg groups, whereas amylase activities increased firstly and then decreased, and the maximum value was achieved in 237.8mg/kg treatment. Both lipase and amylase activities were abruptly decreased in the highest dietary vitamin E group.

Table 5-11   Effects of dietary vitamin E on digestive enzymes activities of intestine of growing sea cucumbers *Apostichopus japonicus* Selenka (unit: U/mg prot)

Digestive enzyme	Groups						P-value
	D1	D2	D3	D4	D5	D6	
Protease	138.46±1.26	139.83±2.33	141.12±0.74	138.07±6.85	144.39±3.00	144.72±6.49	0.862
Lipase	35.95±0.67^{ab}	37.90±0.28^{b}	36.97±0.61^{b}	36.93±0.43^{b}	36.12±0.77^{ab}	34.03±0.98^{a}	0.014
Amylase	0.72±0.02^{a}	0.78±0.02^{ab}	0.79±0.02^{ab}	0.99±0.11^{c}	0.90±0.01^{bc}	0.75±0.01^{ab}	0.000

Notes: Values (mean ± SD) (*n*=3) with the different letters in the same row are significantly different at $P < 0.05$.

Effects of dietary vitamin E on antioxidant enzymes activities of intestine of growing sea cucumbers are listed in Table 5-12. Total superoxide dismutase (T-SOD) activities were evaluated firstly and then dropped, and MDA contents decreased with the increasing of vitamin E levels, whereas catalase (CAT) activities were not affected by dietary vitamin E.

Table 5-12   Effects of dietary vitamin E on antioxidant enzymes activities of intestine of growing sea cucumbers *Apostichopus japonicus* Selenka

Antioxidant enzyme	Groups						P-value
	D1	D2	D3	D4	D5	D6	
T-SOD /(U/mg prot)	156.26±15.06^{a}	195.45±7.08^{b}	233.51±6.20^{c}	210.13±20.78^{bc}	205.56±16.52^{bc}	193.39±3.48^{b}	0.002
CAT /(U/mg prot)	21.43±0.19	24.76±0.03	24.78±0.53	26.63±0.31	25.58±0.04	25.47±0.90	0.075
MDA /(nmol/mg prot)	7.97±0.10^{d}	7.47±0.22^{c}	3.70±0.01^{b}	2.31±0.07^{a}	2.15±0.03^{a}	2.07±0.07^{a}	0.000

Notes: Values (mean ± SD) (*n*=3) with the different letters in the same row are significantly different at $P < 0.05$.

Effects of dietary vitamin E on metabolic enzymes activities of intestine of growing sea cucumbers are listed in Table 5-13. Both alkaline phosphatase (AKP) and acid phosphatase (ACP) activities increased firstly and decreased afterwards with the increasing of dietary vitamin E levels, and reached the maximum at 237.8mg/kg and 237.8mg/kg groups, respectively. Pyruvate kinase (PK) activities increased firstly and decreased afterwards with dietary vitamin E. Lactate dehydrogenase (LDH) activities increased with dietary vitamin E, but there were no differences among adding groups. Glutamic oxaloacetic transaminase (GOT) activities fell firstly and evaluated afterwards, and got minimum values at 237.8mg/kg group.

Table 5-13　Effects of dietary vitamin E on metabolic enzymes activities of intestine of growing sea cucumbers *Apostichopus japonicus* Selenka (unit: U/g prot)

Metabolic enzyme	Groups						P-value
	D1	D2	D3	D4	D5	D6	
AKP	1339.38±49.43[a]	2518.78±193.05[b]	2633.13±164.60[b]	4106.01±206.34[d]	3862.38±50.48[d]	3135.39±184.58[c]	0.000
ACP	1083.39±53.73[a]	1173.44±23.69[b]	1358.85±57.40[c]	1635.89±104.85[d]	1532.80±16.62[d]	1455.49±6.23[c]	0.000
PK	110.51±7.30[a]	130.73±11.65[bc]	146.30±1.10[c]	147.55±8.99[c]	114.98±3.79[ab]	104.97±5.90[a]	0.004
LDH	12.36±1.06[a]	14.41±0.46[ab]	14.54±0.11[ab]	14.81±0.47[b]	15.89±1.32[b]	15.46±1.31[b]	0.036
GOT	1.95±0.10[e]	1.84±0.03[d]	1.26±0.07[b]	1.10±0.02[a]	1.63±0.07[c]	1.81±0.07[d]	0.000

Notes: Values (mean ± SD) ($n=3$) with the different letters in the same row are significantly different at $P < 0.05$.

5.9　Discussions

5.9.1　Growth and body wall compositions

Many researches have illustrated that dietary optimum vitamin E can stimulate the growth performances of aquatic animals (Huang and Lin, 2015; Wassef et al., 2015). Meanwhile, growth can be inhabited by dietary overdose vitamin E (Huang et al., 2003). In this trial, weight gain rate (WGR) and specific growth rate (SGR) presented the trend of first rising and then keeping stable, which were in agreement with hybrid tilapia, *Oreochromis niloticus*×*Oreochromis aureus* (Huang et al., 2003) and grass carp, *Ctenopharyngodon idellus* (Li et al., 2014). With WGR as an evaluating indicator, dietary optimum vitamin E for growing sea cucumbers was 187.2mg/kg diet. It can be concluded from Ko et al. (2009), Wang et al. (2015b) and our trial that the requirement of vitamin E increased with the increasing of animal's size and age, which was contrast to the protein requirement of common dentex, *Dentex dentex* L. (Skalli et al., 2004). There was no vessel-like transporting system in sea cucumbers, and lipids were exposed to oxygen directly, so that more antioxidants were needed for protecting against the damage caused by free radicals, meanwhile, this may be the reason why the dietary lipid requirement for sea cucumbers was much lower than fishes (Seo and Lee, 2011; Liao et al., 2017).

Besides the size and age of animals, it is important to consider the interactions with other feed components, such as Vitamin C (Yildirim-Aksoy et al., 2008; Gao, 2013; Izquierdo et al., 2019), selenium (Zhang et al., 2011; Mahdi et al., 2017) and probably other minerals necessary for activity of the antioxidant enzymes, when measuring the vitamin E requirement (Hamre, 2011). So the contents of vitamin C and selenium were determined in this trial, and there were no differences among all groups.

Crude protein and crude ash contents were not affected by dietary vitamin E levels, which was in accordance with previous research on meager (*Argyrosomus regius*) (Rodríguez et al., 2017). Chaiyapechara et al. (2003) reported that protein and lipid contents in fillet of rainbow trout (*Oncorhynchus mykiss*) were only affected by the dietary lipid levels, and no related with dietary vitamin E contents or the interaction between them. But carcass crude protein of rohu (*Labeo rohita*) evaluated and crude lipid of meager decreased with dietary vitamin E levels. Trenzado et al. (2009) illustrated that lipid deposition occurred with dietary deficiency of HUFA, independent of vitamin E contents. It can be concluded from previous reports that lipid deposition has nothing to do with vitamin E content, and it was hard to attribute the lipid deposition to vitamin E supplementation, considering the fact that there were no differences among the sea cucumbers fed diets containing 81.2mg/kg to 314.6mg/kg vitamin E. More findings and theories were needed to explain this phenomenon.

Meanwhile, it is very interesting that there were similarities and differences in the body wall composition among the reports on the requirement of vitamin E of sea cucumbers (Ko et al., 2009; Wang et al., 2015b). Crude protein contents were lower and crude ash contents were higher in Seo et al.'s reports than our trial, but the crude lipid contents were elevated by dietary vitamin E in both trials. Vitamin E contents increased with the increasing of dietary vitamin E in Seo et al.'s reports, but there was a plateau between dietary 48.5mg/kg to 196.4mg/kg vitamin E, which is in agreement with this study (Seo and Lee, 2011). Growth performances of all three trials increased firstly and then reached a plateau in a certain range. The biggest differences among these three trials were the amount of additions and the growth stage of experimental animals. Generally, the nutrients requirements of fish decreased with increasing size and age, but the requirement of vitamin E for sea cucumbers increased with increasing size.

Hydroxyproline and selenium were also determined in this trial, and both of them increased with dietary vitamin E levels. Hydroxyproline is one of non-essential amino acids that is derived from proline. It is created by the interaction of vitamin C and proline. Shiau and Hsu (2002) reported that high supplementation level of ascorbate could spare vitamin E in diets for hybrid tilapia. Vitamin E also could downregulate activity of the enzyme gulonolactone oxidase, involved in vitamin C synthesis in sturgeon (Moreau and Dabrowski, 2003). Vitamin E could spare vitamin C or improve the efficiency of vitamin C in sea cucumbers. Besides that, vitamin E and selenium can prevent the collagen degradation in vivo (Åsman et al., 1994). Selenium is one of the elements that can play an important role in removing free radicals. Vitamin E protects fatty acids and selenium protects the thiol group of membranes in red blood cells (Chitra et al., 2014). Recent reports on yellowtail kingfish (*Seriola lalandi*) showed that fillet Se and vitamin E were only significantly responsive to dietary Se and vitamin E, respectively (Le et al., 2014). It may be the reason why selenium deposition rate in body wall was slow down in this trial, but more evidence was needed to prove it.

5.9.2 Digestive, antioxidant and metabolic enzymes

Digestive enzymes are very important for the digestion of food, which can promote the hydrolysis of sugar, lipid and protein and break them into micro-molecule. There is no evidence that vitamin E is involved in the regulation of digestive enzymes secretion, but the activity of many enzymes is regulated by reversible oxidation and reduction, which may be affected by vitamin E (Hamre, 2011), and more researches are needed to elucidate the function of vitamin E.

Reactive oxygen species (ROS) are generated as by-products of cellular metabolism. It may cause damage to cellular macromolecules such as lipids, protein, and DNA may ensue when cellular production of ROS overwhelms its antioxidant capacity (Apel and Hirt, 2004). Total superoxide dismutase (T-SOD) and catalase (CAT) are important components of antioxidant enzyme system in organism and their activities reflect the body's antioxidant capacities. Malondialdehyde (MDA) results from lipid peroxidation of polyunsaturated fatty acids. The production of this aldehyde is used as a biomarker to measure the level of oxidative stress in an organism (Draper and Hadley, 1990). There were numerous reports that vitamin E can stimulate the antioxidant capacity of aquatic animals (Liu et al., 2007; Li et al., 2013a; Lu et al., 2016; Shahkar et al., 2018; Wang et al., 2019). T-SOD activities increased significantly firstly and then decreased slightly afterwards with dietary vitamin E levels, which were in accordance with T-SOD activities change in body wall of juvenile sea cucumbers (Wang et al., 2015b). There were no definite trend on the variations of CAT activities, and CAT activities were not affected by dietary vitamin E levels in this study. Generally, there was a positive correlation between SOD and CAT (Liu et al., 2007; Lu et al.,

2016; Shahkar et al., 2018). But it can be concluded that the antioxidant effect of vitamin E on superoxide was better than that of peroxide in sea cucumbers. In other words, vitamin E improved the capability of antioxidant by improving superoxide resistance.

Phosphatase is one kind of enzyme which can remove phosphate groups from the substrate molecule by hydrolyzing phosphate monoesters, and generating phosphate ions and free hydroxyl groups. The increasing of phosphatase activity reflected the increasing of phosphorus metabolic activity in cells. Optimum dietary vitamin E can stimulate the activities of AKP, which increased the digestion of nutrients (Uauy et al., 1990), such as lipid and carbohydrate (illustrated in Table 5-11).

Pyruvate kinase and lactate dehydrogenase are two enzymes which are involved in glycolysis, and pyruvate kinase is one of key rate-limiting enzymes (Wang et al., 2007). Increased activities of two enzymes reflected the increased utilization of glucose by sea cucumbers, which were in accordance with the changes of amylase.

Glutamic oxaloacetic transaminase (GOT) is one of the most important aminotransferases, which is closely related to the metabolism of amino acids in the body. It seemed that dietary vitamin E depressed the metabolism of amino acids, but more evidences are needed to prove that. It can be concluded from the changes of metabolic enzymes that optimum dietary vitamin E can stimulate the activities of digestion and utilization of enzymes involved in lipid and carbohydrate.

5.10 Conclusions

Vitamin E can improve antioxidant capability by increasing the ability to resist superoxide rather than peroxide. Optimum dietary vitamin E for growing sea cucumbers was 165.2-187.2mg/kg diet.

PART III: Optimal dietary phosphorus requirement for juvenile sea cucumber *Apostichopus japonicus*

5.11 Introduction

Japanese sea cucumber (*Apostichopus japonicus*), belonging to the phylum Echinodermata, class Holothuroidea, is mainly distributed off the coasts of the North Pacific Ocean. It is an economically important species as a source of seafood and ingredient in traditional medicine (Song et al., 2023). Recently, the increasing market demand and prices for 'bêche-de-mer' or 'trepang' led to overexploitation of wild stocks worldwide and stimulated the development of commercial aquaculture of *Apostichopus japonicus* (Purcell et al., 2012). Traditionally, *Apostichopus japonicus* is farmed on the ocean bottom with natural food resources. This aquaculture method has a low environmental impact, but it has low yield, high risk and long breeding cycle. Some effective culture models, such as indoor industrial culture, pond culture and floating raft culture, have been developed, providing greater worldwide production for the sea cucumbers trade (Yuan et al., 2006). According to the recent FAO report, global aquaculture production of *Apostichopus japonicus* has more than doubled since 2005, increasing from 57,200 tonnes to 201,500 tonnes in 2020, with production growing at an annual average rate of 16.8% since then (FAO, 2022). In China, *Apostichopus japonicus* aquaculture has become one of the most profitable marine culture industries, and the annual production has reached 196,500 tons in 2020, accounting for approximately 97.5% of the global production according to data from the China Fishery Statistical Yearbook.

The increase in *Apostichopus japonicus* aquaculture output is largely dependent upon the supply of aquaculture feed including natural seaweed and artificial feed. As a deposit feeder, sea cucumbers mainly feed on sedimentary organic matter including bacteria, protozoa, diatoms, and detritus of plants or animals in nature (Yingst, 1976). However, natural food is relatively insufficient in intensive aquaculture practice, and artificial feed is generally supplemented to *Apostichopus japonicus* to meet its requirement (Slater et al., 2009; Yuan et al., 2006). Therefore, information about the quantitative nutrient requirement is required for the formulation of artificial feed. In the last twenty years, efforts towards evaluation of quantitative requirement focused on several nutrients using different nutritional assessment parameters. These studies demonstrate that sea cucumbers have a low protein (11%-17%) and lipid (1.3%) requirement (Seo and Lee, 2011; Liao et al., 2014; Bai et al., 2016) but high carbohydrate requirement (48.56%-49.30%) (Xia et al., 2015a). In addition, several studies indicated that sea cucumbers also required specific vitamins and amino acids for maximizing its growth, including ascorbic acid (100-105.3mg/kg; Okorie et al., 2008), α-tocopherol (23.1-44.0mg/kg; Ko et al., 2009), riboflavin (9.73-17.9mg/kg; Okorie et al., 2011), lysine (0.58%-0.72%; Liu et al., 2017a), methionine (0.76%-1.19%; Li et al., 2020). Up to now, there have been no reports published on phosphorus requirement of *Apostichopus japonicus*.

Phosphorus is not only a vital structural component of cell membranes and nucleic acids, but also participates in many biological processes that are essential for the survival, growth, and development of aquatic animals (Sugiura et al., 2004, 2011; Witten et al., 2016). In general, aquatic animals can obtain most of the minerals they require (calcium, sodium, and potassium) directly from the water (especially in sea water). Phosphorus, however, is generally found at low concentration in natural waters (Lall, 2003). Therefore, absorption of sufficient amounts of phosphorus from water is unlikely, making a dietary source essential for most aquatic animals (Chavez-Sanchez et al., 2000). A few of studies have assessed dietary phosphorus requirements for aquatic animals: 1.59%-1.6% for crab *Portunus trituberculatus* (Zhao et al., 2021), 1.0%-1.2% for abalone *Haliotis discus hannai* (Tan et al., 2002), 1.0% for tiger shrimp *Penaeus monodon* (Ambasankar et al., 2006), 0.86%-0.9% for Japanese seabass, *Lateolabrax japonicas* (Zhang et al., 2006), 0.96% for haddock *Melanogrammus aeglefinus* (Roy and Lall, 2003), 0.76% for yellow catfish *Pelteobagrus fulvidraco* (Luo et al., 2010), and 0.50%-0.87% for Siberian sturgeon *Acipenser baerii* (Xu et al., 2011). These results indicate dietary phosphorus requirements vary across species, and the phosphorus requirement of invertebrate is higher than vertebrate.

Some evidences show that sea cucumbers can obtain available phosphorus from sediment by facilitating the transformation of organic phosphorus to inorganic phosphorus in nature (Hou et al., 2018). However, in intensive cultivation such as land-based culture and floating cage culture, most of the phosphorus intake comes from diets including algal debris or artificial feed. Any excess of phosphorus in the diet above the minimum requirement for sea cucumbers will be excreted. The excess of this element in the effluents of aquaculture systems leads to eutrophication and an adverse effect on the aquatic ecosystems. Therefore, it is critical to know precisely the dietary requirement of phosphorus in order to minimize excess phosphorus in feed without risking phosphorus deficiency in sea cucumbers. In a fiberglass aquaculture system, we estimated the phosphorus requirement of sea cucumbers by investigating the effect of dietary phosphorus on growth, diet utilization, whole body composition, digestion, and metabolism as well as oxidation resistance.

5.12 Methods and materials

5.12.1 Preparation of coated sodium dihydrogen phosphate

Carrageenan was used to coat sodium dihydrogen phosphate to reduce phosphorus loss in water due to its low solubility loss (Slater et al., 2011). The coated sodium dihydrogen phosphate was prepared using the methods of Li et al. (2020) with some modifications. In brief, sodium dihydrogen phosphate was dissolved in distilled water and then equivalent carrageenan was added, following a complete mixing into paste. The mixture was heated in a water bath (95℃) and stirred continuously for 30min. The coated sodium dihydrogen phosphate was freeze-dried, ground (80 mesh), and then stored at –20℃.

5.12.2 Diet preparation

Six experimental diets were formulated to contain different levels of phosphorus (0.24%, 0.37%, 0.51%, 0.62%, 0.77%, and 0.89%) by adding coated sodium dihydrogen phosphate (Table 5-14), which were named as D1, D2, D3, D4, D5, D6 group. Seaweed powder (crude protein, 19.5%; crude fat, 0.82%; carbohydrate, 47.8%), fish meal (crude protein, 67.3%; crude fat, 8.4%), shrimp powder (crude protein, 58%; crude fat, 14%), and yeast powder (crude protein, 53%; crude fat, 3.6%) were ground and sieved (80 mesh). All dry ingredients were weighed in the proportions as presented in Table 5-14, and then mixed thoroughly (20min) in a Patterson-Kelley twin shell® Batch V-mixer (Patterson-Kelley Co., Inc., East Stroudsburg, PA). The coated sodium dihydrogen phosphate and vitamin premix were added by the progressive enlargement method. Subsequently, fish oil and distilled water were appended to the dry ingredients until homogenous in mixer, and then cold-extruded and sliced into strips (1cm×0.5cm×0.08cm) (MZLP400, Anyang Jimke Energy Machinery Co., Ltd). These strip feeds were heated at 90℃ for 30min, air-dried until moisture was lower than 6%, then kept in vacuum-packed bags and stored at –20℃.

Table 5-14 Ingredients and proximate composition of the experimental diets

Items	Groups					
	D1	D2	D3	D4	D5	D6
Ingredients (dry basis)						
Seaweed powder /%	23.5	23.5	23.5	23.5	23.5	23.5
Fish meal /%	8	8	8	8	8	8
Shrimp powder /%	3	3	3	3	3	3
Yeast powder /%	0.5	0.5	0.5	0.5	0.5	0.5
Fish oil /%	1	1	1	1	1	1
Carrageenan /%	5	5	5	5	5	5
Microcrystalline cellulose /%	8	7	6	5	4	3
Sodium dihydrogen phosphate /%	0	1	2	3	4	5

Continued

Items	Groups					
	D1	D2	D3	D4	D5	D6
Sea mud /%	50	50	50	50	50	50
Vitamin premix /%	1	1	1	1	1	1
Proximate composition (Measured value)						
Crude protein /%	12.44	12.45	12.63	12.78	12.52	12.66
Crude fat /%	1.94	1.92	1.98	1.90	1.91	1.92
Ash /%	56.17	56.03	56.24	56.17	56.26	56.14
Total phosphorus /%	0.24	0.37	0.51	0.62	0.77	0.89
Energy /(kJ/kg)	10.65	10.70	10.72	10.74	10.75	10.78

5.12.3 Feeding trial

The feeding trial was conducted in a hatchery (Dongying, China). Sea cucumbers were obtained from Shandong Anyuan Aquaculture Co. Ltd (Penglai, China). After acclimation to the experimental conditions for 2 weeks, 900 sea cucumbers with an average initial weight of 9.99g±0.02g were randomly assigned into 18 fiberglass tanks (L=100cm, W=50cm, H=80cm). Each diet was randomly fed to sea cucumbers in triplicate tanks for 10 weeks. The feeding amount was 10% of wet weight daily (recalculated and adjusted every two weeks) in order to collect enough feces for apparent digestibility and fecal production rate as soon as possible. Feces and uneaten feed residues were collected by siphoning before the next feeding and dried at 60℃ to a constant weight for further analysis (Yuan et al., 2006). Some intact feces were picked out for further digestibility analysis. The weight of uneaten feed was assessed by the leaching ration of diets in water (Shi et al., 2015). During the trail, the water temperature was maintained at (18.0±0.5)℃ by seawater source heat pump system (Zhuoren Air Conditioning Equipment Co., Ltd, Shandong, China). The pH was maintained at 7.1±0.1 by adjusting the water exchange rate (300%/d), and dissolved oxygen was kept higher than 6.0mg/L by oxygen pump (2HB520-7HH57, Weisida Electromechanical Co., Ltd, Huzhou, China). These water quality parameters were monitored daily during the trial.

5.12.4 Sample collection and growth calculation

At the end of the feeding trial, sea cucumbers in each tank were bulk-weighed. Then, 13 sea cucumbers from each tank were weighed individually and dissected. Intestines were weighed and quickly frozen in liquid nitrogen. Pooled intestines of each replicate tank were homogenated with physiological saline solution (0.7% NaCl) and then centrifuged at 8,000r/min for 10min. The supernatant was divided into aliquots of 400ul in 1.5ml centrifugal tube and stored at −80℃ until the enzyme activity analysis.

The growth parameters and diet utilization were calculated according to the following formulas:

Weight gain (WG, g) = final weight (g) − initial weight (g);

Specific growth rate (SGR, %/d) = (ln final weight − ln initial weight)/days of experiment × 100;

Daily feed intake (DFI, %/d) = dry weight of consumed feed(g) /[(initial weight + final weight)/2 ×days of experiment] × 100;

Daily phosphorus intake (DPI, %/d)= Daily feed intake × phosphorus content;

Fecal production rate (FPR, %/d)= Fecal production /[days×(initial weight+ final weight)/2];

Feed efficiency (FE, %) = weight gain (g) / feed supplied (g) × 100;

Survival rate (SR, %) = final sea cucumber number / initial sea cucumber number × 100;

Apparent digestibility coefficient of dry material (AD_m, %) = (1 − dietary acid − insoluble ash content/fecal acid − insoluble ash content) × 100;

Apparent digestibility coefficients of protein (AD_p, %), fat (AD_f, %), energy (AD_e, %), and phosphorus (AD_{pi}, %) =100−100×(fecal nutrient content × dietary acid-insoluble ash content)/(dietary nutrient content ×fecal acid-insoluble ash content).

5.12.5 Proximate composition analysis

Proximate compositions of diets, body walls and feces were analyzed according to the standard methods of AOAC (1990). Moisture was determined by drying samples at 105℃ for 2h in oven (Binder FD-S56, BINDER GmbH, German). Crude protein ($N×6.25$) was determined using Kjeldahl nitrogen analyzer (Kjeltec 8100, FOSS Analytical Co., Ltd, Denmark) following a acid digestion (DT 220 Digestor, FOSS Analytical Co., Ltd, China). Crude fat was analyzed by the ether extraction method using fat analyzer (SOX-406, Jinan Hanon Instruments Co., Ltd, China). Crude ash was determined using a muffle furnace (Linder/blue M1100, Thermo Fisher Scientific Co., Ltd, China) at 550℃ for 6h. Total energy was measured with a automatic calorimeter (IKA C6000, Aika Instrument and Equipment Co., Ltd, Guangzhou). Acid-insoluble ash was determined according to the method of Khalil et al. (2021). Total phospholipids were separated from total lipid with a silica gel-based solid-phase extraction colum and then quantitation was performed by hydrophilic interaction HPLC coupled to evaporative light-scattering detection using a quaternary separation method (Ferraris et al., 2020). The samples of whole body and diets were digested with HNO_3 in microwave oven (ZUOT-SYS-WBL Shanghai Satian Precision Co., Ltd, China), and total phosphorus contents were analyzed using the vanadomolybdophosphoric acid method with a spectrophotometer (U-3900H, Hitachi Co., Ltd, Japan) set at a wavelength of 420nm (Tanner et al., 1999).

5.12.6 Activity analyses of intestinal enzymes

The activities of acid and alkaline proteases were assayed using the Folin-Ciocalteu's reagent according to methods of Cui et al. (2015). Lipase activity was determined according to the method of Massadeh and Sabra (2011) using p-nitrophenyl palmitate (pNPP) (Sigma, USA) as substrate. Amylase activity was analyzed by measuring absorbance value change of starch-iodine reaction solution at the wavelength of 660nm according to the method of Al-Qodah et al. (2007). The ALP and ACP activities were estimated by mixing crude enzyme extract with a reaction of disodium phenyl phosphate buffer at pH 10.5 and 4.9, and followed by estimation of absorbance of the resultant chromogenic solution at 520nm, respectively (Kind and King, 1954).

Superoxide dismutase (SOD) activity was measured with a microplate reader according to the method of Peskin and Winterbourn (2000). One unit (U/mg prot) of SOD activity is defined as the amount of enzyme in 1mg tissue protein that inhibits the rate of reduction of cytochrome c by 50% in a coupled system, using xanthine and xanthine oxidase at pH 7.8 at 37℃. Catalase (CAT) activity was determined using a spectrophotometric assay of hydrogen peroxide based on formation of its stable complex with ammonium molybdate as described in detail by Góth (1991). Glutathione peroxidase (GSH-Px) activity was measured by DTNB (5,5'-dithiobis-(2-nitrobenzoic acid)) method as described by Fukuzawa and Tokumura (1976). One unit (U/mg prot) of GSH-Px activity is defined as the amount of enzyme that catalyzes the oxidation by H_2O_2 of 1.0μmol of reduced glutathione to oxidized glutathione per minute at pH 7.0 at 37℃. Malondialdehyde (MDA) content was determined using the acid extraction TBA (thiobarbituric acid) method as reported by Lynch and Frei (1993) and results were expressed as nmol MDA per mg of tissue protein (nmol/mg prot). GSH content was determined spectrophotometrically according to the method of Drukarch et al. (1996), and the content was expressed as μmol GSH per gram of tissue protein (μmol/g prot).

Protein content was determined by the method of Lowry et al. (1951).

5.12.7　Statistical analysis

The Software SPSS 12.0 microcomputer software package (SPSS, Chicago, IL, USA) was used for all statistical evaluations. A homogeneity test for variance was conducted. All data were subjected to one-way analysis of variance (ANOVA) followed by Tukey's test. Differences were regarded as significant when $P<0.05$. Data are expressed as mean and standard deviation with pooled SE. Nonlinear regression analysis was used to describe the relationship between growth, diet utilization and whole-body phosphorus content.

5.13　Results

5.13.1　Growth performance

As summarized in Table 5-15, there was no difference in survival rate (SR) among dietary treatments ($P>0.05$). Final weight (FW), weight gain (WG), and specific growth rate (SGR) significantly increased with dietary phosphorus level increasing from 0.24% (D1) to 0.62% (D4), but decreased thereafter ($P<0.05$). The higher FW, WG and SGR were observed in *Apostichopus japonicus* fed 0.51% and 0.62% dietary phosphorus than in those fed other diets ($P<0.05$). *Apostichopus japonicus* receiving diet containing 0.24% phosphorus showed the lowest WG and SGR ($P<0.05$). Application of quadratic regression analyses to the WG and SGR provided an estimate of 0.63% dietary phosphorus for optimum growth, with the predicted of the maximum WG (11.39g) and SGR (1.09%/d) (Figure 5-5).

Table 5-15　Growth performance of *Apostichopus japonicus* fed experimental diets with different phosphorus levels

Items	Groups						Pooled SE
	D1	D2	D3	D4	D5	D6	
IW /g	9.99±0.02	9.99±0.05	10.01±0.05	9.94±0.19	10.00±0.00	10.01±0.04	0.018
FW /g	16.47±0.38^a	19.31±0.13^b	21.02±0.25^c	21.60±0.93^d	20.14±0.30^c	19.56±0.19^b	0.408
WG /g	6.47±0.37^a	9.32±0.12^b	11.01±0.27^d	11.65±0.85^d	10.14±0.30^c	9.55±0.20^{bc}	0.408
SGR /(%/d)	0.71±0.03^a	0.94±0.01^b	1.06±0.02^d	1.11±0.05^d	1.00±0.02^c	0.96±0.02^{bc}	0.031
SR /%	87.78±3.85	92.22±6.94	91.11±6.94	90.00±3.33	91.11±3.33	90.00±5.77	1.094

Notes: Values in the same row with different superscript letters show significant difference ($P<0.05$).

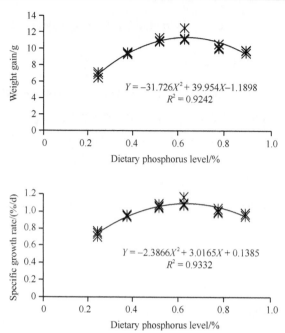

Figure 5-5 Quadratic regression analyses of weight gain and specific growth rate in sea cucumbers fed diets with dietary phosphorus level ($n=3$). The predicted dietary phosphorus level is 0.63% for maximum growth.

5.13.2 Diet utilization

Diet utilization was presented in Table 5-16. Apparent digestibility coefficient of protein (AD_p) and apparent digestibility coefficient of fat (AD_f) were not affected by dietary phosphorus levels ($P>0.05$). Daily feed intake (DFI) and fecal production rate (FPR) increased to the peak values (2.55%/d and 2.08%/d, respectively) as dietary phosphorus increased from 0.24% (D1) to 0.51% (D3), and then both decreased ($P<0.05$). However, apparent digestibility coefficient of dry material (AD_m) and apparent digestibility coefficient of energy (AD_e) followed an opposite tendency with the lowest AD_m (17.56%) and AD_e (25.85%) were observed in *Apostichopus japonicus* fed 0.51% (D3) phosphorus. Feed efficiency (FE) significantly increased to 0.41 with dietary phosphorus increasing up to 0.62%, and then reached a plateau within 0.41-0.43. Daily phosphorus intake (DPI) gradually increased while apparent digestibility coefficient of phosphorus (AD_{pi}) decreased with the increase of dietary phosphorus ($P<0.05$). The quadratic regression analysis was applied to illustrate the DFI response, predicting dietary phosphorus of 0.55% for maximum DFI. The piecewise regression analyses were applied to FE and AD_{pi}, locating the breakpoint (0.57% and 0.66%) of dietary phosphorus levels (Figure 5-6).

Table 5-16 Diet utilization of *Apostichopus japonicus* fed diets with different phosphorus levels

	Groups						Pooled SE
	D1	D2	D3	D4	D5	D6	
DFI /(%/d)	2.12±0.05^a	2.47±0.10^c	2.55±0.06^c	2.54±0.05^c	2.26±0.04^b	2.16±0.04^{ab}	0.051
FPR /(%/d)	1.65±0.17^{ab}	2.01±0.06^c	2.08±0.06^c	2.04±0.11^c	1.77±0.08^b	1.57±0.08^a	0.052
FE	0.33±0.01^a	0.37±0.02^a	0.40±0.01^b	0.41±0.02^{bc}	0.43±0.00^c	0.43±0.00^c	0.009
DPI /(10^{-2}%/d)	0.51±0.01^a	0.91±0.04^b	1.30±0.03^c	1.58±0.03^d	1.74±0.03^e	1.92±0.04^f	0.118
AD$_{pi}$ /%	33.70±1.18^c	32.52±0.62^{bc}	31.15±0.72^b	30.92±2.14^b	26.51±0.78^a	25.05±1.44^a	0.800
AD$_m$ /%	18.74±1.58^{ab}	18.26±1.25^{ab}	17.56±1.52^a	19.16±0.91^{ab}	19.98±1.48^{ab}	20.73±0.68^b	0.359
AD$_p$ /%	22.76±1.36	22.06±1.27	23.35±0.77	23.78±1.99	22.51±2.08	22.83±1.79	0.347
AD$_f$ /%	34.24±3.70	32.92±1.52	31.78±1.42	33.50±1.18	34.09±2.45	34.63±1.32	0.480
AD$_e$ /%	29.47±0.83^{bc}	28.51±0.98^{ab}	25.85±0.71^a	27.99±0.87^{ab}	30.90±3.19^{bc}	32.35±2.23^c	0.557

Notes: Values in the same row with different superscript letters show significant difference ($P<0.05$).

5.13.3 Activities analysis of intestinal enzymes

As summarized in Table 5-17, the activities of acid protease(ACPT), alanine transaminase(ALT), hexokinase(HK), and citrate-synthase(CS) were not affected by dietary phosphorus($P > 0.05$). The activities of alkaline protease(AKPT), aspartate transaminase(AST), and phosphoenolpyruvate carboxykinase(PEPCK) decreased significantly

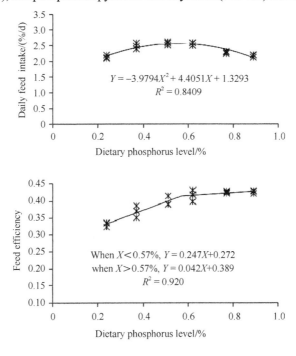

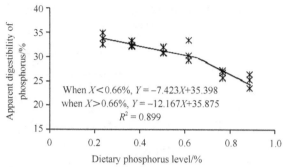

Figure 5-6 The quadratic regression analysis of daily feed intake and piecewise regression analyses of feed efficiency and apparent digestibility of phosphorus in sea cucumbers fed diets with dietary phosphorus level ($n=3$), respectively. The predicted dietary phosphorus levels are 0.55% for the maximum feed intake, 0.57% for the optimal feed efficiency, and 0.66% for optimal phosphorus digestibility.

but pyruvate kinase (PK) activity increased ($P<0.05$) with dietary phosphorus level. The activities of alkaline phosphatase (ALP), phosphofructokinase (PFK), and succinate dehydrogenase (SDH) increased firstly and then decreased, responding to the increasing dietary phosphorus. The highest activities of ALP and SDH were observed at dietary phosphorus of 0.62% and the highest PFK activity was observed at dietary phosphorus of 0.51%. *Apostichopus japonicus* fed 0.51%-0.62% phosphorus had similar ALP activities ($P>0.05$), and *Apostichopus japonicus* fed 0.51%-0.77% phosphorus had similar SDH activities ($P>0.05$), which were significantly higher than those fed other diets ($P<0.05$). The amylase activity increased to 21.32U/mg prot with dietary phosphorus increasing up to 0.62%, but decreased slightly thereafter.

Table 5-17 Intestinal enzymes of sea cucumbers fed diets with different phosphorus levels

Items	D1	D2	D3	D4	D5	D6	Pooled SE
Digestive enzymes							
ACPT /(U/mg prot)	0.56±0.03	0.55±0.06	0.59±0.10	0.49±0.02	0.51±0.02	0.55±0.03	0.013
AKPT /(U/mg prot)	243.60±21.78^c	220.63±6.20^b	199.25±4.86^b	156.35±9.73^a	150.88±12.25^a	138.83±12.26^a	4.317
Amylase /(U/mg prot)	15.97±0.75^a	17.92±0.64^{ab}	19.60±1.29^{bc}	21.32±2.63^c	20.84±1.82^{bc}	19.71±1.79^{bc}	0.546
Lipase /(U/g prot)	4.97±0.83^a	7.61±0.47^b	6.95±0.16^b	7.35±0.57^b	7.44±0.93^b	6.97±1.47^b	0.273
Metabolic enzymes /(U/g prot)							
ALP	52.20±1.80^a	68.08±1.73^b	83.41±1.02^{cd}	84.30±2.64^d	80.60±2.02^c	66.68±1.56^b	2.807
AST /(×10^1)	72.82±5.51^d	63.73±2.50^{cd}	53.66±4.69^b	56.26±6.74^{bc}	46.87±4.20^{ab}	39.94±6.73^a	2.800

Continued

Items	\multicolumn{6}{c}{Groups}	Pooled SE
	D1	D2	D3	D4	D5	D6	
ALT /($\times 10^2$)	38.58±2.44	37.65±3.88	39.49±2.57	40.81±2.18	38.57±3.78	37.83±0.96	61.512
HK	67.72±3.42	67.40±4.86	65.70±4.73	63.45±5.02	64.21±1.90	62.91±3.93	0.878
PK	32.62±5.12^a	33.52±2.28^a	33.83±4.04^a	42.46±1.83^b	43.76±2.77^b	45.72±5.40^b	1.518
PFK	8.09±2.07^a	9.21±0.86^{ab}	12.01±1.18^b	11.96±2.80^b	10.90±1.18^{ab}	10.57±0.68^{ab}	0.417
PEPCK	42.16±6.59^c	40.83±7.24^c	36.12±6.13^{bc}	26.39±4.89^{ab}	26.16±5.22^{ab}	24.35±1.56^a	2.087
CS	5.32±0.44	5.51±1.31	4.88±0.65	4.84±0.53	5.14±0.41	4.96±0.33	0.149
SDH	5.16±0.81^a	5.31±0.34^a	5.90±0.34^b	6.35±0.35^b	6.25±0.63^b	5.23±0.20^a	0.153

Notes: Values in the same row with different superscript letters show significant difference ($P<0.05$).

5.13.4 Proximate composition of whole body

As presented in Table 5-18, there were no significant differences in the contents of crude protein, crude fat, and crude ash of whole body among all groups ($P>0.05$). The contents of phospholipid and whole-body phosphorus increased with the dietary phosphorus increasing up to 0.51% ($P<0.05$), and fluctuated within 0.56%-0.59% and 1.96%-2.02mg/g respectively as dietary phosphorus further increased ($P>0.05$). A piecewise regression analysis was used to describe the relationship between whole-body phosphorus content and dietary phosphorus level, with the predicted optimal phosphorus level of 0.55% for the whole-body phosphorus deposition (Figure 5-7).

Table 5-18 Proximate composition of sea cucumbers fed diets with different phosphorus levels

Items	Groups						Pooled SE
	D1	D2	D3	D4	D5	D6	
Crude protein /%	45.42±0.65	44.70±0.43	44.36±0.38	44.61±0.58	45.06±0.19	45.51±0.67	0.126
Crude fat /%	4.04±0.09	4.06±0.12	4.05±0.08	4.07±0.09	4.13±0.08	4.16±0.08	0.021
Phospholipid /%	0.46±0.02^a	0.52±0.02^b	0.56±0.01^c	0.58±0.01^c	0.58±0.01^c	0.59±0.01^c	0.011
Crude ash /%	35.15±0.45	34.90±0.12	35.04±0.13	35.24±0.23	34.28±0.37	35.35±0.10	0.065
Phosphorus /(mg/g)	1.78±0.05^a	1.88±0.03^b	1.96±0.03^c	2.00±0.02^c	2.01±0.02^c	2.02±0.03^c	0.022

Notes: Values in the same row with different superscript letters show significant difference ($P<0.05$).

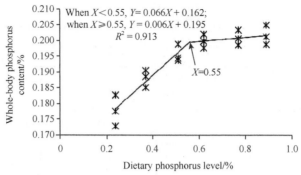

Figure 5-7　A piecewise regression analysis of whole-body phosphorus content in sea cucumbers fed diets with dietary phosphorus level ($n=3$). The predicted dietary phosphorus level is 0.55% for the optimal whole-body phosphorus deposition.

5.13.5　Intestinal oxidation resistance

As presented in Table 5-19, the intestinal of catalase (CAT) activity and malondialdehyde (MDA) content were not affected by dietary phosphorus ($P>0.05$). The activities of glutathione peroxidase (GSH-Px) and superoxide dismutase (SOD) showed a quadratic response to the the incremental increase of dietary phosphorus ($P<0.05$), with the highest activities of GSH-Px (33.52U/mg prot) and SOD (5.26U/mg prot) occurring at dietary phosphorus levels of 0.62% and 0.51% respectively. The contents of glutathione (GSH) and glutathione oxidized (GSSG) exhibited an ascending trend with dietary phosphorus increasing from 0.24% to 0.51%, and then GSSG followed by a significant reduction ($P<0.05$). The contents of GSH and GSSG were highest in *Apostichopus japonicus* fed 0.51% phosphorus, and lowest in those fed 0.24% phosphorus ($P<0.05$). The ratio of GSH/GSSG was significantly elevated in *Apostichopus japonicus* fed diets containing 0.77% and 0.89% phosphorus as compared to those fed other diets ($P<0.05$).

Table 5-19　The intestinal oxidation resistance of sea cucumbers fed diets with different phosphorus levels

Items	Groups						Pooled SE
	D1	D2	D3	D4	D5	D6	
SOD /(U/mg prot)	4.38±0.30^a	4.93±0.09^{bc}	5.26±0.16^c	5.24±0.13^c	4.96±0.31^{bc}	4.67±0.17^{ab}	0.085
CAT /(U/mg prot)	6.11±1.00	5.97±0.72	6.00±1.68	5.86±0.35	5.89±0.14	6.03±0.20	0.108
GSH-Px /(U/mg prot)	26.03±1.26^a	30.72±2.29^{bc}	33.36±1.30^{cd}	33.52±2.38^d	30.74±1.54^{bc}	27.51±2.80^{ab}	0.863
GSH /(μmol/g prot)	27.36±2.05^a	31.75±1.65^b	35.70±2.20^{bc}	34.79±1.00^c	34.47±1.14^{bc}	34.31±1.49^{bc}	0.691

Continued

Items	Groups						Pooled SE
	D1	D2	D3	D4	D5	D6	
GSSG /(μmol/g prot)	73.88±1.32^a	83.78±10.53^b	107.44±13.34^c	91.34±12.79^{bc}	79.87±7.71^{ab}	85.21±17.55^{bc}	4.232
GSH/GSSG	0.37±0.03^a	0.35±0.01^a	0.34±0.03^a	0.35±0.01^a	0.45±0.01^b	0.47±0.04^b	0.013
MDA /(nmol/mg prot)	0.83±0.02	0.80±0.07	0.94±0.09	0.95±0.13	0.89±0.10	0.84±0.14	0.024

Notes: Values in the same row with different superscript letters show significant difference ($P<0.05$).

5.14 Discussions

5.14.1 Effects of dietary phosphorus on growth of *Apostichopus japonicus*

During the feeding trial, *Apostichopus japonicus* grew from 9.99g to 16.47-21.60g with an acceptable SGR (0.71-1.11%/d), as compared to the growth data from other studies (Liu et al., 2009; Song et al., 2016; Xia et al., 2017; Li et al., 2020). These positive growth responses fit quadratic regression models, and thus confirm that phosphorus deficiency or excess could hinder the growth of *Apostichopus japonicus* as reported in a previous study on tiger shrimp (Ambasankar et al., 2006). Therefore, an appropriate supply of dietary phosphorus is necessary for *Apostichopus japonicus* in aquaculture. In addition, the estimated dietary phosphorus requirement from these quadratic regression analyses appears lower than the published reports on fish (0.72%-1.57%) (Roy and Lall, 2003; Yang et al., 2006; Tang et al., 2012; Wang et al., 2021), shrimp (1%-2%) (Deshimaru and Yone, 1978; Kanazawa et al., 1984; Davis et al., 1993; Ambasankar et al., 2006), crab (1.59%-1.68%) (Zhao et al., 2021) and abalone (1%-1.2%) (Tan et al., 2002), implicating that sea cucumbers may have a specific phosphorus acquisition mechanism. This is proved by a recent physiological findings indicating the phosphatase exhibited comparable activity levels in the respiratory tree segments of sea cucumbers *Isostichopus badionotus* (Martínez-Milián et al., 2021). Another evidence from an early ^{32}P isotope study also revealed that sea cucumbers absorbed a limited amount of phosphorus from the environment through the integumentary surface and respiratory tree (Ahearn, 1968). The phosphorus obtained by this additional way can not completely meet the growth needs of *Apostichopus japonicus*, but reduce its dietary requirement.

5.14.2　Effects of dietary phosphorus on diet utilization of *Apostichopus japonicus*

Feeding response to dietary phosphorus varies across species, and phosphorus deficiency symptom generally manifests as appetite loss and low feed efficiency (Sugiura et al., 2000, 2004; Roy and Lall, 2003). Therefore, in the present study feeding diet containing 0.24% phosphorus resulted in low daily feed intake and feed efficiency. However, it was noted that with dietary phosphorus increasing, DFI and FPR showed a quadratic variation while FE showed a piecewise linear variation, fitting different regression models. This indicated that dietary phosphorus exerted different effects on the DFI, FPR and FE. Aquatic animal can control the nutritional equilibrium by altering their feeding response including feeding intake and feed efficiency, as reported on gibel carp *Carassius auratus gibelio* (Xie et al., 2017) and sea cucumbers *Australostichopus mollis* (Zamora and Jeffs, 2011). In the present study, the feeding response result indicated that *Apostichopus japonicus* could sense body phosphorus status and control phosphorus intake by regulating feed intake. According to quadratic regression analysis of DFI, dietary phosphorus of 0.55% was optimal for maximum feed intake. In addition, the change point of slope estimated by a piecewise linear regression analysis of FE represented a change of body phosphorus status, suggesting that *Apostichopus japonicus* feeding >0.57% dietary phosphorus could allocate more phosphorus for supporting growth than that fed low phosphorus.

Apparent digestibility is not only as an estimate of digestive efficiency. Generally, there is a trade-off between intake and digestive efficiency. Higher diet intake equates to faster digesta transit, which can result in a lower digestive efficiency (German, 2011). Therefore, the reduction of AD_m and AD_e of *Apostichopus japonicus* in the present study may result from high feed intake, which agree with the report on turtle *Pelodiscus sinensis* (Wang et al., 2022). In addition, apparent digestibility was also indicator of the balance of nutrient deposition and excretion. In the present study, AD_{pi} decreased proportionately with dietary phosphorus increasing, which was consistent with the findings on largemouth bass *Micropterus salmoides* (Wang et al., 2021). However, a chang point (0.66% dietary phosphorus) was detected by a piecewise linear regression model in the relationship via a change in slope. This change point represents the threshold of dietary phosphorus, above which body phosphorus status of *Apostichopus japonicus* is altered and the proportion of fecal phosphorus becomes higher. This agrees with the studies on rainbow trout (Coloso et al., 2003) and tiger shrimp (Ambasankar et al., 2006), and suggests that phosphorus excretion occurs when the phosphorus intake is above the requirement level. Therefore, dietary phosphorus should be controlled within the range of 0.57%-0.66%.

5.14.3 Effects of dietary phosphorus on intestinal enzymes of *Apostichopus japonicus*

The whole digestion process primarily relies on the types and activities of digestive enzymes (Sveinsdóttir et al., 2006; Nazemroaya et al., 2015). Several studies reported that *Apostichopus japonicus* could modulate their digestive enzyme activities in response to different diet qualities, suggesting its digestive flexibility (Liao et al., 2015; Wen et al., 2016a; Song et al., 2016). In the present study, dietary phosphorus stimulated the activities of lipase and amylase, which agreed with the results reported on coho salmon *Oncorhynchus kisutch* (Xu et al., 2021), and red swamp crayfish *Procambarus clarkia* (Xu et al., 2022). However, intestinal protease exhibited a descending trend as dietary phosphorus increased, suggesting that dietary phosphorus reduced protein digestion. Induction and secretion of phosphatase is one of the important adaptive responses of the animal to low phosphorus status. Studies with abalone (Tan et al., 2002) and turtle *Pelodiscus sinensis* (Wang et al., 2022) demonstrated that the tissue activity of alkaline phosphatase is negatively correlated with dietary phosphorus levels. However, the present study showed that dietary phosphorus-stimulated intestinal phosphatase activity as dietary phosphorus level increased up to 0.62%. This stimulatory effect may be associated with the increased feeding intake because feed intake is considered as a major driver of alkaline phosphatase activity diatoms (Lallès, 2019), and the debris from plants or animals.

In line with the above mentioned protease, the activities of two intestinal transaminases of *Apostichopus japonicus* decreased by dietary phosphorus in the present study. Similar finding was also reported on catfish *Silurus asotus* (Yoon et al., 2015), suggesting that dietary phosphorus reduced oxidation of amino acids. On the other hand, dietary phosphorus significantly stimulated the activities of PK and PFK but reduced PEPCK activity, which agreed with the recent findings on blunt snout bream *Megalobrama amblycephala* (Yang et al., 2021) and swimming crab *Portunus trituberculatus* (Zhao et al., 2021), indicating an enhanced glycolysis coupling with a depressed gluconeogenesis. In addition, the membrane-bound succinate dehydrogenase (SDH) contributes to the establishment of the mitochondrial membrane potential and ATP synthesis (van Vranken et al., 2014). This elevated SDH activity in *Apostichopus japonicus* fed 0.37%-0.62% phosphorus suggested that appropriate amount of dietary phosphorus facilitated the tricarboxylic acid cycle and energy production. Taken collectively, dietary phosphorus altered *Apostichopus japonicus* intestinal metabolism allowing more energy produced for growth, which may explain the growth result.

5.14.4 Effects of dietary phosphorus on proximate composition of *Apostichopus japonicus*

The whole-body phosphorus content has been commonly used as an indicator of dietary phosphorus status. Signs of phosphorus deficiency were generally characterized by low whole-body phosphorus (Tang et al., 2012; Braga et al., 2016). Our results showed the content of whole-body phosphorus significantly increased in response to the increasing dietary phosphorus, which was consistent with the research results on yellow catfish *Pelteobagrus fulvidraco* (Tang et al., 2012), pejerrey fingerlings *Odontesthes bonariensis* (Rocha et al., 2014), stinging catfish *Heteropneustes fossilis* (Zafar and Khan, 2018), largemouth bass (Wang et al., 2021) and swimming crab (Zhao et al., 2021). The piecewise regression analysis located the breakpoint of whole-body phosphorus at a dietary phosphorus of 0.57%, lower than the growth requirement of 0.63%. This may be attributed to a fact that a dynamic "phosphorus pool" exists in organisms, playing an important role in controlling phosphorus homeostasis (Berndt et al., 2005). Sea cucumbers preferentially fill the "phosphorus pool" with exogenous phosphorus and then allocate for growth.

Phospholipid is an essential component of the cell membrane and predominates in the composition of lipids in sea cucumbers. The increase in phospholipids represents an important adaptive strategy for marine invertebrates in resilience to environmental stress (Li et al., 2019; Hu et al., 2022; Zhao et al., 2022). In the present study, dietary phosphorus prompted the biosynthesis of phospholipids, and thus explained the increased phospholipids content. This contradicted the findings of another study on the growth of Japanese flounder (Uyan et al., 2007), and implicated that sea cucumbers may develop phospholipid synthesis ability. More researches are needed to verify this speculation.

5.14.5 Effects of dietary phosphorus on oxidation resistance of *Apostichopus japonicus*

Reactive oxygen species (ROS) are byproducts of normal mitochondrial metabolism and homeostasis (Zorov et al., 2014). However, excessive ROS can induce cell damage (oxidative stress) (Liemburg-Apers et al., 2015). Sea cucumbers, similarly to other invertebrates, are endowed with efficient ROS-scavenging mechanisms (Du et al., 2013; Yu et al., 2016). Superoxide dismutase (SOD) and glutathione peroxidase (GSH-Px) are two major antioxidant enzymes, involved in response of *Apostichopus japonicus* to environmental stress and nutritional stimulation (Wang et al., 2015a; Huo et al., 2018; Hou et al., 2019). Dietary phosphorus deficiency downregulated the mRNA levels and activities of antioxidant enzymes in fish (Chen et al., 2017). In the present study, the activities of SOD and GSH-Px as well as GSSH concentration were significantly elevated by the moderate level (0.37%-0.77%) of dietary phosphorus, which agreed with the findings in juvenile Jian carp (Feng et al., 2013), grass carp (Wen et al., 2015) and juvenile snakehead (Shen et al., 2017). This suggests that appropriate amount of dietary phosphorus promotes syntheses of SOD and GSH-Px, protecting the intestine from free radicals and keeping MDA at a low level. Therefore, dietary phosphorus levels should be controlled within the range of 0.37%-0.77% without negative impacts on *Apostichopus japonicus*.

5.15 Conclusions

In summary, it is necessary to provide appropriate dietary phosphorus in *Apostichopus japonicus* aquaculture, though *Apostichopus japonicus* has a relatively low requirement. Dietary phosphorus deficiency hindered the growth of sea cucumbers, but excessive phosphorus increased fecal phosphorus excretion which exerted a potential negative impact on the aquaculture environment. Considering the results in the current study, the optimal dietary phosphorus for sea cucumbers was 0.57%-0.63%. This provides an important reference for feed formulators to develop nutritionally balanced commercial diets that promote optimal growth and health of sea cucumbers with the minimal impact on the environment. In addition, the present result in terms of diet utilization provides a reference for calculating the maximum tolerable daily or weekly feed consumption/intake, to reduce feed waste. In the future, a strategy for improving phosphorus utilization in commercial feeds for sea cucumbers is necessary, not only for economic but also for environmental reasons.

PART IV: Application of hydrolyzed soybean meal in feed of sea cucumber *Apostichopus japonicus*

5.16 Introduction

Sea cucumber *Apostichopus japonicus* is an economically important aquatic animal and has long been exploited as an important fishery resource in China, Russia,

Japan, Republic of Korea and Democratic People's Republic of Korea (Sloan, 1984; Yang et al., 2005) for its high quality meat and medicinal purposes. Wild stocks have been heavily fished in many areas around the world, and as a result of overfishing, wild stock of sea cucumbers has dropped in recent years (Choo, 2008). To satisfy growing demand for sea cucumber products, sea cucumber aquaculture boosted and many sea cucumber farming patterns have been developed, such as farming in shrimp ponds, in offshore ponds, and in cement pond in green house. Global aquaculture production for *Apostichopus japonicus* increased from 34,100 tonnes in 2003 to 193,941 tonnes in 2013, with an annual increasing rate of 1.05% (FAO). However, the rapid development of sea cucumbers culture imposes great pressure on artificial feed production (Song et al., 2016).

As a typical deposit-feeding species, sea cucumbers *Apostichopus japonicus* preferentially inhabit areas of coarse sediments and complex bottom topography that contain large flourishing algae, the rich detritus of which provides sea cucumbers with their main organic nutrients (Gavrilova and Sukhin, 2011). It ingests sediment bearing organic matter, including bacteria, protozoa, diatoms, and the debris from plants or animals. Sea cucumbers have a clear preference for seaweed especially *Sargassum thunbergii* (Xia et al., 2012b), thus powdered macroalgae is generally used as the main component of formulated feeds (Liu et al., 2010). However, the rapid development of sea cucumbers feed industry has led to the massive macroalgal powder input, causing overexploitation of several seaweed species and increase of feed cost. Therefore, finding alternative feed ingredients to reduce the use of seaweed powder in feed is an important challenge to increase the sustainability of sea cucumbers culture.

Hydrolyzed soybean meal (HSBM) is a mixture of amino acids, oligopeptides and polypeptides, which allow a broader range of nitrogen utilization. Compared with soybean meal, enzymatic hydrolysis of soybean meal improves its functional and nutritional properties (Hrčková et al., 2002). Short-chain peptides from this hydrolyzed protein have a higher nutritive value and may be utilized more efficiently than an equivalent mixture of free amino acids (Grimble et al., 1986). Although there is no published study on the replacement of seaweed with hydrolyzed soybean meal in the artificial feed of the sea cucumbers, but recent studies indicated that appropriate inclusion of hydrolyzed soy protein could improve growth and health of flatfish (Mamauaga et al., 2011; Song et al., 2014).This study aimed to elucidate the utilization of hydrolyzed soybean meal by sea cucumbers and determine the maximum and optimal quantity of seaweed powder replacement by hydrolyzed soybean meal in diets of sea cucumbers without any negative effects on their growth and diet utilization.

5.17 Materials and methods

5.17.1 Diet formula

HSBM was obtained using soybean meal (SBM) as a substrate according to the method of Song et al. (2014). Compound enzymes (SunHY Biotechnology Co., Ltd., Wuhan, China) containing neutrase (100,000U/g), flavourzyme (50,000U/g) and non-starch polysaccharide enzymes (xylanase, 8000U/g; β-glucanase, 2000U/g; β-mannase,150U/g; cellulose, 300U/g) were employed in the hydrolysis. Protein solubility was determined according to the method of Mamauaga et al. (2011). Protein solubility at pH 7.0 was 58.33% and molecular weight of peptides (Table 5-20) was analysed by China National Analytical Center (Guangzhou, China).

Table 5-20 The information of hydrolyzed soybean meal(HSBM)and soybean meal(SBM)(%)

	HSBM	SBM
Water soluble nitrogen (total nitrogen; pH 7.0, 22℃)	44.6	10.24
Molecular mass distribution of water-soluble nitrogen (total nitrogen)		
<1,000Da	21.81	3.68
1,000-3,000Da	14.00	5.15
3,000-5,000Da	2.05	0.23
5,000-10,000Da	0.45	0.08
>10,000Da	6.29	1.10

Six iso-energetic (11.39kJ/g ± 0.08kJ/g DM) and isoprotein (13.62g/kg±0.05g/kg DM) diets were formulated by replacing 0%, 20%, 40%, 60%, 80% and 100% protein deprived from seaweed powder with graded HSBM inclusion levels of 0% (control), 4.05%, 8.18%, 12.27%, 16.36% and 20.45% based on an equivalent N basis, which were named as D1, D2, D3, D4, D5, D6 group. To test efficiency of HSBM in promoting growth of sea cucumbers *Apostichopus japonicus* in comparison with SBM, 12.27% SBM (S12.27%) was included in seventh diet to replace 60% seaweed powder instead of 12.27% HSBM, which was named as D7 group. Fish oil was used to balance the lipid levels throughout the seven diets. Chromic oxide (0.4% dry weight) was added in experimental diets as an inert marker for digestibility measurement. All ingredients and proximate composition of experimental diets were shown in Table 5-21.

Table 5-21 Ingredients and proximate composition of experimental diets

Items	Groups						
	D1	D2	D3	D4	D5	D6	D7
Ingredients (dry basis)							
Seaweed powder /%	50	40	30	20	10	0	20
Hydrolyzed soybean meal /%	0	4.09	8.18	12.27	16.36	20.45	0
Soybean meal /%	0	0	0	0	0	0	12.27
Microcrystalline cellulose /%	0	5.90	11.80	17.70	23.60	29.50	17.70
Fish meal /%	5	5	5	5	5	5	5
Fish oil /%	0	0.01	0.02	0.03	0.04	0.05	0.03
Wheat meal /%	3	3	3	3	3	3	3
Potato starch /%	6	6	6	6	6	6	6
Alginate /%	6	6	6	6	6	6	6
Sea mud /%	27.6	27.6	27.6	27.6	27.6	27.6	27.6
Minerals[①] /%	1	1	1	1	1	1	1
Vitamins[②] /%	1	1	1	1	1	1	1
Chromic oxide /%	0.4	0.4	0.4	0.4	0.4	0.4	0.4
Proximate composition							
Crude protein /%	13.54	13.56	13.54	13.55	13.53	13.54	13.55
Crude fat /%	0.88	0.87	0.87	0.87	0.88	0.87	0.88
Crude ash /%	46.61	43.76	42.87	40.39	38.24	36.06	39.97
Carbohydrate /%	35.95	37.24	39.19	41.23	43.07	45.52	41.16
Energy /(kJ/g)	11.38	11.31	11.37	11.46	11.51	11.43	11.28
Amino acids composition (dry material)							
Aspartic acid /%	1.38	1.36	1.35	1.36	1.27	1.26	1.35
Threonine /%	0.63	0.60	0.59	0.56	0.51	0.48	0.57
Serine /%	0.69	0.68	0.68	0.68	0.64	0.63	0.69
Glutamic acid /%	2.16	2.24	2.38	2.50	2.48	2.58	2.53
Glycine /%	0.84	0.77	0.73	0.72	0.65	0.61	0.73
Alanine /%	0.86	0.65	0.59	0.56	0.47	0.42	0.56
Cysteine /%	0.16	0.17	0.18	0.19	0.21	0.22	0.19
Valine /%	0.62	0.60	0.58	0.56	0.53	0.51	0.58
Methionine /%	0.03	0.02	0.02	0.02	0.02	0.02	0.02
Isoleucine /%	0.42	0.42	0.41	0.41	0.39	0.39	0.42
Leucine /%	0.97	0.92	0.92	0.91	0.82	0.80	0.92
Tyrosine /%	0.69	0.80	0.85	0.88	0.89	0.89	0.88
Phenylalanine /%	0.60	0.59	0.59	0.57	0.54	0.53	0.59
Lysine /%	0.52	0.56	0.59	0.62	0.64	0.67	0.63
Histidine /%	0.13	0.15	0.18	0.21	0.23	0.25	0.22
Arginine /%	0.60	0.60	0.63	0.67	0.63	0.65	0.67

Continued

Items	Groups						
	D1	D2	D3	D4	D5	D6	D7
Proline /%	0.59	0.58	0.57	0.59	0.58	0.60	0.60
Leucine /%	0.97	0.92	0.92	0.91	0.82	0.80	0.92

Notes: ① Mineral mixture: $MgSO_4 \cdot 7H_2O$, 3568.0mg/kg; $NaH_2PO_4 \cdot 2H_2O$, 25,568.0mg/kg; KCl, 3020.5mg/kg; $KAl(SO_4)_2$, 8.3mg/kg; $CoCl_2$, 28.0mg/kg; $ZnSO_4 \cdot 7H_2O$, 353.0mg/kg; Ca-lactate, 15,968.0mg/kg; $CuSO_4 \cdot 5H_2O$, 9.0mg/kg; KI, 7.0mg/kg; $MnSO_4 \cdot 4H_2O$, 63.1mg/kg; Na_2SeO_3, 1.5 mg/kg; $C_6H_5O_7Fe \cdot 5H_2O$, 1533.0mg/kg; NaCl, 100.0mg/kg; NaF, 4.0mg/kg.

② Vitamin mixture: retinol acetate, 38.0mg/kg; cholecalciferol, 13.2mg/kg; α-tocopherol, 210.0mg/kg; thiamin, 115.0mg/kg; riboflavin, 380.0mg/kg; pyridoxine HCl, 88.0mg/kg; pantothenic acid, 368.0mg/kg; niacin acid, 1030.0mg/kg; biotin, 10.0mg/kg; folic acid, 20.0mg/kg; vitamin B_{12}, 1.3mg/kg; inositol, 4000.0mg/kg; ascorbic acid, 500.0mg/kg.

The experimental diets were prepared by thoroughly mixing the dry ingredients, blending them with the fish oil and water (30%), and then forcing the resulting paste through a 2mm die using a Hobart (AE200) mincer/8 processor (Hobart Corporation, Troy, OH, USA). The moist pellets were air-dried to approximately 5% moisture at room temperature, and then stored at –20℃ until used.

5.17.2 Experimental procedure

Sea cucumbers were obtained from Shandong Anyuan Aquaculture Co., LTD (Penglai, China). The feeding trial was conducted in a thermoregulated flow-through system equipped with 21 fiberglass tanks of 400L capacity. A polyethylene corrugated plate was used as an artificial shelter at the bottom of each tank. Prior to the trial, sea cucumbers were acclimated to the experimental facilities for two weeks. Then, 840 sea cucumbers (initial weight of 29.40g ± 0.19g) were randomly allocated to 21 tanks in equal numbers (n=40) to form seven groups in triplicate. The total weight of sea cucumbers per tank was approximately same. Each group was fed one of the seven experimental diets at feeding amount of 2% total wet weight, by hand, once daily (16:00). During the trial, temperature was maintained at (19.0±1.0)℃, salinity was maintained at 31±0.5, and photoperiod was maintained on a 12∶12 light∶dark schedule. Oxygen concentration was kept higher than 6mg/L. Every other day a thorough cleaning of tank bottom was performed through siphoning with a pipe, and then feces was collected using tweezers every 2h and dried for calculating the amount of feces. The feeding trial lasted for 45 days.

5.17.3 Sampling and growth performance analysis

At the end of the feeding trial, all sea cucumbers were fasted for 48h before harvest. Total number and mean body weight of sea cucumbers in each tank were measured to calculate percent weight gain (PWG), specific growth rate (SGR) and survival. Fourteen sea cucumbers from each tank were randomly collected and dissected for organ and tissue sampling. Body wall and intestine were weighed and recorded separately, then frozen in liquid nitrogen for further analysis.

The variables of growth performance, diet utilization and organ coefficients were calculated as follows:

Percent weight gain (PWG, %) = (final weight – initial weight)/initial weight × 100;

Specific growth rate (SGR, %/d) = (ln final weight – ln initial weight)/days × 100;

Daily feed intake (DFI, %/d) = dry feed weight/[(initial weight + final weight)/2 × days];

Feed efficiency (FE, %) = feed intake/weight gain × 100;

Apparent digestibility coefficients of dry materials (ADC_{dm}, %), crude protein (ADC_{cp}, %) and crude fat (ADC_{cf}, %) = [1–(dietary chromic oxide content × faecal nutrient content)/(faecal chromic oxide content × dietary nutrient content)] × 100;

Survival rate (SR, %)= final sea cucumber number/initial sea cucumber number × 100;

Viscera-body wall ratio (VBR, %) = viscera weight/body wall weight × 100;

Intestine-body wall ratio (IBR, %) = intestine weight/body wall weight × 100;

Intestine-viscera ratio (IVR, %) = intestine weight/viscera weight × 100.

5.17.4 Analysis of digestive and metabolic enzymes activities

To measure the activities of digestive enzymes and metabolic enzymes, fourteen body walls and intestines of sea cucumbers from each tank were homogenated respectively in 4 volumes (W/V) of ice-cold Tris-HCl buffer solution (50mmol/L, pH 7.6) and then centrifuged at 4000 ×g, 4℃ for 10min. The supernatants were collected and kept frozen at –80℃ until activity assay. The activity of acid protease, alkaline protease, lipase, amylase, cellulose and ATPase were assayed using commercial kits purchased from Nanjing Jiancheng Bioengineering Institute (Nanjing, China) following the manufacturer's instructions. One unit (U/mg prot) of acid protease is defined as the amount of enzyme of 1mg tissue homogenate that hydrolyzed the haemoglobin to form 1μg tyrosine equivalent per minute at 37℃, pH 2.0. One unit (U/mg prot) of alkaline protease is defined as the amount of enzyme of 1mg tissue homogenate that hydrolyzed benzoyl-L-arginine-4-nitroanilide to 1μmol 4-nitroanilide per minute at 37℃, pH 8.1. One lipase unit (U/mg prot) is defined as the amount of enzyme of 1mg tissue homogenate that hydrolyzed 1μmol of triglyceride per minute at 37℃, pH 7.5. One amylase unit (U/mg prot) is defined as the amount of enzyme of 1mg tissue homogenate that liberated 1.0mg of maltose from starch in 3min at 20℃, pH 6.9. One cellulase unit (U/mg prot) is defined as the amount of enzyme of 1mg tissue homogenate that liberated 1.0μmol of glucose from cellulose in one hour at 37℃, pH 5.0. One unit (U/L) of GPT (or GOT) was defined as the amount of enzyme that generated 1.0μmol of glutamate (or pyruvate) per minute at 37℃, pH 7.4 (or pH 7.5). One ATPase unit (U/mg prot) is defined as the amount of enzyme that catalyzes the conversion of 1μmol of ATP to ADP in one minute at 30℃ in a total reaction volume of 50μl.

5.17.5 Analysis of antioxidant enzymes activities

To measure the activities of antioxidant enzymes, fourteen body walls and intestines of sea cucumbers from each tank were homogenated respectively in 4 volumes (*W/V*) of ice-cold Tris-HCl buffer solution (50mmol/L, pH 7.6) and then centrifuged at 4000 ×*g*, 4℃ for 10min. The supernatants were collected for activity analysis. Superoxide dismutase (SOD) was measured spectrophotochemically by the ferricytochrome c method using xanthine/xanthine oxidase as the source of superoxide radicals. One unit of SOD activity was defined as the amount of enzyme necessary to produce a 50% inhibition of the ferricytochrome reduction rate measured at 550nm (Peng et al., 2013). Catalase (CAT) activity was determined using a colorimetric assay based on the yellow complex with molybdate and hydrogen peroxide (Góth, 1991). One unit (U/ml) of CAT activity was defined as the quantity of enzyme that liberated 1μmol of hydrogen peroxide per minute at 37℃, pH 6.8. Glutathione peroxidase (GSH-Px) activity was determined using the commercial kit purchased from Nanjing Jiancheng Bioengineering Institute. One unit (IU/ml) of GSH-Px activity was defined as the amount of enzyme that consumed 1μmol of NADPH per minute at 37℃, pH 7.2. To normalize enzyme activity, total protein content of the supernatant was determined by the Bradford Assay using bovine serum albumin as a standard (Bradford, 1976). Free malondialdehyde (MDA) content of 1g tissue protein was determined by thiobarbituric acid (TBA) method using the commercial kit purchased from Nanjing Jiancheng Bioengineering Institute.

5.17.6 Chemical analysis

Fourteen body walls of sea cucumbers from each tank were freeze-dried for proximate composition analysis. Protein (*N*×6.25) was determined by the Kjeldahl digestion method, crude lipid was determined by the Soxhlet extraction method, crude ash was determined by combustion at 550℃ (AOAC, 2000), and energy was determined by an adiabatic bomb calorimeter (IKA®C 6000, Janke& Kunkel KG.IKA-werk, German). Minerals and chromium concentrations were determined with an inductively coupled plasma atomic emission spectrophotometer (ICP-OES; VISTAMPX, VARIAN) after perchloric acid digestion.

The body wall (approximate 20mg protein) was hydrolyzed with 6ml of 6mol/L HCl at 110℃ for 22h in an evacuated sealed tube to determine amino acids composition. The hydrolysate was dried under nitrogen gas to remove HCl, re-dissolved in 0.1mol/L HCl loading buffer, and filtered through a 0.22μm polyethersulfone ultrafiltration membrane.

The filtrate was loaded on a high-performance liquid chromatography system (LC1200, Agilent Technologies Inc., PaloAlto, CA, USA) equipped with an Agilent ZORBAX Eclipse Plus C18 column (150 μm×5μm). Signals of 16 amino acids were detected after derivatization with o-phthaldialdehyde. Asparagine, glutamine, proline, and tryptophan were not within the determination range. The HPLC conditions followed the protocol for the Agilent ZORBAX Eclipse Plus C18 column.

5.17.7 Statistical analysis

Data are compared by LSD test with one-way analysis of variance (ANOVA) after tests for normality and heterogeneity of variances using SPSS16.0 (SPSS Inc., Chicago, USA). Differences were considered statistically significant at $P < 0.05$. Data were presented as mean±SE, mean square within groups, F and P values. Quadratic re-gression models of PWG and SGR corresponding to HSBM inclusion level were established in Office Excel 2007 (Microsoft Corp., Redmond, WA, USA).

5.18 Results

5.18.1 Growth performance of sea cucumbers fed experimental diets for 45 days

Growth performance of sea cucumbers fed experimental diets for 45 days was presented in (Table 5-22). There was no significant difference in survival rate among all groups ($P>0.05$). Sea cucumbers fed diets with HSBM inclusion level from 4.05% to 12.27% (replacing 20% to 60% seaweed) had higher final weight, PWG and SGR than those fed the control diet ($P<0.05$). However, the highest HSBM inclusion level (20.45%) decreased final weight by 17.29%, PWG by 66.96% and SGR by 64.62% as

Table 5-22 Growth performance, survival and organ coefficients of sea cucumber fed with experimental diets for 45 days

	\multicolumn{6}{c}{Dietary HSBM levels}	MSE within group	Pooled SE	F	P-value
	0	4.05%	8.18%	12.27%	16.36%	20.45%	S12.27%				
\multicolumn{11}{l}{Growth performance}
Initial weight (g)	29.30±0.10	29.17±0.26	29.75±0.26	29.50±0.21	29.36±0.58	29.26±0.19	29.43±0.20	0.256	0.100	0.451	0.832
Final weight (g)	39.38±1.10^b	43.57±1.33^c	46.50±1.55^c	46.02±0.83^c	36.85±0.60^b	32.57±1.45^a	32.61±0.77^a	3.925	1.278	26.802	0.000
PWG (%)	34.41±4.21^b	49.31±3.66^c	56.27±4.37^c	56.00±3.07^c	25.53±5.46^b	11.37±5.67^a	10.79±1.96^a	41.111	4.246	28.371	0.000
SGR (%/d)	0.65±0.07^b	0.89±0.06^c	0.99±0.06^c	0.99±0.04^c	0.51±0.01^b	0.23±0.11^a	0.23±0.04^a	0.012	0.071	27.972	0.000
Survival (%)	90.83±4.53	87.50±2.89	85.00±2.89	85.83±4.64	87.17±3.83	84.17±2.20	83.33±2.20	22.321	1.133	0.543	0.767
\multicolumn{11}{l}{Diet utilization}
DFI (%/d)	0.36±0.01^b	0.40±0.01^{bc}	0.42±0.02^c	0.42±0.03^c	0.43±0.00^c	0.36±0.02^b	0.20±0.01^a	0.001	0.018	22.466	0.000
FE (%)	183.74±24.47^{bc}	218.88±16.70^c	230.56±17.16^c	234.11±26.32^c	117.73±17.52^{ab}	68.70±36.25^a	114.86±20.57^{ab}	1544.980	15.480	8.520	0.001
ADCdm (%)	7.11±0.36^b	7.96±0.52^{bc}	8.63±0.27^{cd}	9.42±0.33^d	12.38±0.74^c	13.22±0.51^f	5.57±0.22^a	0.034	0.518	685.646	0.000
ADCcp (%)	44.29±0.24^b	43.27±0.87^b	51.35±0.30^c	65.00±2.39^d	70.87±0.38^e	71.43±0.12^e	40.38±0.22^a	83.239	3.102	192.328	0.000
APCcf (%)	3.28±0.05	3.30±0.08	3.26±0.07	3.33±0.06	3.35±0.04	3.27±0.08	3.23±0.05	0.005	0.016	0.904	0.521
\multicolumn{11}{l}{Organ coefficient}
VBR (%)	10.83±0.75^c	10.37±0.78^{bc}	8.11±0.57^b	6.56±0.48^{ab}	6.13±0.39^a	5.12±0.38^a	6.29±0.47^a	13.383	0.250	14.676	0.000
IBR (%)	3.35±0.17^b	3.26±0.18^b	3.51±0.20^b	3.11±0.12^b	2.58±0.14^a	2.59±0.12^a	2.31±0.16^a	1.537	0.075	1.387	0.220
IVR (%)	30.40±2.32^a	33.77±2.19^{ab}	39.21±2.15^{bc}	48.10±2.41^d	47.62±2.17^d	47.69±2.58^d	42.95±2.68^{cd}	212.615	0.966	9.532	0.000

compared to the control treatment. Sea cucumbers fed diet with HSBM level of 16.36% had a comparable final weight, PWG and SGR compared to those fed the control diet ($P>0.05$). Two quadratic regression models were adopted to describe well the relationship between PWG, SGR and HSBM inclusion level in Figure 5-8, with estimated optimal HSBM level of 8.03%-8.15% (replacing 39.25%-39.86% seaweed). Sea cucumbers fed diets with HSBM levels from 8.18% to 16.36% have higher DFI ($P<0.05$), while sea cucumbers fed diets with HSBM level of 4.05% and 20.45% have similar DFI values ($P>0.05$) compared with that of the control group. No significant difference was detected in FE of sea cucumbers fed diets with HSBM levels from 0% to 16.36% ($P>0.05$), but a significant reduction was detected at the HSBM level of 20.45% ($P<0.05$).

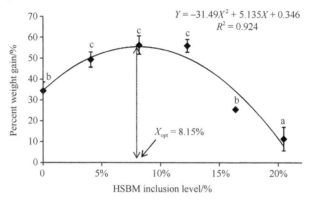

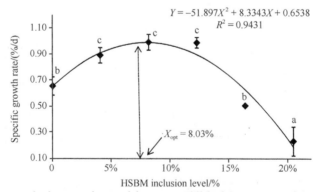

Figure 5-8　Two quadratic regression models were established on percent weight gain and specific growth rate in response to hydrolyzed soybean meal (HSBM) inclusion level. Different letters above the diamond denote significant difference between diet treatments ($P<0.05$). According to these two models, the optimal HSBM inclusion level is 8.03%-8.15% in current diet formula.

ADC_{dm} and ADC_{cp} gradually increased with the increasing HSBM level and sea cucumbers fed diets with HSBM levels from 8.18% to 20.45% have higher ADC_{dm} and

ADC_{cp} than those fed the control diet ($P<0.05$). In addition, VBR and IBR of sea cucumbers showed a decreased trend with the increasing HSBM inclusion level whereas IVR followed the opposite trend. Dietary medium to high HSBM levels (8.18% to 20.45%) decreased significantly VBR and IVR of sea cucumbers ($P<0.05$). Sea cucumbers fed diet with HSBM level of 4.09% have higher IBR than those fed diet with the HSBM level of 20.45% ($P<0.05$), but they showed no difference compared to other treatments ($P>0.05$). Sea cucumbers fed diet with SBM level of 12.27% had lower final weight, PWG, SGR, DFI, FE, ADC_{dm} and ADC_{cp} than those fed diet with HSBM level of 12.27% ($P<0.05$).

5.18.2 Activities of digestive and metabolic enzymes of sea cucumbers fed experimental diets for 45 days

Activities of intestinal lipase and cellulose remained unchanged regardless of HSBM inclusion ($P>0.05$) (Table 5-23). Acid protease activity was higher in sea cucumbers fed diet with HSBM level of 8.18% versus those fed diet with HSBM level of 20.45% ($P<0.05$), although both of them showed no difference compared to the control treatment ($P>0.05$). Alkaline protease activity was comparable in sea cucumbers fed diets with HSBM level of 0% and 4.09%, whereas it increased significantly in those fed diets with HSBM level from 8.18% to 12.27% ($P<0.05$) and then reduced by 41.80% in those fed diet with HSBM level of 20.45%.

Amylase activity increased with increasing HSBM level and diets with HSBM level of 8.18% significantly increased amylase activity compared with the control diet ($P<0.05$). Sea cucumbers fed diet with SBM level of 12.27% have higher amylase activity but lower alkaline protease activity than those fed diet with HSBM level of 12.27% ($P<0.05$). Intestinal GPT and GOT activities showed a similar trend that they increased to a peak value with 8.18% HSBM inclusion and then decreased with further increases of HSBM inclusion. Their activities were significantly elevated in sea cucumbers fed diets with HSBM levels from 4.09% to 12.27% ($P<0.05$), but intestinal GPT activity was depressed in those fed diets with HSBM level of 16.36% ($P<0.05$). In body wall, GPT activity was not affected by the low to medium HSBM levels (4.09% to 12.27%) ($P>0.05$) whereas reduced by the highest HSBM inclusion level (20.45%) ($P<0.05$). The highest HSBM inclusion level also significantly reduced the ATPase activity of intestine and body wall ($P<0.05$). GOT activity of body wall showed no significant change among all groups ($P>0.05$). Diet with SBM level of 12.27%

Table 5-23 Activities of digestive and metabolic enzymes of sea cucumbers fed experimental diets for 45 days (unit: U/mg prot)

Items	Groups							MSE within groups	Pooled SE	F	P-value
	D1	D2	D3	D4	D5	D6	D7				
Digestive enzymes											
Acid protease	5.75±0.18	5.96±0.22	5.99±0.46	5.57±0.36	4.83±0.19	4.90±0.66	4.86±0.17	0.395	0.158	2.108	0.117
Alkaline protease	705.95±23.63[bc]	795.52±70.57[c]	994.98±66.19[d]	968.57±14.77[d]	625.65±23.74[ab]	523.08±22.98[a]	530.10±30.15[a]	5202.243	42.239	21.674	0.000
Lipase	0.43±0.03	0.45±0.02	0.48±0.02	0.41±0.03	0.45±0.01	0.47±0.02	0.48±0.03	0.002	0.009	1.447	0.266
Amylase	0.11±0.01[a]	0.14±0.02[a]	0.17±0.01[b]	0.18±0.01[b]	0.20±0.01[b]	0.31±0.02[d]	0.26±0.01[c]	0.003	0.014	40.122	0.000
Cellulase	1.68±0.06	1.75±0.03	1.73±0.03	1.76±0.04	1.81±0.03	1.80±0.09	1.73±0.04	0.007	0.018	0.855	0.550
Metabolic enzymes of intestine											
GPT	31.00±0.44[c]	35.72±2.03[d]	40.23±1.74[e]	37.31±0.98[de]	24.56±1.39[ab]	21.80±0.89[a]	27.19±1.68[bc]	5.931	1.501	24.264	0.000
GOT	2.97±0.11[a]	4.81±0.12[d]	5.07±0.18[d]	3.63±0.10[c]	3.35±0.05[abc]	3.17±0.17[ab]	3.38±0.11[bc]	0.047	0.176	44.204	0.000
ATPase	2.61±0.12[b]	2.84±0.14[b]	2.97±0.22[b]	2.64±0.12[b]	2.72±0.14[b]	2.16±0.11[a]	1.95±0.02[a]	0.053	0.087	7.675	0.001
Metabolic enzymes of body wall											
GPT	44.69±1.31[bc]	46.22±1.05[c]	45.69±0.81[c]	45.17±1.23[bc]	41.98±0.36[ab]	39.45±0.84[a]	40.68±0.80[a]	2.773	0.632	7.758	0.001
GOT	2.60±0.14	2.83±0.15	2.74±0.10	2.88±0.18	3.01±0.07	2.44±0.14	2.48±0.11	0.072	0.061	1.326	0.309
ATPase	0.71±0.04[d]	0.75±0.02[cd]	0.79±0.02[d]	0.77±0.01[cd]	0.68±0.02[b]	0.48±0.02[a]	0.53±0.01[a]	0.002	0.026	26.871	0.000

Notes: Values are presented as means±SE (n=3), with different superscript letters indicating significant difference ($P<0.05$). GPT, Glutamic-pyruvic transaminase; GOT, Glutamic oxaloacetic transaminase; ATPase, Adenosine triphosphatase.

significantly suppressed ATPase activity of body wall compared to diet with HSBM level of 12.27% ($P<0.05$).

5.18.3 Oxidation resistance of tissues of sea cucumbers fed experimental diets for 45 days

Intestinal MDA content had no changes with HSBM inclusion ($P>0.05$) (Table 5-24). Intestinal SOD activity significantly increased in sea cucumbers fed diets with HSBM level from 4.09% to 8.18% while GSH-Px activity significantly increased in sea cucumbers fed diets with HSBM level from 4.09% to 12.27% ($P<0.05$). In body wall, SOD and GSH-Px activities decreased with increasing dietary HSBM levels and both of them showed significant declines at high inclusion levels (16.36% to 20.45%) ($P<0.05$). No significant difference was observed in CAT activity and MDA content of body wall among all treatments ($P>0.05$). Sea cucumbers fed diet with SBM level of 12.27% have lower SOD and GSH-Px activities of intestine and body wall than those fed diet with HSBM level of 12.27% ($P<0.05$).

5.18.4 Body composition of sea cucumbers fed experimental diets for 45 days

Body composition of sea cucumbers fed experimental diets for 45 days was presented in Table 5-25. Amino acid, Crude protein and crude fat contents of body wall were not affected by HSBM inclusion level ($P>0.05$). However, the contents of crude ash, Mg, Ca and Se of body wall significantly increased in sea cucumbers fed diet with HSBM level of 8.18% ($P<0.05$). Fe content of body wall significantly increased in sea cucumbers fed diets with HSBM level from 8.18% to 20.45%. The contents of crude ash and Ca of body wall were lower in sea cucumbers fed diet with SBM level of 12.27% compared with those fed diet with HSBM level of 12.27% ($P<0.05$).

5.19 Discussions

Sea cucumbers have a low protein (13.5%) and fat (0.19%-1.38%) requirement (Liao et al., 2014, 2015) but a high carbohydrate (48.56%-49.30%) requirement (Xia et al., 2015a), which may be related to their feeding habit of long-term feeding benthic algae because most algae that sea cucumbers preferred contain low protein and fat (Table 5-25). Seaweed is traditionally considered the favourite food of sea cucumbers, such as *Sargassum thunbergii*, *Laminaria japonica*, *Sargassum polycystum* and so on. However, recent studies have shown considerable success in partial (25%-40%) or total

Table 5-24　Oxidation resistance of sea cucumbers fed experimental diets for 45 days

Items	Groups							MSE within groups	Pooled SE	F	P-value
	D1	D2	D3	D4	D5	D6	D7				
				Intestine							
SOD/(U/mg prot)	5.50±0.20^{ab}	6.36±0.15^c	6.29±0.13^c	5.93±0.19^{bc}	6.05±0.17^{bc}	5.49±0.23^{ab}	5.35±0.14^a	0.091	0.101	5.487	0.004
GSH-Px/(U/mg prot)	14.38±0.28^b	16.59±0.15^d	16.25±0.16^d	15.11±0.22^c	14.23±0.16^{ab}	14.47±0.17^{bc}	13.57±0.33^a	0.146	0.240	25.238	0.000
CAT/((U/mg prot)	3.90±0.15^{abc}	4.01±0.11^{bc}	4.18±0.15^c	4.07±0.10^{bc}	3.89±0.17^{abc}	3.55±0.15^a	3.69±0.16^{ab}	0.057	0.063	2.516	0.073
MDA/(nmol/g prot)	1.67±0.15	1.48±0.32	1.63±0.20	1.55±0.14	1.54±0.08	1.63±0.34	1.42±0.16	0.037	0.038	0.471	0.819
				Body wall							
SOD/(U/mg prot)	8.28±0.37^d	8.14±0.20^d	8.03±0.41^d	7.59±0.36^{cd}	6.70±0.44^b	6.75±0.60^{bc}	5.59±0.59^a	0.228	0.223	12.846	0.000
GSH-Px/(U/mg prot)	3.68±0.28^c	3.81±0.33^c	3.74±0.21^c	3.69±0.12^c	3.02±0.09^b	2.38±0.33^a	2.84±0.39^{ab}	0.080	0.127	11.919	0.000
CAT/((U/mg prot)	3.52±0.68	3.83±0.52	3.27±0.64	3.61±0.28	3.34±0.27	3.43±0.36	3.54±0.28	0.093	0.068	1.116	0.401
MDA/(nmol/g prot)	0.74±0.04	0.73±0.05	0.74±0.6	0.85±0.6	0.65±0.8	0.81±0.11	0.71±0.10	0.016	0.026	0.585	0.737

Notes: Values are presented as means±SE (n=3), with different superscript letters indicating significant difference (P<0.05). SOD, Superoxide Dismutase; GSH-Px, Glutathione peroxidase; CAT, Catalase; MDA, Malondialdehyde.

Table 5-25 Nutrition composition of body wall of sea cucumbers fed experimental diets for 45 days

	Dietary HSBM levels						MSE within groups	Pooled SE	F	P-value
	0%	4.05%	8.18%	12.27%	16.36%	20.45%	S12.27%				
	Proximate composition (dry material)									
Crude protein/%	51.80±0.17	51.38±0.12	51.50±0.23	51.26±0.10	51.62±0.11	51.65±0.07	51.47±0.10	0.058	0.058	1.685	0.197
Crude fat/%	6.43±0.11	6.34±0.12	6.41±0.10	6.45±0.10	6.44±0.05	6.35±0.13	6.39±0.05	0.029	0.032	0.173	0.980
Crude ash/%	33.55±0.07[ab]	33.53±0.14[ab]	34.19±0.20[c]	34.02±0.15[bc]	33.20±0.17[a]	33.07±0.21[a]	33.17±0.17[a]	0.081	0.104	6.915	0.001
	Amino acid contents (dry material)									
Aspartic acid/%	5.04±0.06	5.02±0.08	5.07±0.07	5.04±0.08	5.23±0.07	5.22±0.06	5.05±0.06	0.015	0.029	1.543	0.23
Threonine/%	2.71±0.07	2.68±0.06	2.61±0.06	2.66±0.05	2.68±0.08	2.68±0.06	2.61±0.11	0.016	0.016	0.293	0.930
Serine/%	2.46±0.05	2.46±0.02	2.45±0.03	2.44±0.09	2.44±0.06	2.46±0.07	2.39±0.06	0.010	0.019	0.183	0.977
Glutamic acid/%	8.26±0.06	8.22±0.10	8.22±0.08	8.23±0.09	8.34±0.14	8.48±0.12	8.29±0.07	0.029	0.037	0.954	0.489
Glycine/%	7.26±0.07	7.37±0.07	7.34±0.07	7.24±0.06	7.36±0.08	7.35±0.12	7.14±0.07	0.019	0.031	1.175	0.373
Alanine/%	3.06±0.07	3.13±0.11	2.98±0.12	3.00±0.16	3.16±0.07	3.13±0.13	2.96±0.11	0.040	0.040	0.518	0.786
Cysteine/%	1.02±0.08	1.02±0.14	0.97±0.07	0.98±0.11	1.02±0.10	0.98±0.10	0.93±0.05	0.027	0.031	0.124	0.991
Valine/%	1.77±0.10	1.83±0.06	1.78±0.14	1.79±0.11	1.80±0.15	1.85±0.12	1.80±0.10	0.040	0.037	0.056	0.999
Methionine/%	0.34±0.03	0.35±0.07	0.39±0.04	0.35±0.09	0.34±0.04	0.36±0.02	0.33±0.07	0.010	0.018	0.128	0.991
Isoleucine/%	1.34±0.05	1.39±0.10	1.39±0.08	1.35±0.07	1.33±0.05	1.33±0.04	1.34±0.07	0.014	0.022	0.161	0.98
Leucine/%	2.55±0.03	2.50±0.03	2.54±0.04	2.52±0.02	2.57±0.08	2.54±0.05	2.49±0.03	0.059	0.045	0.088	0.977
Tyrosine/%	1.54±0.21	1.42±0.12	1.49±0.04	1.52±0.17	1.39±0.10	1.54±0.22	1.52±0.20	0.077	0.053	0.140	0.988
Phenylalanine/%	1.41±0.04	1.39±0.02	1.46±0.05	1.42±0.01	1.40±0.02	1.45±0.01	1.39±0.02	0.026	0.030	0.092	0.996
Lysine/%	1.84±0.18	1.57±0.05	1.69±0.07	1.84±0.14	1.84±0.13	1.81±0.11	1.79±0.05	0.041	0.041	0.667	0.671
Histidine/%	0.60±0.11	0.71±0.13	0.55±0.04	0.59±0.09	0.60±0.05	0.61±0.04	0.59±0.06	0.019	0.027	0.377	0.882
Arginine/%	3.39±0.10	3.48±0.09	3.53±0.07	3.52±0.12	3.48±0.07	3.54±0.10	3.35±0.12	0.028	0.034	0.581	0.740

Continued

	Dietary HSBM levels						MSE within groups	Pooled SE	F	P-value
	0%	4.05%	8.18%	12.27%	16.36%	20.45%	S12.27%				
	Mineral contents									
Na/(g/kg)	53.45±0.23	52.70±0.49	53.43±0.26	53.06±0.24	53.43±0.11	53.39±0.18	53.31±0.10	0.204	0.101	1.143	0.388
Mg/(g/kg)	7.68±0.15^a	7.85±0.1^{ab}	8.20±0.18^b	7.83±0.13^{ab}	7.70±0.11^a	7.68±0.13^a	7.45±0.12^a	0.058	0.065	2.741	0.056
K/(g/kg)	6.20±0.10	6.17±0.11	6.46±0.15	6.35±0.07	6.25±0.18	6.45±0.02	6.26±0.12	0.041	0.044	1.010	0.457
Ca/(g/kg)	7.30±0.16^{ab}	7.69±0.06^{bc}	7.81±0.30^c	7.57±0.07^{bc}	7.02±0.13^a	7.07±0.08^a	7.05±0.12^a	0.069	0.084	4.794	0.007
Mn/(mg/kg)	4.57±0.10	4.31±0.17	4.27±0.15	4.66±0.12	4.42±0.14	4.36±0.15	4.65±0.08	0.054	0.054	1.432	0.271
Fe/(mg/kg)	4.17±0.07^a	4.31±0.14^{ab}	4.54±0.12^{ab}	4.62±0.18^b	4.65±0.12^b	4.65±0.10^b	4.45±0.13^{ab}	0.036	0.042	1.024	0.450
Cu/(mg/kg)	1.52±0.06	1.34±0.04	1.32±0.12	1.45±0.13	1.48±0.07	1.45±0.11	1.29±0.10	0.027	0.036	0.912	0.514
Zn/(mg/kg)	26.54±0.22^{ab}	27.98±0.14^{ab}	28.71±0.19^b	28.04±1.77^{ab}	28.16±0.22^{ab}	27.63±0.21^{ab}	26.04±0.10^a	1.421	0.293	1.910	0.149
Se/(mg/kg)	3.36±0.07^{ab}	3.58±0.05^b	3.90±0.10^c	3.14±0.11^a	3.39±0.09^{ab}	3.34±0.06^{ab}	3.40±0.04^b	0.018	0.055	9.543	0.000

Notes: Values are presented as means ± SE (n=3), with different superscript letters indicating significant difference (P<0.05).

Replacement of seaweed with terrestrial plant source (Xia et al., 2015b; Wu et al., 2015; Yu et al., 2015b). Present study indicates that up to 16.36% HSBM can be added to replace 80% of seaweed powder to achieve the satisfactory growth performance of sea cucumbers. These results are better than the maximum seaweed powder replacement level (25%-40%) reported by Xia et al. (2015a) with SBM and Wu et al. (2015) with corn meal and extruded SBM, suggesting that sea cucumbers can tolerate dietary high HSBM level well. Furthermore, sea cucumbers fed diets with 4.09%-16.36% HSBM grew better than those fed seaweed powder-based diet, indicating that sea cucumbers utilize appropriate levels of HSBM more efficiently than the seaweed powder. These findings agreed with previous conclusions that appropriate inclusion level of soy protein hydrolysates improved the growth performance and diet utilization of flatfish (Song et al., 2014; Mamauaga et al., 2011). Based on present results, it is concluded that there is an optimal nitrogen supply pattern which supported rapid growth of sea cucumbers and this pattern may be achieved by adding 8.03%-8.15% HSBM (replacing 39.25%-39.86% seaweed) as predicted by the PWG and SGR models.

Several studies demonstrated that feed intake rate of deposit feeders was related to dietary energy and protein levels (McBride et al., 1998; Otero-Villanueva et al., 2004). In present study, all sea cucumbers received diets with equal energy and protein level, thus palatability was speculated to be a key factor affecting the feed intake. The daily feed intake results suggest that HSBM can be an effective feed attractant at the inclusion level of 8.18%-16.36%. The improved palatability may be associated with free amino acid released from soy protein by hydrolysis of flavourzyme (Ma et al., 2013). Although sea cucumbers fed diet with 16.36% HSBM have a similar DFI compared with those fed diets with 4.05%-12.27% HSBM, their significantly decreased FE suggested a poor utilization occurred. This poor utilization may be due to high metabolic loss from excessive nitrogen intake (Moughan et al., 2005; Hodgkinson et al., 2000) and eventually induced a low conversion efficiency. This mechanism provides a reason for the decreased growth in sea cucumbers fed diets with high inclusion of HSBM. In addition, apparent digestibility coefficients of dry material and protein correlate positively to HSBM inclusion level indicating that sea cucumbers can digest HSBM well. This finding agreed with those reports on other animals with soy protein and marine protein hydrolysates (Jin et al., 2015; Khosravi et al., 2015; Zhou et al., 2011b). However, at highest HSBM inclusion levels of 20.45% (replacing 100% seaweed), high digestibility may not compensate for insufficient nutrients supply due to low feed intake and feed efficiency, and thus cannot support a rapid growth. Present study also indicated sea cucumbers fed diet containing 12.75% SBM exhibited a poor growth result compared to those fed diet containing 12.75% HSBM. Since the difference between SBM and HSBM is not only in peptides content but also in partially digested protein content, the present results are difficult to explain that which kind of matter can promote growth of sea cucumbers. However, two hypotheses are proposed based on present result and need further study. First, the presence in the digestive tract of high amounts of partially digested protein may have accelerated the amino acid absorption rate and increase nitrogen utilization. Second, soy peptides benefit cell growth and protein synthesis by modulating signal transduction (Lee et al., 2012), promoting the development of intestinal epithelial cells (Jiang et al., 2009) and bone growth (Lv et al., 2013), facilitating transport of special amino acids (Kitagawa et al., 2013).

Organ coefficient is generally considered as an indicator of organ development which is affected by nutritional status of animals (Azarm and Lee, 2014). In present study, viscera-body wall ratio decreased gradually with the increasing HSBM level while intestine-viscera ratio followed the opposite trend. These changes suggested that HSBM selectively promoted the growth of body wall and intestine of sea cucumbers. Similar phenomenon was also observed in our previous study with starry flounder fed diets containing soy protein hydrolysate (Song et al., 2014). In addition, diets with high inclusion levels of HSBM (16.36%-20.45%) reduced intestine-body wall ratio which indicated that the intestinal atrophy occurred probably in response to the poor feeding intake.

Diet-related plasticity of digestive enzyme activity can be explained by the "adaptive modulation hypothesis" which states that variation in diet should confer upon animals the ability to modulate their digestive enzyme activity accordingly (Karasov, 1992). The activity of several digestive enzymes has been detected in the gastrointestinal tract of sea cucumbers including pepsin, lipase, amylase (Sun et al., 2015) and alkaline protease activity (Fu et al., 2005a, 2005b). A recent study demonstrated that intestinal protease activity of sea cucumbers varied responding to changes in dietary nutritional composition (Liao et al., 2015). In current study, it was found that compared to control group, dietary addition of HSBM at low and moderate levels of 4.09%-12.27% (replacing 20%-60% seaweed) increased in alkaline protease activity, but higher inclusion level (16.36%-20.45%) depressed its activity. This may be due to low protein ingestion which reduces the concentration of substrate for alkaline protease in intestine. However, amylase activity increased gradually with the increasing

HSBM levels, implicating that HSBM inclusion improved the digestion of starch by sea cucumbers. Endogenous amylase activity could be induced by appropriate dietary carbohydrate level (Xia et al., 2015a, 2015b) or non-starch polysaccharide enzymes (Li et al., 2009). Similarly, non-starch polysaccharide enzymes employed in this study could destroy cellulose of cell wall and release more starch, which may be responsible for the increased amylase activity.

It is generally accepted that intestine and body wall are main metabolic tissues of sea cucumbers (Byrne, 2001), where most metabolic enzymes are highly expressed to regulate nutrient metabolism (Zang et al., 2012; Xia et al., 2015a, 2015b; Dong et al., 2011). GPT and GOT are two cytoplasmic enzymes aiding gluconeogenesis through transamination of glucogenic amino acid to meet the energy demand under specific conditions (Ramaswamy et al., 1999). Inhibition of GOT and GPT may be due to a reduction in metabolic activities and blockage of protein metabolism (Sanchez et al., 2007). In present study, sea cucumbers fed diets with HSBM levels from 4.09% to 12.27% (replacing 20%-60% seaweed) have higher GPT and GOT activity of intestine as well as GPT activity of body wall compared to the control treatment, indicating that they efficiently regulate transamination and gluconeogenesis in response to the increased nitrogen nutrient influx resulting from fast absorption of soy peptides. This enhanced metabolism may trigger a series of energy-production reactions to support their fast growth as observed in current growth results. However, several metabolic enzymes activities including ATPase and GPT activities of these two organs all showed significant reductions at the highest HSBM level of 20.45%, which were consistent with changes of alkaline protease activity. The reduced digestion and metabolism may lead to a poor growth of sea cucumbers fed the diet without seaweed powder.

Free radicals generated by endogenous metabolic processes may cause oxidative damage resulting in cell death and tissue damage (Kehrer, 1993). The primary antioxidant protection is provided by the enzymes SOD, CAT and GPX. The combined action of these enzymes keeps intracellular steady state levels of superoxide anion and hydrogen peroxide low. Although there is no relevant research on application of plant peptide in sea cucumber diet, dietary supplementation of plant protein hydrolysates at certain level has been reported to enhance oxidation resistance of fish (Khosravi et al., 2015; Song et al., 2014) and other animals (Nazeer et al., 2012; Han et al., 2014; Wang et al., 2014). Present study showed that dietary intake of low and medium levels of HSBM (4.09%-12.27%) could reinforce intestinal oxidant status by increasing SOD and GSH-Px activities. The enhanced antioxidase system may respond to the elevated reactive oxygen species which result from high nitrogen metabolism and protect enterocytes from oxygen free radicals. SOD and GSH-Px activities of body wall were not affected by low and medium level of HSBM whereas they were significantly reduced at higher level of 16.36% to 20.45% (replacing 80%-100% seaweed) indicating that excessive hydrolysates depressed antioxidant capacity of these organs. Combined with these oxidation resistance results, HSBM should be included in the diet at a level of <12.27%, so as not to negatively affect the health of sea cucumbers.

The minerals are responsible for skeletal formation, maintenance of colloidal systems, regulation of acid-base equilibrium and for biologically important compounds such as hormones and enzymes (Watanabe et al., 1997). Present results showed that inclusion of HSBM (8.18%-16.36%) (replacing 40%-80% seaweed) improved bioavailability of most minerals as reflected by an increased deposition of ash and minerals in body wall. Similar findings were also found with crucian carp fed cottonseed meal hydrolysate (Gui et al., 2010). A possible reason is that peptides derived from plant protein bind with supplemented minerals to form soluble complexes and promote their uptake (Lv et al., 2013).

5.20 Conclusions

In conclusion, present study indicated that up to 16.36% HSBM can be included to effectively replace up to 80% seaweed powder in diet without retarding the growth of sea cucumbers with the optimal inclusion level restricted to 8.03%-8.15%. Appropriate inclusion level of HSBM could promote growth and improve diet utilization, metabolism and minerals deposition. Further investigation should be conducted on the elucidation of the underlying mechanisms of nutrient metabolism from the molecular level in response to HSBM-rich diets in sea cucumbers.

主要参考文献

艾庆辉, 严晶, 麦康森. 2016. 鱼类脂肪与脂肪酸的转运及调控研究进展. 水生生物学报, 40(4): 859-868.

安振华, 董云伟, 董双林. 2008. 日粮水平对周期性变温模式下刺参生长和能量收支的影响. 中国海洋大学学报, 38(5): 739-743.

白雪峰. 2004. 木聚糖酶在饲料中的应用进展. 中国畜牧兽医, 31(12): 11-13.

白燕, 王维新. 2012. 刺参肠道蛋白酶、淀粉酶、脂肪酶与纤维素酶活性的测定方法. 饲料工业, 33(20): 28-32.

白阳, 徐玮, 汪东风, 等. 2016. 不同分子量壳聚糖对刺参(Apostichopus japonicus Selenka)生长和免疫功能的影响. 渔业科学进展, (37): 93-99.

蔡琨, 苏东海, 陈静, 等. 2012. 大豆低聚糖的生理功能研究进展. 中国食物与营养, 18(12): 56-61.

蔡英华. 2006. 几种大豆抗营养因子对牙鲆(Paralichthys olivaceus)生长和消化生理的影响. 青岛: 中国海洋大学.

曹功明, 卢建雄, 马桂兰, 等. 2010. 甘露寡糖对动物肠道环境及生长性能影响的研究进展. 江苏农业科学, 38(1): 226-227, 345.

曹新勇. 2013. 酶制剂在饲料中的合理使用. 农村养殖技术, (10): 59

常杰, 牛化欣, 张文兵. 2011. 刺参免疫系统及其免疫增强剂评价指标的研究进展. 中国饲料, (6): 8-12.

常杰. 2010. 对虾和刺参敏感免疫学指标的筛选和评价. 青岛: 中国海洋大学.

常忠岳, 衣吉龙, 慕康庆. 2003. 影响刺参生长及成活的因素. 河北渔业, (2): 32-33, 36.

陈超, 陈京华. 2012. 牛磺酸、晶体氨基酸对大菱鲆摄食、生长和饲料利用率的影响. 中国农学通报, 28(23): 108-112.

陈娟, 王安, 张淑云, 等. 2010. 饲粮钙和维生素D3水平对肉仔鸡生长性能、屠宰性能和肉品质的影响. 动物营养学报, 22(6): 1544-1550.

陈书秀, 刘学迁, 王青岩, 等. 2014. 微绿球藻作为刺参幼体饵料的可行性研究. 河北渔业, (7): 53-55.

成永旭. 2005. 生物饵料培养学(第二版). 北京: 中国农业出版社.

程会昌, 霍军, 宋予振, 等. 2006. 木聚糖酶在黄河鲤鱼饲料中的应用. 安徽农业科学, 34(13): 3077-3079.

程柳, 李静. 2016. 3,5-二硝基水杨酸法测定山楂片中还原糖和总糖含量. 轻工科技, 32(3): 25-28.

迟淑艳, 周歧存, 周健斌, 等. 2006. β-葡聚糖对奥尼罗非鱼生长性能及抗嗜水气单胞菌感染的影响. 中国水产科学, 13(5): 767-774.

丛大鹏, 咸洪泉. 2009. 我国海藻的开发利用价值及产业化生产. 中国渔业经济, 27(1): 93-97.

崔凤霞, 薛长湖, 李兆杰, 等. 2006. 仿刺参胶原蛋白的提取及理化性质. 水产学报, 30(4): 549-553.

崔洪斌. 2000. 大豆生物活性物质的开发与应用. 中国食物与营养, 6(1): 15-17.

邓必阳, 张展霞. 1999. 鲨鱼软骨营养成分分析及其评价. 营养学报, 21(1): 104-108.

邓岳松. 2005. 耐高温酶制剂对草鱼生长的影响. 内陆水产, 30(6): 45-46.

董晓弟, 潘如佳, 王长海. 2013. 海地瓜、黑乳参和乌皱辐肛参营养成分对比. 现代食品科技, 29(12): 2986-2990.

董晓亮, 李成林, 赵斌, 等. 2013. 低盐胁迫对刺参非特异性免疫酶活性及抗菌活力的影响. 渔业科学进展, 34(3): 82-87.

董学前, 张艳敏, 张永刚, 等. 2017. 复合酶法综合提取海带中褐藻糖胶与海藻酸的研究. 中国食品添加剂, 4(9): 171-176.

董育红, 封涛, 张振兰, 等. 2003. 螺旋藻的营养成分分析. 食品研究与开发, 24(3): 70-71.

董云伟, 牛翠娟, 杜丽. 2001. 饲料蛋白水平对罗氏沼虾 (*Macrobrachium rosenbergii*) 生长和消化酶活性的影响. 北京师范大学学报(自然科学版), 37(1): 96-99.

窦江丽, 谭成玉, 白雪芳, 等. 2008. 棉子糖及水苏糖对兔外周血中性粒细胞功能的影响. 精细与专用化学品, 16(3-4): 26-27.

杜以帅. 2010. 酶解海藻产物对刺参(*Apostichopus japonicus*)肠道菌群和免疫相关因子的影响. 青岛: 中国海洋大学.

段鸣鸣, 王春芳, 谢从新. 2014. 维生素 D_3 对黄颡鱼幼鱼抗氧化能力及免疫功能的影响. 淡水渔业, 44(3): 80-84.

段培昌, 张利民, 王际英, 等. 2009. 新型蛋白源替代鱼粉对星斑川鲽幼鱼生长、体成分和血液学指标的影响. 水产学报, 33(5): 797-804.

段树丽. 2021. 当归对血虚大鼠 Na^+-K^+-ATP 酶和 $Ca^{2+}-Mg^{2+}$-ATP 酶的影响. 甘肃畜牧兽医, 51(4): 50-53.

方微, 单玉萍, 李峰, 等. 2011. 木聚糖酶的作用机理及其在饲料中的应用. 中国饲料, (21): 21-24.

冯仁勇. 2006. 维生素 A 对幼建鲤消化能力和免疫功能的影响. 雅安: 四川农业大学.

冯秀妮, 张文兵, 麦康森, 等. 2006. 饲料中维生素 B_6 对皱纹盘鲍幼鲍蛋白质代谢的影响. 海洋科学, 30(10): 52-60.

付京花, 张文兵, 麦康森, 等. 2006. 维生素 D 对皱纹盘鲍生长和体组织抗氧化反应的影响. 高技术通讯, 16(12): 1306-1311.

付雪艳. 2004. 海参 (*Apostichopus japonicus*) 消化蛋白酶的初步研究. 青岛: 中国海洋大学.

付燕红, 王庆玉, 付学军, 等. 2019. 海带中褐藻糖胶不同提取工艺. 食品工业, 40(8): 49-53.

干懿洁, 丁树哲. 2006. 丙酮酸的抗氧化作用. 中国临床康复, 10(8): 141-143.

高春生, 范光丽, 李建华, 等. 2006a. 纤维素酶对草鱼生长性能和饲料消化率及体成分的影响. 中国农学通报, 22(10): 473-475.

高春生, 刘忠虎, 肖传斌, 等. 2006b. 木聚糖酶对草鱼生长性能和消化率的影响. 饲料研究, (8): 48-49.

高晓莉, 岳淑芹, 王丽敏, 等. 2003. 重金属对鲢鱼肝组织 SOD 和 CAT 活性影响研究简报. 河北农业大学学报, 26(4): 130.

谷珉. 2010. 仿刺参(*Apostichopus japonicus*)免疫增强剂的体外筛选与养殖试验的评估. 青岛: 中国海洋大学.

关莹, 薛敏, 王伟. 2021. 饲用酶制剂在水产动物中应用的最新研究进展. 中国渔业质量与标准, 11(1): 61-67.

郭娜, 董双林, 刘慧. 2011. 几种饲料原料对刺参幼参生长和体成分的影响. 渔业科学进展, 32(1): 122-128.

郭娜, 姚子昂, 于国友, 等. 2019. 海带酶解产物对海参生长及其免疫相关因子的影响. 中国酿造, 38(4): 160-164.

郭鹏, 王际英, 李宝山, 等. 2022. 棉子糖对刺参幼参生长、生理指标及糖代谢的影响. 水产学报, 46(10): 1940-1949.

郭勤. 2005. 海参精氨酸激酶的折叠与结构研究. 北京: 清华大学.

韩承义, 林培华, 谢河中, 等. 2011. 南方海区刺参养殖关键技术研究. 现代渔业信息, 26(9): 31-33.

韩诗雯, 高加涛, 桑若杰, 等. 2019. 水苏糖调节肠道功能及作用机制研究. 食品科技, 44(4): 281-284.

韩秀杰, 李宝山, 王际英, 等. 2019. 仿刺参幼参对缬氨酸最适需求量的研究. 水产学报, 43(3): 628-638.

郝继浦. 2014. 利用海带渣和啤酒糟生产刺参饵料的研究. 哈尔滨: 哈尔滨工业大学.

郝林华, 孙丕喜, 石红旗, 等. 2007. 牛蒡寡糖对大菱鲆生长和免疫机能的影响. 海洋科学进展, 25(2): 208-214.

郝甜甜, 王际英, 李宝山, 等. 2014. 饲料中添加糖萜素对大菱鲆幼鱼生长、免疫及热休克蛋白70含量的影响. 浙江大学学报: 农业与生命科学版, 40(3): 338-347.

何四旺, 许国焕, 吴月嫦, 等. 2003. 低聚异麦芽糖和低聚果糖对罗非鱼生长和非特异性免疫的影响. 中国饲料, (23): 14-15.

何伟. 2008. 吡哆醇对幼建鲤消化吸收能力和免疫能力的影响. 雅安: 四川农业大学.

何晓庆. 2010. 玉米DDGS在奥尼罗非鱼(*Oreochromis niloticus*♀×*O. aureus*♂)饲料中的应用研究. 广州: 华南农业大学.

何亚男. 2006. 两种寡糖对大鼠肠黏膜结构及黏膜免疫相关细胞影响的研究. 北京: 中国农业大学.

何云, 崔金忠, 郑素玲, 等. 2008. 木聚糖酶在水产养殖中的应用研究进展. 广东农业科学, 35(10): 100-101.

侯传宝. 2005. 刺参育苗单胞藻培育技术. 齐鲁渔业, 22(9): 50.

胡静, 林英庭. 2012. 半乳甘露寡糖在动物生产中的研究进展. 饲料博览, (9): 29-31.

胡友军, 周安国, 杨凤, 等. 2002. 饲料淀粉糊化的适宜加工工艺参数研究. 饲料工业, 23(12): 5-8.

黄爱霞, 孙丽慧, 陈建明, 等. 2018. 饲料亮氨酸水平对幼草鱼生长、饲料利用及体成分的影响. 饲料工业, 39(2): 26-32.

黄峰, 施培松, 文华, 等. 2008a. 外源酶对草鱼鱼种生长及饲料表观消化率的影响. 安徽农业科学, 36(3): 1057-1059.

黄峰, 张丽, 周艳萍, 等. 2008b. 木聚糖酶在制粒前后的热稳定性及对异育银鲫生长的影响. 粮食与饲料工业, (1): 29-31.

黄峰, 张丽, 周艳萍, 等. 2008c. 外源木聚糖酶对异育银鲫生长、超氧化物歧化酶及溶菌酶活性的影响. 淡水渔业, 38(1): 44-48.

黄峰, 张丽, 周艳萍, 等. 2008d. 复合酶制剂对异育银鲫生长、SOD 和溶菌酶活性的影响. 华中农业大学学报, 27(1): 96-100.

黄利娜, 梁萌青, 张海涛, 等. 2013. 饲料中添加不同水平维生素 A 对大菱鲆亲鱼繁殖性能的影响. 渔业科学进展, 34(4): 62-70.

黄亮华, 李浩洋, 李彬, 等. 2014. 裂壶藻对刺参生长、免疫及消化酶的影响. 渔业科学进展, 35(3): 91-97.

黄文庆, 王国霞, 罗志锋, 等. 2012. 玉米 DDGS 替代豆粕对草鱼生长性能血清化指标和免疫指标的影响. 饲料工业, 33(16): 33-36.

黄燕华, 王国霞, 黄文庆, 等. 2009. 酶制剂对南美白对虾幼虾生长、体组成及非特异免疫的影响. 第六届饲料安全与生物技术专业委员会大会暨第三届全国酶制剂在饲料工业中应用学术与技术研讨会. 北京: 中国农业科学技术出版社, 175-180.

黄云, 胡毅, 肖调义, 等. 2012. 双低菜粕替代豆粕对青鱼幼鱼生长及生理生化指标的影响. 水生生物学报, 36(1): 41-48.

江晓路, 杜以帅, 王鹏, 等. 2009. 褐藻寡糖对刺参体腔液和体壁免疫相关酶活性变化的影响. 中国海洋大学学报(自然科学版), 39(6): 1188-1192

姜令绪, 杨宁, 李建, 等. 2007. 温度和pH对刺参(*Apostichopus japonicus*)消化酶活力的影响. 海洋与湖沼, 38(5): 476-480.

姜婷婷. 2013. 外源添加木聚糖酶对幼建鲤生长性能、消化吸收能力、肠道菌群和免疫功能的影响. 雅安: 四川农业大学.

姜燕, 王印庚, 薛太山, 等. 2012. 刺参池塘养殖系统中发酵饲料的制作与投喂. 渔业科学进展, 33(1): 66-71.

蒋明, 吴凡, 文华, 等. 2009. 饲料中添加不同水平的维生素 D_3 对草鱼幼鱼生长和体成分的影响. 淡水渔业, 39(5): 38-42.

蒋明. 2007. 草鱼幼鱼对维生素 A、D 和 K 需要量的研究. 武汉: 华中农业大学.

解俊美, 王安. 2012. 饲粮维生素 D 添加水平对蛋雏鸭免疫及抗氧化功能的影响. 动物营养学报, 24(9): 1819-1824.

黎德兵, 邵珊珊, 张龚炜, 等. 2015. 饲料中维生素 D_3 添加水平对黄鳝生长性能及免疫功能的影响. 动物营养学报, 27(4): 1145-1151.

李爱杰. 1996. 水产动物营养与饲料学. 北京: 中国农业出版社.

李宝山, 王际英, 王成强, 等. 2019. 仿刺参幼参对饲料中维生素 B6 需求量的研究. 水产学报, 43(12): 2545-2553.

李宝山, 王丽丽, 王际英, 等. 2023. 刺参幼参对维生素 A 最适需求量. 水产学报, 47(8): 126-134.

李宝山, 张利民, 张德瑞, 等. 2017. 发酵豆粕替代藻粉对刺参(*Apostichopus japonicus*)生长及体组成的影响. 渔业科学进展, 38(5): 130-139.

李成军, 雷帅. 2016. 刺参吊笼养殖技术. 现代农业, (8): 13.

李春艳, 常亚青. 2006. 海参的营养成分介绍. 科学养鱼, (2): 71-72.

李丹彤, 常亚青, 陈炜, 等. 2006. 獐子岛野生刺参体壁营养成分的分析. 大连水产学院学报, 21(3): 278-282.

李丹彤, 范友敏, 常亚青, 等. 2014. 海带和条斑紫菜凝集素提取液对刺参体壁主要免疫酶活性的影响. 大连海洋大学学报, 29(6): 607-612.

李二超, 陈立侨, 顾顺樟, 等. 2009. 水产饲料蛋白源营养价值的评价方法. 海洋科学, 33(7): 113-117

李二超, 曾嶒, 禹娜, 等. 2010. 饲料蛋白质和维生素 B_6 对低盐度下凡纳滨对虾生长和转氨酶活力的影响. 动物营养学报, 22(3): 634-639.

李桂梅, 解绶启, 雷武, 等. 2010. 异育银鲫幼鱼对饲料中缬氨酸需求量的研究. 水生生物学报, 34(6): 1157-1165.

李桂梅. 2009. 异育银鲫幼鱼对饲料苏氨酸、亮氨酸、缬氨酸和异亮氨酸需求量的研究. 武汉: 中国科学院水生生物研究所.

李行先, 陆清儿, 赵芸, 等. 2005. 丁(鱼岁)不同发育阶段鱼体营养成分和氨基酸组成的比较分析. 淡水渔业, 35(2): 23-24.

李红岗, 郭德姝, 李辉. 2010. 木聚糖酶添加量对黄河鲤血糖水平及增重率的影响. 河南水产, (2): 33-34.

李华磊. 2014. 玉米酒糟及其可溶物对免疫抑制肉鸡肉品质、抗氧化性能和免疫功能的影响. 咸阳: 西北农林科技大学.

李继业. 2007. 养殖刺参免疫学特征与病害研究. 青岛: 中国海洋大学.

李君华, 刘佳亮, 曹学彬, 等. 2016. 饲料中添加低聚木糖对仿刺参幼参生长性能、肠道消化酶活力和免疫力的影响. 动物营养学报, 28(8): 2534-2541.

李猛, 廖梅杰, 常青, 等. 2017. 不同浒苔型饲料对幼刺参(*Apostichopus japonicus*)生长、消化及非特异性免疫的影响. 渔业科学进展, 38(5): 122-129.

李猛, 廖梅杰, 王印庚, 等. 2015. 浒苔添加比例与微生物发酵对幼刺参生长、消化和非特异性免疫的影响. 动物营养学报, 27(10): 3270-3278.

李平凡, 钟彩霞. 2012. 淀粉糖与糖醇加工技术. 北京: 中国轻工业出版社.

李清, 肖调义, 毛华明. 2005. 生物活性肽对鲤鱼血液生理生化指标的影响. 长江大学学报: 自然科学版, 2(5): 27-29.

李淑霞. 2010. 海参蛋白酶与淀粉酶性质的研究. 大连: 大连工业大学.

李素红, 梁萌清, 孙慧玲, 等. 2012. 饲料中适宜蛋白质和氨基酸水平对刺参生长的影响. 渔业科学进展, 33(5): 59-63.

李卫芬, 占秀安, 陆清儿. 1998. 中华鳖不同发育阶段氨基酸组成的研究. 江西农业大学学报, 20(3): 3.

李晓丽, 吕林, 解竞静, 等. 2013. 锰在鸡肠道中吸收的特点、影响因素及分子机制. 动物营养学报, 25(3): 486-493.

李旭, 章世元, 陈四清, 等. 2013. 四种饲料原料对刺参生长、体成分及消化生理的影响. 饲料工业, 34(8): 36-40.

李旭, 章世元, 陈四清, 等. 2014. 四种饲料原料对刺参蛋白质代谢及免疫功能的影响. 广东饲料, 23(10): 29-32.

李燕. 2010. 鲈鱼和大黄鱼支链氨基酸与组氨酸营养生理的研究. 青岛: 中国海洋大学.

李治龙, 孟良玉. 2010. 发酵食品工艺. 北京: 中国计量出版社.

李忠清, 夏斌, 王际英, 等. 2016. 青、白刺参(*Apostichopus japonicus*)体壁营养成分的比较分析. 渔业科学进展, 37(3): 101-107.

梁永, 陈中平, 杨洪森. 2012. 木聚糖酶在饲料应用上的研究进展. 饲料研究, 35(4): 30-31.

廖国周, 柴仕名. 2005. 支链氨基酸营养研究进展. 饲料博览, (4): 7-10.

廖明玲. 2014. 刺参幼参脂肪及 n-3 系列高度不饱和脂肪酸需求的研究. 大连: 大连海洋大学.

林少珍, 刘伟成, 单乐州. 2009. 酶制剂在水产养殖饲料中的应用. 水产科学, 28(12): 798-800.

刘爱君, 冷向军, 李小勤, 等. 2009. 甘露寡糖对奥尼罗非鱼(*Oreochromis niloticus× O. aureus*)生长、肠道结构和非特异性免疫的影响. 浙江大学学报: 农业与生命科学版, 35(3): 329-336.

刘财礼, 王际英, 李宝山, 等. 2022. 仿刺参幼参对亮氨酸最适需求量的研究. 水产学报, 46(2): 238-249.

刘晨光, 刘成圣, 刘万顺, 等. 2000. 海洋生物酶的研究和应用. 海洋科学, 24(7): 24-26.

刘存歧, 王伟伟, 张亚娟. 2005. 水生生物超氧化物歧化酶的酶学研究进展. 水产科学, 24(11): 49-52.

刘红英, 薛长湖, 李兆杰, 等. 2002. 海带岩藻聚糖硫酸酯测定方法的研究. 青岛海洋大学学报, 32(2): 236-240.

刘镜格, 陈晓琳, 李岿然, 等. 2005. 实验微粒饲料中花生四烯酸含量对牙鲆(*Paralichthys divaceus*)仔稚鱼生长、存活的影响. 海洋与湖沼, 36(5): 418-422.

刘凯, 麦康森, 艾庆辉, 等. 2010. 军曹鱼幼鱼对吡哆醇的需要量. 水产学报, 34(2): 307-314.

刘凯, 张玲, 曹香林, 等. 2008. 木聚糖酶在鲤鱼小麦基础饲料中的应用研究. 水利渔业, 28(4): 55-57.

刘凯, 张玲, 聂国兴. 2012. 小麦基础饲料添加木聚糖酶对鲤鱼体成分及消化酶活力的影响. 饲料工业, 33(14): 37-40.

刘梅, 朱曦露, 苏艳秋, 等. 2016. 微藻和动物性生物饵料在水产养殖中的应用研究. 海洋与渔业, (4): 56-57.

刘梦梅, 陈娇娇, 朱文欢, 等. 2017. 维生素 A 对养成期草鱼生长性能以及骨骼中钙磷含量变化的影响. 水生生物学报, 41(1): 101-107.

刘梦坛, 李超伦, 孙 松. 2010. 两种甲藻和两种硅藻脂肪酸组成的比较研究. 海洋科学, 34 (10): 77-82.

刘青, 赵恒寿. 2007. 鱼类常用免疫指标及其检测技术. 渔业现代化, 34(3): 28-30, 27.

刘伟成, 冀德伟, 周志明, 等. 2010. 木聚糖酶添加对条石鲷幼鱼能量收支的影响. 宁波大学学报(理工版), 23(3): 6-10.

刘小芳, 薛长湖, 王玉明, 等. 2011. 乳山刺参体壁和内脏营养成分比较分析. 水产学报, 35(4): 587-593.

刘小芳. 2014. 刺参营养成分的地域性差异分析及其磷脂的活性研究. 青岛: 中国海洋大学.

刘长琳, 王有廷, 秦搏, 等. 2015. 蓬莱玉参(*Apostichopus* sp.)体壁的营养成分分析及评价. 渔业科学进展, 36(5): 111-118.

路凯. 2015. 花鲈对亮氨酸、异亮氨酸和色氨酸需求量的研究. 青岛: 中国海洋大学.

罗莉, 王亚哥, 李芹, 等. 2010. 草鱼幼鱼对缬氨酸需要量的研究. 动物营养学报, 22(3): 616-624.

吕永智. 2018. 葡萄糖感应机制与肠道内分泌调控的研究进展. 黑龙江畜牧兽医, (1): 80-82.

马得友. 2013. 基于高通量测序的刺参白化发生和子代体色分离研究. 青岛: 中国科学院研究生院(海洋研究所).

马晶晶, 王际英, 孙建珍, 等. 2014. 饲料中 DHA/EPA 值对星斑川鲽幼鱼生长、体组成及血清生理指标的影响. 水产学报, 38(2): 244-256.

马同江, 周清凯, 蔡云见. 1982. 海参的药理作用及应用. 海洋药物, 2(2): 9-13

马文强, 冯杰, 刘欣. 2008. 微生物发酵豆粕营养特性研究. 中国粮油学报, 23(1): 121-124

马悦欣, 许珂, 王银华, 等. 2010. κ-卡拉胶寡糖对仿刺参溶菌酶、碱性磷酸酶和超氧化物歧化酶活性的影响. 大连海洋大学学报, 25(3): 224-227

马跃华, 胡守义. 2006. 免疫多糖投喂海参幼体试验. 河北渔业, 151(7): 22-23.

明红, 刘涌涛, 杜习翔, 等. 2006. 木聚糖酶对尼罗罗非鱼生长及血脂血糖水平的影响. 新乡医学院学报, 23(6): 556-558.

明建华. 2007. 酶制剂在水产动物饲料中的应用. 饲料工业, 28(10): 17-19.

聂国兴, 明红, 张玲, 等. 2006. 外源木聚糖酶对尼罗罗非鱼消化器官消化酶活力及分布的影响. 华北农学报, 21(4): 123-130.

聂国兴, 明红, 郑俊林, 等. 2007a. 木聚糖酶对尼罗罗非鱼血液生理生化指标的影响. 大连水产学院学报, 22(5): 361-365.

聂国兴, 王俊丽, 周洪琪, 等. 2007b. 饲料中添加木聚糖酶对尼罗罗非鱼生长及血清激素水平的影响. 中国水产科学, 14(2): 251-256.

聂国兴, 王俊丽, 朱命炜, 等. 2007c. 小麦基础饲料添加木聚糖酶对尼罗罗非鱼肠道食糜粘度和绒毛、微绒毛发育的影响. 水产学报, 31(1): 54-61.

聂品. 1997. 鱼类非特异性免疫研究的新进展. 水产学报, 21(1): 69-73.

牛娟娟, 宋扬. 2009. 刺参粘多糖的生物活性物质作用研究进展. 华北煤炭医学院学报, (11): 651-652.

欧阳亮, 李亮, 蔡源锋. 2008. 垂直型 SDS-PAGE 分析发酵豆粕中蛋白类抗营养因子的研究. 饲料工业, 29(21): 43-45.

彭士明, 施兆鸿, 孙鹏, 等. 2012. 饲料组成对银鲳幼鱼生长率及肌肉氨基酸、脂肪酸组成的影响. 海洋渔业, 34(1): 51-56.

彭素晓. 2017. 海带渣的酶解工艺优化及在对虾养殖中的应用. 上海: 上海海洋大学.

鹏翔, 邵庆均. 2010. 维生素 D 在水产动物中的研究进展. 饲料工业, 31(22): 47-49.

强俊, 王辉, 李瑞伟, 等. 2009. 低聚木糖对奥尼罗非鱼幼鱼生长、体成分和消化酶活力的影响. 淡水渔业, 39(6): 63-68.

秦搏, 常青, 陈四清, 等. 2015. 饲料中浒苔添加量以及处理方法对幼刺参生长、消化率、消化酶和非特异性免疫酶的影响. 水产学报, 39(4): 547-556.

邱燕. 2010. 三种微生态制剂对草鱼(*Ctenopharyngodon idellus*)生长性能、生理机能及肠道黏膜的影响. 苏州: 苏州大学.

任和, 占秀安. 2006. 水产动物氨基酸营养研究进展. 饲料研究, 29(2): 41-43.

任红立, 汪晶晶, 宋建楼, 等. 2016. 功能性低聚糖的研究进展. 饲料博览, (4): 35-39.

任庆印, 潘鲁青. 2013. 刺参消化酶性质与活性分布的研究. 海洋湖沼通报, (2): 51-56.

阮同琦, 赵祥颖, 刘建军. 2008. 木聚糖酶及其应用研究进展. 山东食品发酵, (1): 42-45.

邵庆均, 苏小凤, 许梓荣. 2004. 饲料蛋白水平对宝石鲈增重和胃肠道消化酶活性影响. 浙江大学学报: 农业与生命科学版, 30(5): 553-556.

沈文英, 胡洪国, 潘雅娟. 2004. 温度和pH值对南美白对虾(*Penaeus vannmei*)消化酶活性的影响. 海洋与湖沼, 35(6): 543-548.

石军, 陈安国. 2002. 木聚糖酶的应用研究进展. 中国饲料, (4): 10-12.

石亚庆, 孙玉轩, 罗莉, 等. 2014. 吉富罗非鱼亮氨酸需求量研究. 水产学报, 38(10): 1778-1785.

时皎皎, 糜漫天, 韦娜, 等. 2012. 不同膳食脂肪酸构成对大鼠肝脏脂代谢相关基因表达的影响. 重庆医学, 71(2): 1957-1966.

司滨. 2017. 壳寡糖对仿刺参养殖的应用效果研究. 大连: 大连海洋大学.

宋志东, 王际英, 王世信, 等. 2009. 不同生长发育阶段刺参体壁营养成分及氨基酸组成比较分析. 水产科技情报, 36(1): 11-13.

苏秀榕, 李太武, 丁明进, 等. 1997. 扇贝营养成分的研究. 海洋科学, (2): 10-11.

孙红梅. 2004. 饥饿对黄颡鱼血液指标及免疫机能的影响. 长春: 吉林农业大学.

孙虎山, 李光友. 2000. 栉孔扇贝血淋巴中超氧化物歧化酶和过氧化氢酶活性及其性质的研究. 海洋与湖沼, 31(3): 259-265.

孙虎山, 李光友. 2002. 硒化卡拉胶和酵母葡聚糖对栉孔扇贝血淋巴中两种水解酶活力的影响. 海洋与湖沼, 33(3): 245-249.

孙纪录, 贾英民, 田洪涛, 等. 2003. 双歧杆菌生长促进因子的研究进展. 食品工业科技, 24(8): 110-112.

孙静秋, 许燕, 张慧绮, 等. 2007. 凡纳对虾体内ACP、AKP酶的细胞化学定位. 复旦学报(自然科学版), 46(6): 947-951.

孙丽娜. 2013. 刺参 *Apostichopus japonicus* (Selenka)消化道再生的组织细胞特征与关键基因分析. 北京: 中国科学院大学.

孙淑洁, 王宝维, 葛文华, 等. 2012. 维生素A对鹅生长性能及血清生化指标的影响. 动物营养学报, 24(1): 78-84.

孙新瑾. 2009. 中华绒螯蟹(Eriocheir sinensis)幼蟹对脂溶性维生素A、D3和K3的营养需要研究. 华东师范大学学报, 40-44.

孙永泰. 2015. 海带粉在饲料中的应用. 江西饲料, (4): 23-25, 32

孙永欣. 2008. 黄芪多糖促进刺参免疫力和生长性能的研究. 大连: 大连理工大学.

索兰弟, 魏建民, 闫素梅. 2002. 日粮锌和维生素A水平及其交互作用对肉仔鸡血清碱性磷酸酶活性的影响. 内蒙古畜牧科学, 23(6): 1-3.

谭北平, 周歧存, 郑石轩, 等. 2003. beta-1, 3/1, 6 葡聚糖制剂对凡纳对虾生长及免疫力的影响. 高技术通讯, 14: 73-77.

谭青. 2017. n-3/n-6HUFA对大菱鲆幼鱼生长、体组成、脂肪酸成分及免疫的影响. 上海: 上海海洋大学.

汤海燕. 2010. 刺海参的育苗与养殖. 特种经济动植物, 13(1): 32-33.

唐宏刚. 2008. 鱼蛋白水解物对大黄鱼生长代谢、肌肉品质、免疫及抗氧化性能的影响. 杭州: 浙江大学, 26.

唐黎, 王吉桥, 许重, 等. 2007. 不同发育期的幼体和不同规格刺参消化道中四种消化酶的活性. 水产科学, 26(5): 275-277.

唐薇, 王庆吉, 张蕾, 等. 2014. 不同藻粉对刺参组织免疫性能和体壁成分的影响. 资源开发与市场, 30(8): 905-907.

唐雪, 徐钢春, 徐跑. 2011. 野生与养殖刀鲚肌肉营养成分的比较分析. 动物营养学报, 23(3): 514-520.

滕怀丽, 黄旭雄, 周洪琪, 等. 2010. 充气方式对盐藻生长、细胞营养成分及氮磷营养盐利用的影响. 水产学报, 34(6): 762-768.

万军利, 麦康森, 艾庆辉. 2006. 鱼类精氨酸营养生理研究进展. 中国水产科学, 13(4): 679-685.

汪将. 2015. 维生素E对环境胁迫下刺参生长和免疫的影响. 大连: 大连理工大学.

王彬, 印遇龙, 黄瑞林, 等. 2006. 半乳甘露寡糖对母猪泌乳性能和血清激素水平的影响. 西南农业大学学报, 28(1): 115-119, 123.

王昌义, 单守水, 徐世艾. 2012. 预消化法酶解饲用海藻粉中非淀粉多糖的初步研究. 饲料工业, 33(2): 44-48.

王玮韡, 瞿明仁. 2006. 浅谈几种常见酶制剂的研究及其应用. 饲料工业, 27(24): 55-59.

王成强, 梁萌青, 徐后国, 等. 2016. 大规格鲈鱼(*Lateolabrax japonicus*)对饲料中花生四烯酸的需求量. 渔业科学进展, 37(5): 46-55.

王成强, 王际英, 李宝山, 等. 2018. 饲料中花生四烯酸含量对刺参生长性能、抗氧化能力及脂肪酸代谢的影响. 中国水产学报, 25(3): 555-566.

王成强. 2016. 饲料花生四烯酸、亚麻酸含量及亚麻酸/亚油酸比值对大规格鲈鱼生长性能、脂肪酸组成和脂肪沉积的影响. 上海: 上海海洋大学.

王冠. 2005. 利用微胶囊技术改善晶体氨基酸添加效果的研究. 上海: 上海水产大学.

王广军, 谢骏, 余德光, 等. 2008. 注射L-精氨酸和环磷酰胺对杂色鲍血清NO水平、NOS活性及免疫指标的影响. 中国水产科学, 15(6): 902-909.

王吉桥, 蒋湘辉, 姜玉声, 等. 2009. 在饲料中添加包膜赖氨酸对仿刺参幼参生长、消化和体成分的影响. 水产科学, 28(5): 241-245.

王吉桥, 蒋湘辉, 赵丽娟, 等. 2007a. 不同饲料蛋白源对仿刺参幼参生长的影响. 饲料博览, 19: 9-13.

王吉桥, 隋晓楠, 顾公明, 等. 2008. 不同饲料搭配及投喂量对仿刺参稚、幼参生长和成活的影响. 水产科学, 27(2): 55-58.

王吉桥, 唐黎, 许重, 等. 2007b. 仿刺参消化道的组织学及其4种消化酶活力的周年变化. 水产科学, 26(9): 481-484.

王吉桥, 张凯, 袁成玉, 等. 2013. 不同比例微藻和微生态制剂对仿刺参幼参生长的影响. 水产科学, 32(9): 524-530.

王吉潭, 李德发, 龚利敏, 等. 2003. 半乳甘露寡糖对肉鸡生产性能和免疫机能的影响. 中国畜牧杂志, 39(2): 5-7.

王际英, 李宝山, 马晶晶, 等. 2010. 褐牙鲆亲鱼野生群体与养殖群体维生素A、C、E含量的比较. 中国水产科学, 17(6): 1250-1256.

王际英, 李宝山, 张德瑞, 等. 2015. 饲料中添加精氨酸对仿刺参幼参生长、免疫能力及消化酶活力的影响. 水产学报, 39(3): 410-420.

王际英, 宋志东, 李培玉, 等. 2014. 饲料添加半乳甘露寡糖对刺参幼参生长、体壁营养组成及免疫力的影响. 中国水产科学, 21(2): 310-319.

王际英, 宋志东, 王世信, 等. 2009. 刺参不同发育阶段对蛋白质需求量的研究. 水产科技情报, 36(5): 229-233.

王建学, 卫育良, 徐后国, 等. 2021. 红鳍东方鲀对 8 种饲料原料的表观消化率. 渔业科学进展, 42(2): 96-103.

王晶, 王加启, 卜登攀, 等. 2009. DDGS 的营养价值及在动物生产中的应用研究进展. 中国畜牧杂志, 45(23): 71-75

王镜岩, 朱圣庚, 徐长法. 2007. 生物化学第三版(下册). 北京: 高等教育出版社.

王俊丽, 于广丽, 刘凯, 等. 2007. 饲料中添加木聚糖酶对尼罗罗非鱼生长性能的影响. 华北农学报, 22(3): 178-182.

王可宝. 2011. 饲料中不同水平维生素 D_3 对团头鲂生产性能、非特异性免疫及抗病原菌感染的影响. 南京: 南京农业大学.

王丽丽, 李宝山, 王际英, 等. 2019. 维生素 D_3 对刺参幼参生长、体组成及抗氧化能力的影响. 渔业科学进展, 40(1): 110-118.

王莉丽, 梅文泉, 陈兴连, 等. 2020. 3, 5-二硝基水杨酸比色法测定大米中水溶性糖含量. 中国粮油学报, 35(9): 168-173.

王美琪, 宋志东, 郭鹏, 等. 2023. 饲料中添加酶解海带粉对刺参幼参生长、体组成、消化代谢和抗氧化能力的影响. 渔业科学进展, 44(3): 176-187.

王鹏, 江晓路, 江艳华, 等. 2006. 褐藻低聚糖对提高大菱鲆免疫机能的作用. 海洋科学, 30(8): 6-9.

王庆吉, 唐薇, 罗亚军, 等. 2014. 日粮中蛋白质含量对刺参生长和免疫的影响. 饲料工业, 35(22): 32-36.

王秋梅. 2007. 微量元素锰在家禽生产中的应用. 畜禽业, (12): 12-14.

王锐, 刘军, 刘辉宇, 等. 2008. 半乳甘露寡糖对异育银鲫幼鱼生长和非特异性免疫的影响. 上海水产大学学报, 17(4): 502-506.

王文娟, 叶元土, 蔡春芳, 等. 2010. 豆粕及其抗营养因子对异育银鲫血清生化和非特异性免疫指标的影响. 中国饲料, (18): 30-33, 41.

王熙涛, 徐永平, 尤建嵩, 等. 2014. 刺参饲料的使用与研究新进展. 饲料工业, 35(21): 65-69.

王晓安, 蒋小满, 郑哲民. 2003. 软体动物的一氧化氮及其合酶的研究进展. 动物学杂志, 38(6): 97-103.

王晓艳, 乔洪金, 黄炳山, 等. 2018. 海藻在刺参养殖中的营养应用研究进展. 海洋渔业, 40(3): 377-384.

王晓艳, 乔洪金, 黄炳山, 等. 2019. 5 种海藻在刺参幼参饲料中的应用研究. 渔业科学进展, 40(3): 160-167.

王兴春. 2014. 饲料对吊笼养殖刺参生长及体成分的影响. 海洋渔业, 36(5): 453-460.

王雪良. 2008. 酵母 β-葡聚糖对中华绒螯蟹免疫功能的影响. 苏州: 苏州大学.

王永辉, 李培兵, 李天, 等. 2010. 刺参的营养成分分析. 氨基酸和生物资源, 32(4): 35-37.

王用黎. 2013. 凡纳滨对虾幼虾对苏氨酸、亮氨酸、色氨酸和缬氨酸需要量的研究. 湛江: 广东海洋大学.

王玥, 孙新瑾, 陈立侨, 等. 2009. 中华绒螯蟹幼蟹对饲料中维生素 B_6 的适宜需求量. 华东师范大学学报(自然科学版), (6): 47-55.

王哲平, 刘淇, 曹荣, 等. 2012. 野生与养殖刺参营养成分的比较分析. 南方水产科学, 8(2): 64-70.

王震, 徐玮, 麦康森, 等. 2016. 饲料缬氨酸水平对军曹鱼鱼体脂肪含量、血浆生化指标和肝脏脂肪代谢基因表达的影响. 水生生物学报, 40(4): 744-751.

魏东, 俞建中. 2014. 微藻在水产养殖和新型饲料开发中的应用. 海洋与渔业·水产前沿, (4): 84-89.

魏玉婷. 2010. 大菱鲆 (*Scophthalmus maximus*) 幼鱼对饲料中蛋氨酸、精氨酸、维生素 A 及维生素 E 需求量的研究. 青岛: 中国海洋大学.

邬佳颖, 陈敏暄, 金天赐, 等. 2021. 双歧杆菌和乳杆菌对水苏糖的利用特性. 食品与发酵工业, 47(24): 13-20.

吴凡, 任春, 文华, 等. 2018. 吉富罗非鱼对饲料中维生素 B_6 的需要量. 动物营养学报, 30(5): 1781-1788.

吴海歌, 于超, 姚子昂, 等. 2008. 鼠尾藻营养成分分析. 大连大学学报, 29(3): 84-85.

吴韬, 张振龙, 蔡春芳, 等. 2015. 果胶和木聚糖对中华绒螯蟹生长性能和消化生理的影响. 动物营养学报, 27(7): 2282-2291.

吴阳. 2012. 寡糖和益生菌对团头鲂生长、消化及免疫抗氧化的影响. 南京: 南京农业大学.

吴永恒, 王秋月, 冯政夫, 等. 2012. 饲料粗蛋白含量对刺参消化酶及消化道结构的影响. 海洋科学, 36(1): 36-41.

武明欣, 王际英, 李宝山, 等. 2015. 酶制剂对刺参生长、体成分、免疫能力及氨氮胁迫下免疫酶活力和热休克蛋白 70 含量的影响. 动物营养学报, 27(4): 1293-1301.

武明欣, 王雅平, 李培玉, 等. 2018. 饲料中添加木聚糖酶对刺参幼参生长、消化和体腔液酶活力的影响. 大连海洋大学学报, 33(3): 329-335.

夏斌, 王际英, 李培玉, 等. 2015. 饲料中不同碳水化合物水平对刺参幼参生长和能量收支的影响. 中国水产科学, 22(4): 645-653.

向怡卉, 苏秀榕, 董明敏. 2006. 海参体壁及消化道的氨基酸和脂肪酸分析. 水产科学, 25(6): 280-282.

肖宝华, 杨小东, 劳赞, 等. 2014. 北方刺参与南方糙海参口感及营养成分比较分析. 水产科技情报, 41(6): 280-289.

肖露, 崔婷, 李莹莹, 等. 2017. 孕期开始的持续维生素 A 缺乏加重 LPS 诱导的仔鼠肠上皮屏障功能障碍. 免疫学杂志, 33(10): 837-843, 849.

谢奉军. 2011. 大黄鱼仔稚鱼氨基酸及脂肪酸营养生理的研究. 青岛: 中国海洋大学.

谢瑾. 2018. 微生物发酵法提高水苏糖纯度的研究. 广州: 华南理工大学.

谢玉英. 2009. 谈活性氧与人类疾病. 现代农业科技, (15): 285-286, 288.

徐志昌, 刘铁斌, 李爱杰. 1995. 中国对虾对维生素 B_2、B_5、B_6 营养需要的研究. 水产学报, 19(2): 97-104.

徐宗法, 毕庶万, 王际英, 等. 1999. 饵料对稚幼参生长变色的影响. 齐鲁渔业, 16(1): 30-33.

严晶, 曹俊明, 王国霞, 等. 2012. 家蝇蛆粉替代鱼粉对凡纳滨对虾肌肉营养成分、氨基酸和肌苷酸含量的影响. 中国水产科学, 19(2): 265-274.

杨宝灵, 姜健, 王冰, 等. 2009. 海胆微量元素的光谱测定. 安徽农业科学, 37(2): 613-615.

杨东升. 2004. 双歧杆菌及寡糖类双歧因子研究. 中国食品添加剂, 75(2): 20-24.

杨耐德, 符广才. 2008. 凡纳滨对虾饲料中发酵豆粕替代鱼粉的研究. 饲料工业, 29(10): 24-26.

杨宁, 郭中帅, 王正丽. 2016. 饲料中添加浒苔对刺参幼参生长、消化酶活性和免疫力的影响. 水产科学, 35(5): 498-503.

杨奇慧, 周歧存, 迟淑艳, 等. 2007. 饲料中维生素A水平对凡纳滨对虾生长、饲料利用、体组成成分及非特异性免疫反应的影响. 动物营养学报, 19(6): 698-705.

杨奇慧, 周小秋. 2005. 维生素A缺乏对建鲤生长性能及免疫功能的影响. 中国水产科学, 12(1): 62-67.

杨霞, 叶金云, 周志金, 等. 2014. 中华绒螯蟹幼蟹对亮氨酸和异亮氨酸的需要量. 水生生物学报, 38(6): 1062-1070.

杨先乐. 1989. 鱼类免疫学研究的进展. 水产学报, 13(3): 271-284.

杨志刚, 陈乃松, 郑剑伟, 等. 2009. 酶解小麦对凡纳滨对虾生长性能的影响. 粮食与饲料工业, (2): 38-40.

姚骏, 张弘, 郭森, 等. 2018. 海带的生物活性及系列产品开发研究进展. 食品研究与开发, 39(8): 198-202.

姚朋波. 2018. 苹果酸酶调控磷酸戊糖途径的分子机制研究. 北京: 清华大学.

姚永峰, 张继红, 方建光, 等. 2014. 温度、饵料质量对不同规格刺参摄食率、吸收率的影响. 水产学报, 38(7): 992-998.

叶金云, 王友慧, 郭建林, 等. 2010. 中华绒螯蟹对赖氨酸、蛋氨酸和精氨酸的需要量. 水产学报, 34(10): 1541-1548.

叶元土, 王永玲, 蔡春芳, 等. 2007. 谷氨酰胺对草鱼肠道L-亮氨酸、L-脯氨酸吸收及肠道蛋白质合成的影响. 动物营养学报, 19(1): 28-32.

叶元土, 王友慧, 林仕梅, 等. 2003. 草鱼肠道对10种必需氨基酸的跨壁运输量. 中国水产科学, 10(4): 311-317.

殷旭旺, 李文香, 白海锋, 等. 2015. 不同海藻饲料对刺参幼参生长的影响. 大连海洋大学学报, 30(3): 276-280.

于东祥, 孙慧玲, 陈四清, 等. 2010. 海参健康养殖技术. 第2版. 北京: 海洋出版社, 46-61.

于瑞海, 周文江, 孔静. 2012. 海参池塘高产稳产养殖技术的探讨. 海洋湖沼通报, 34(3): 25-28.

于世浩, 何伯峰, 赵倩, 等. 2009. 海参营养与饲料研究现状. 饲料研究, 32(10): 53-54.

于艳梅. 2010. 魔芋甘露寡糖对黄颡鱼的益生功能研究. 武汉: 华中农业大学.

余致远. 2015. 两种饲料对海上筏式吊笼养殖仿刺参生长的影响. 福建水产, 37(1): 73-78.

袁成玉. 2005. 海参饲料研究的现状与发展方向. 水产科学, 24(12): 54-56.

袁秀堂, 杨红生, 王丽丽, 等. 2007. 夏眠对刺参(*Apostichopus japonicus* (Selenka))能量收支的影响. 生态学报, 27(8): 3155-3161.

臧元奇, 田相利, 董双林, 等. 2012. 氨氮胁迫对刺参免疫酶活性及热休克蛋白表达的影响. 中国海洋大学学报, 6(42): 60-66.

臧元奇. 2012. 刺参(*Apostichopus japonicus* Selenka)对典型环境胁迫的生理生态学响应及其机制研究. 青岛: 中国海洋大学.

詹冬梅, 李成林, 菅玉霞, 等. 2004. 絮凝浓缩藻膏投喂刺参饵料效果的研究. 中国水产, (z1): 213-215.

展学孔, 周海妹, 马小花, 等. 2011. 海参多糖提取新工艺. 中国实验方剂学杂志, 17(15): 40-42.

张傲. 2020. 东北林蛙油中糖和脂类及其代谢相关因子表达的研究. 北京: 北京林业大学.

张彩霞, 陈文, 黄艳群, 等. 2010. 限饲对哈巴德肉鸡肠道结构的影响. 江西农业大学学报, 32(4): 677-682.

张春晓, 麦康森, 艾庆辉, 等. 2008. 饲料中添加外源酶对大黄鱼和鲈氮磷排泄的影响. 水生生物学报, 32(2): 231-236.

张德瑞, 张利民, 马晶晶, 等. 2016. 配合饲料中添加玉米 DDGS 对刺参(*Apostichopus japonicus*)生长、体组成及免疫指标的影响. 渔业科学进展, 37(6): 115-122.

张佳明. 2007. 鲈鱼和大黄鱼微量元素——锌、铁的营养生理研究. 青岛: 中国海洋大学.

张杰. 2010. 刺海参育苗及养殖技术. 农村实用技术, (2): 58-59.

张俊杰, 段蕊, 许可, 等. 2010. 海带工业中海带渣应用的研究及展望. 水产科学, 29(10): 620-623.

张丽. 2016. 维生素 A 对生长中期草鱼生产性能、肠道、机体和鳃健康以及肌肉品质的作用及作用机制. 雅安: 四川农业大学.

张丽靖, 齐莉莉, 杨郁. 2008. 纳豆菌发酵对豆粕脲酶活性的影响. 大豆科学, 27(4): 669-671.

张玲, 聂国兴, 周洪琪. 2006. 木聚糖酶对鲫鱼生长性能和小肠绒毛的影响. 浙江海洋学院学报: 自然科学版, 25(2): 133-137.

张璐, 艾庆辉, 麦康森, 等. 2008. 肽聚糖对鲈鱼生长和非特异性免疫力的影响. 中国海洋大学学报, 38(4): 551-554.

张璐, 李静, 麦康森, 等. 2016. 饲料中不同维生素 D 含量对鲈鱼幼鱼生长性能和钙磷代谢的影响. 动物营养学报, 28(5): 1402-1411.

张璐, 李静, 谭芳芳, 等. 2015. 饲料中不同维生素 A 含量对花鲈生长和血清生化指标的影响. 水产学报, 39(1): 88-96.

张璐, 麦康森, 艾庆辉, 等. 2006. 饲料中添加植酸酶和非淀粉多糖酶对大黄鱼生长和消化酶活性的影响. 中国海洋大学学报, 36(6): 923-928.

张琴. 2010. 刺参(*Apostichopus japonicus* Selenka)高效免疫增强剂的筛选与应用. 青岛: 中国海洋大学.

张淑云, 王安. 2010. 钙和维生素 D 对生长肉鸡免疫及抗氧化功能的影响. 动物营养学报, 22(3): 579-585.

张桐, 徐奇友, 许红, 等. 2011. 饲料中维生素 D_3 对松浦镜鲤幼鱼体成分和血清碱性磷酸酶的影响. 华北农学报, 26(S1): 258-263.

张永亮, 张浩江, 谢水波, 等. 2009. 藻类吸附重金属的研究进展. 铀矿冶, 28(1): 31-37.

张宇鹏, 田燚, 商艳鹏, 等. 2017. 镉对刺参幼参体内金属硫蛋白含量及其变化规律的影响. 大连海洋大学学报, 32(2): 178-182.

张圆琴, 徐后国, 曹林, 等. 2017. 饲料中花生四烯酸对发育前期大菱鲆亲鱼性类固醇激素合成的影响. 水产学报, 41(4): 588-601.

张圆圆, 王连生. 2020. 水产动物亮氨酸营养研究进展. 动物营养学报, 32(12): 5516-5523.

章超桦, 吴红棉, 洪鹏志, 等. 2000. 马氏珠母贝肉的营养成分及其游离氨基酸组成. 水产学报, 24(2): 180-184.

赵斌, 胡炜, 李成林, 等. 2015. 低盐环境对 3 种规格刺参(*Apostichopus japonicus*)幼参生长与消化酶活力的影响. 渔业科学进展, 36(1): 91-96.

赵斌, 胡炜, 李成林, 等. 2016. 两种甘薯饲料原料的营养成分及其对仿刺参(*Apostichopus japonicus*)摄食与生长的影响. 渔业科学进展, 37(1): 80-86.

赵鹤凌, 杨红生, 赵欢, 等. 2012. 刺参虾青素基因的克隆及不同体色个体间表达差异的分析. 海洋科学, 36(3): 22-28.

赵红霞, 曹俊明, 吴建开, 等. 2007. 军曹鱼幼鱼对饲料中精氨酸的需要量. 华南农业大学学报, 28(4): 87-90.

赵红霞. 2009. 玉米 DDGS 的研究及在水产饲料中应用. 现代渔业信息, 24(1): 12-14.

赵丽丽, 包焕玲, 姜永新, 等. 2019. 低聚果糖对仿刺参生长及生长免疫因子的影响. 水产科学, 38(2): 145-151.

赵芹, 王静凤, 薛勇, 等. 2008. 3 种海参的主要活性成分和免疫调节作用的比较研究. 中国水产科学, 15(1): 154-159.

赵瑞祯, 戴继勋, 刘秋明, 等. 2007. 紫菜单细胞活饵料的稚参育苗研究. 中国海洋大学学报: 自然科学版, 37(s2): 91-94.

赵瑞祯, 戴继勋, 王娟, 等. 2012. 大型海藻资源在刺参饲育中的可持续应用研究. 饲料研究, 35(10): 64-67.

郑海羽, 饶道专, 陈高峰, 等. 2008. 保护性开发南麂列岛铜藻 *Sargassaum horneri*(Turn.)Ag. 资源的思考. 现代渔业信息, 23(10): 25-26.

郑慧, 李彬, 荣小军, 等. 2014. 盐度和溶解氧对刺参非特异性免疫酶活性的影响. 渔业科学进展, 35(1): 118-124.

钟广贤. 2013. 玉米 DDGS 对草鱼、鲤鱼生长性能及体成分的影响. 长沙: 湖南农业大学.

钟国防, 周洪琪. 2005a. 木聚糖酶和复合酶制剂 PS 对尼罗罗非鱼生长性能、非特异性免疫能力的影响. 海洋渔业, 27(4): 286-291.

钟国防, 周洪琪. 2005b. 木聚糖酶和复合酶制剂 PS 对尼罗罗非鱼生长性能、消化率以及肌肉营养成分的影响. 浙江海洋学院学报, 24(4): 324-329.

钟鸣, 胡超群. 2016. 海参养殖饲料学研究进展. 饲料工业, 37(18): 58-64.

周凡. 2011. 饲料赖氨酸和精氨酸对黑鲷幼鱼生长影响及其拮抗作用机理研究. 杭州: 浙江大学.

周恒永. 2011. 大口黑鲈对饲料中精氨酸需求量的研究. 上海: 上海海洋大学.

周金敏, 张伟, 程时军. 2010. 非淀粉多糖酶在鱼类饲料中的应用研究. 饲料研究, 33(7): 68-71.

周率, 冉照收, 于珊珊, 等. 2016. 几种微藻饵料对缢蛏稚贝脂质营养组成的影响. 生物学杂志, 33(3): 52-56.

周歧存, 麦康森. 2004. 皱纹盘鲍维生素 D 营养需要的研究. 水产学报, 28(2): 155-160.

周歧存, 王用黎, 黄文文, 等. 2015. 凡纳滨对虾幼虾的缬氨酸需求量. 动物营养学报, 27(2): 459-468.

周玮, 田甲申, 黄俊鹏, 等. 2010. 不同生长阶段仿刺参肠道内含物及消化酶活性的变化. 大连海洋大学学报, 25(5): 460-464.

朱峰. 2009. 仿刺参 *Apostichopus japonicus* 胚胎发育和主要系统的组织学研究. 青岛: 中国海洋大学.

朱建新, 刘慧, 冷凯良, 等. 2007. 几种常用饵料对稚幼参生长影响的初步研究. 海洋水产研究, 28(5): 48-53.

朱伟, 麦康森, 张百刚, 等. 2005, 刺参稚参对蛋白质和脂肪需求量的初步研究. 海洋科学, 29(3):

54-58.

朱伟. 2001. 皱纹盘鲍(*Haliotis discus hannai* Ino.)B 族维生素营养生理及营养需要的研究. 青岛: 中国海洋大学.

左然涛, 李敏, 秦宇博, 等. 2017. 饲料中 DHA 含量对刺参成参生长及其体壁营养成分的影响. 大连海洋大学学报, 32(2): 172-177.

左然涛, 麦康森, 徐玮, 等. 2015. 脂肪酸对鱼类免疫系统的影响及调控机制研究进展. 水产学报, 39(7): 1079-1088.

左然涛. 2013. 饲料脂肪酸调控大黄鱼免疫力和脂肪酸代谢的初步研究. 青岛: 中国海洋大学.

曾呈奎. 1962. 中国经济海藻志. 北京：科学出版社.

Abdel-Latif H M R, Soliman A A, Sewilam H, et al. 2020. The influence of raffinose on the growth performance, oxidative status, and immunity in Nile tilapia (*Oreochromis niloticus*). Aquaculture Reports, 18: 100457.

Abidi S F, Khan M A. 2004. Dietary valine requirement of Indian major carp, Labeo rohita (Hamilton) fry. Journal of Applied Ichthyology, 20(2): 118-122.

Abidi S F, Khan M A. 2007. Dietary leucine requirement of fingerling Indian major carp, *Labeo rohita* (Hamilton). Aquaculture Research, 38(5): 478-486.

Abidi S F, Khan M A. 2011. Total sulphur amino acid requirement and cystine replacement value for fingerling rohu, *Labeo rohita*: effects on growth, nutrient retention and body composition. Aquaculture Nutrition, 17: e583-e594.

Abimorad E G, Favero G C, Castellani D, et al. 2009. Dietary supplementation of lysine and/or methionine on performance, nitrogen retention and excretion in pacu *Piaractus mesopotamicus* reared in cages. Aquaculture, 295: 266-270.

Ahearn G A. 1968. A comparative study of P32 uptake by whole animals and isolated body regions of the sea cucumber *Holothuria atra*. The Biological Bulletin, 134(3): 367-381.

Ahmed I, Khan M A. 2006. Dietary branched-chain amino acid valine, isoleucine and leucine requirements of fingerling Indian major carp, *Cirrhinus mrigala* (Hamilton). British Journal of Nutrition, 96(3): 450-460.

Ahmed I. 2014. Dietary amino acid L-methionine requirement of fingerling India catfish, *Heteropneustes fossilis* (Bloch-1974) estimated by growth and haemato-biochemical parameters. Aquaculture Research, 45: 243-258.

Ai Q H, Mai K S, Zhang L, et al. 2007. Effects of dietary β-1, 3-glucan on innate immune response of large yellow croaker, *Pseudosciaena crocea*. Fish and Shellfish Immunology, 22(4): 394-402.

Ala F S, Hassanabadi A, Golian A. 2019. Effects of dietary supplemental methionine source and betaine replacement on the growth performance and activity of mitochondrial respiratory chain enzymes in normal and heat-stressed broiler chickens. Journal of Animal Physiology and Animal Nutrition, 103: 87-99.

Alam S, Teshima S, Koshio S, et al. 2002. Arginine requirement of juvenile Japanese flounder *Paralichthys olivaceus* estimated by growth and biochemical parameters. Aquaculture, 205(1-2): 127-140.

Alami-Durante H, Bazin D, Cluzeaud M, et al. 2018. Effect of dietary methionine level on muscle growth mechanisms in juvenile rainbow trout (*Oncorhynchus mykiss*). Aquaculture, 483: 273-285.

Al-Qodah Z, Daghstani H, Goepll P, et al. 2007. Determination of kinetic parameters of α-amylase producing thermophile *Bacillus sphaericus*. African Journal of Biotechnology, 6(6): 699-706.

Ambasankar K, Ali S A, Dayal J S. 2006. Effect of dietary phosphorus on growth and its excretion in tiger shrimp *Penaeus monodon*. Asian Fisheries Science, 19: 21-26.

Anggraeni M S, Owens L. 2000. The haemocytic origin of lymphoid organ spheroid cells in the penaeid prawn *Penaeus monodon*. Diseases of Aquatic Organisms, 40(2): 85-92.

AOAC. 1990. "Official methods of analysis, " in Association of Official Analytical Chemists, 15th, D. C, Washington.

AOAC. 1999. Official Methods of Analysis of the Association of Official Analytical Chemists. Association of Official Analytical Chemists, Arlington, VA, USA.

AOAC. 2000. Official Methods of Analysis of the Association of Official Analytical Chemists, Gaithersburg, Maryland, USA (1018 pp.).

AOAC. 2003. Official Methods of Analysis of the Association of Official Analytical Chemists. Association of Official Analytical Chemists, Arlington, VA, USA.

Apel K, Hirt H. 2004. Reactive oxygen species: metabolism, oxidative stress, and signal transduction. Annual Review of Plant Biology, 55: 373-399.

Archer S Y, Meng S F, Shei A, et al. 1998. p21^{WAF1} is required for butyrate-mediated growth inhibition of human colon cancer cells. Proceedings of the National Academy of Sciences of the United States of America, 95(12): 6791-6796.

Ásman B, Wijkander P, Hjerpe A. 1994. Reduction of collagen degradation in experimental granulation tissue by vitamin E and selenium. Journal of Clinical Periodontology, 21(1): 45-47.

Atalah E, Hernándezcruz C M, Benítezsantana T, et al. 2011. Importance of the relative levels of dietary arachidonic acid and eicosapentaenoic acid for culture performance of gilthead seabream (*Sparus aurata*) larvae. Aquaculture Research, 42(9): 1279-1288.

Azarm H M, Lee S M. 2014. Effects of partial substitution of dietary fish meal by fermented soybean meal on growth performance, amino acid and biochemical parameters of juvenile black sea bream, *Acanthopagrus schlegelii*. Aquaculture Research, 45(6): 994-1003.

Azaza M S, Khiari N, Dhraief M N, et al. 2013. Growth performance, oxidative stress indices and hepatic carbohydrate metabolic enzymes activities of juvenile Nile tilapia, *Oreochromis niloticus* L., in response to dietary starch to protein ratios. Aquaculture Research, 46(1): 14-27.

Babalola T O O, Adebayo M A, Apata D F, et al. 2009. Effect of dietary alternative lipid sources on haematological parameters and serum constituents of *Heterobranchus longifilis* fingerlings. Tropical Animal Health and Production, 41(3): 371-377.

Bagnyukova T V, Storey K B, Lushchak V I. 2003. Induction of oxidative stress in *Rana ridibunda* during recovery from winter hibernation. Journal of Thermal Biology, 28(1): 21-28.

Bai Y, Zhang L, Xia S, et al. 2016. Effects of dietary protein levels on the growth, energy budget, and physiological and immunological performance of green, white and purple color morphs of sea cucumber, *Apostichopus japonicas*. Aquaculture, 450: 375-382.

Baker H. 2006. Comparative species utilization and toxicity of sulfur amino acids. The Journal of Nutrition, 136(6Suppl): 1670S-1675S.

Balbas J, Hamid N, Liu T T, et al. 2015. Comparison of physicochemical characteristics, sensory properties and volatile composition between commercial and New Zealand made wakame from *Undaria pinnatifida*. Food Chemistry, 186(5): 168-175.

Balcázar J L, de Blas I, Ruiz-Zarzuela I, et al. 2006. The role of probiotics in aquaculture. Veterinary Microbiology, 114(3-4): 173-186.

Bannister J V, Bannister W H, Rotilio G. 1987. Aspects of the structure, function, and applications of superoxide dismutase. Critical Reviews in Biochemistry, 22(2): 111-180.

Barnett B J, Cho C Y, Slinger S J. 1982. Relative biopotency of dietary ergocalciferol and cholecalciferol and the role of and requirement for vitamin D in rainbow trout (*Salmo gairdneri*). The Journal of Nutrition, 112(11): 2011-2019.

Bayne C J, Gerwick L. 2001. The acute phase response and innate immunity of fish. Developmental and Comparative Immunology, 25(8-9): 725-743.

Bedford M R, Classen H L. 1992. Reduction of intestinal viscosity through manipulation of dietary rye and pentosanase concentration is effected through changes in the carbohydrate composition of the intestinal aqueous phase and results in improved growth rate and food conversion efficiency of broiler chicks. The Journal of Nutrition, 122(3): 560-569.

Bellamy D. 1961. The endogenous citric acid-cycle intermediates and amino acids of mitochondria. Biochemical Journal, 82: 218.

Berge G E, Bakke-McKellep A M, Lied E. 1999. *In vitro* uptake and interaction between arginine and lysine in the intestine of Atlantic salmon (*Salmo salar*). Aquaculture, 179(1-4): 181-193.

Berndt T J, Schiavi S, Kumar R. 2005. "Phosphatonins" and the regulation of phosphorus homeostasis. American Journal of Physiology. 289(6): F1170-F1182.

Bertheussen K. 1982. Receptors for complement on echinoid phagocytes. II. Purified human complement mediates echinoid phagocytosis. Developmental and Comparative Immunology, 6(4): 635-642

Bessonart M, Izquierdo M S, Salhi M, et al. 1999. Effect of dietary arachidonic acid levels on growth and survival of gilthead sea bream (*Sparus aurata* L.) larvae. Aquaculture, 179(1-4): 265-275.

Blier P U, Pelletier D, Dutil J D. 1997. Does aerobic capacity set a limit on fish growth rate? Reviews in Fisheries Science, 5: 323-340.

Bornet F R J, Brouns F. 2002. Immune-stimulating and gut health-promoting properties of short-chain fructo-oligosaccharides. Nutrition Reviews, 60: 326-334.

Bradford M M. 1976. A dye binding assay for protein. Anal. Biochem., 72: 248-254.

Braga W F, Araújo J G, Martins G P, et al. 2016. Dietary total phosphorus supplementation in goldfish diets. Latin American Journal of Aquatic Research, 44(1): 129-136.

Brauge C, Médale F, Corraze G. 1994. Effect of dietary carbohydrate levels on growth, body composition and glycaemia in rainbow trout, *Oncorhynchus mykiss*, reared in seawater. Aquaculture, 123(1-2): 109-120.

Brosnan J T, Brosnan M E, Bertolo R F P, et al. 2007. Methionine: a metabolically unique amino acid.

Livestock Science, 112(1-2): 2-7.

Brot N, Weissbach H. 2000. Peptide methionine sulfoxide reductase: biochemistry and physiological role. Peptide Science, 55(4): 288-296.

Brouwers M C G J, Jacobs C, Bast A, et al. 2015. Modulation of glucokinase regulatory protein: A double-edged sword? Trends in Molecular Medicine, 21(10): 583-594.

Buentello J A, Gatlin Ⅲ D M. 2000. The dietary arginine requirement of channel catfish (*Ictalurus punctatus*) is influenced by endogenous synthesis of arginine from glutamic acid. Aquaculture, 188(3-4): 311-321.

Byrne M. 2001. The morphology of autotomy structures in the sea cucumber *Eupentacta quinquesemita* before and during evisceration. The Journal of Experimental Biology, 204: 849-863.

Campbell K, Vowinckel J, Keller M A. 2016. Methionine metabolism alters oxidative stress resistance via the pentose phosphate pathway. Antioxidants and Redox Signaling, 10: 543-547.

Canicattí C. 1990. Lysosomal enzyme pattern in *Holothuria polii coelomocytes*. Journal of Invertebrate Pathology, 56(1): 70-74.

Cárdenas W, Dankert J R. 1997. Phenoloxidase specific activity in the red swamp crayfish *Procambarus clarkii*. Fish and Shellfish Immunology, 7(5): 283-295.

Carter C G, Houlihan D F, Buchanan B, et al. 1994. Growth and feed utilization efficiencies of seawater Atlantic salmon, *Salmon salar* L., fed a diet containing supplementary enzymes. Aquaculture Research, 25(1): 37-46.

Castell J D, Bell J G, Tocher D R, et al. 1994. Effects of purified diets containing different combinations of arachidonic and docosahexaenoic acid on survival, growth and fatty acid composition of juvenile turbot (*Scophthalmus maximus*). Aquaculture, 128(3-4): 315-333.

Chaiyapechara S, Casten M T, Hardy R W, et al. 2003. Fish performance, fillet characteristics, and health assessment index of rainbow trout (*Oncorhynchus mykiss*) fed diets containing adequate and high concentrations of lipid and vitamin E. Aquaculture, 219: 715-738.

Chang J, Zhang W B, Mai K S, et al. 2010. Effects of dietary β-glucan and glycyrrhizin on non-specific immunity and disease resistance of the sea cucumber (*Apostichopus japonicus* Selenka) challenged with *Vibrio splendidus*. Journal of Ocean University of China, 9(4): 389-394.

Chavez-Sanchez M C, Martinez-Palacios C A, Martinez-Perez G, et al. 2000. Phosphorus and calcium requirements in the diet of the American cichlid *Cichlasoma urophthalmus* (Günther). Aquaculture Nutrition, 6(1): 1-9.

Chen B A, Leng X J, Li X Q. 2008. Study on the effect of crystalline or coated amino acids for *Cyprinus carpio*. Acta hydrobiologica sinica, 32(5): 774-778 (In Chinese with English abstract).

Chen C, Zhu W, Wu F, et al. 2016. Quantifying the dietary potassium requirement of subadult grass carp (*Ctenopharyngodon idellus*). Aquaculture Nutrition, 22(3): 541-549.

Chen H, Simar D, Ting J H Y, et al. 2012a. Leucine improves glucose and lipid status in offspring from obese dams, dependent on diet type, but not caloric intake. Journal of Neuroendocrinology, 24(10): 1356-1364.

Chen J H, Ren Y C, Wang G D, et al. 2018a. Dietary supplementation of biofloc influences growth performance, physiological stress, antioxidant status and immune response of juvenile sea cucumber *Apostichopus japonicas* (Selenka). Fish and Shellfish Immunology, 72: 143-152.

Chen J N, Li X Q, Han D Y, et al. 2018b. Comparative study on the utilization of crystalline methionine and methionine hydroxy analogue calcium by Pacific white shrimp (*Litopenaeus vannamei* Boone). Aquaculture Research, 49: 3088-3096.

Chen K, Jiang W, Wu P, et al. 2017. Effect of dietary phosphorus deficiency on the growth, immune function and structural integrity of head kidney, spleen and skin in young grass carp (*Ctenopharyngodon idella*). Fish and Shellfish Immunology, 63: 103-126.

Chen L P, Huang C H. 2015. Estimation of dietary vitamin A requirement of juvenile soft-shelled turtle, *Pelodiscus sinensis*. Aquaculture Nutrition, 21(4): 457-463.

Chen L X, Li P, Wang J J, et al. 2009. Catabolism of nutritionally essential amino acids in developing porcine enterocytes. Amino acids, 37(1): 143-152.

Chen Y J, Tian L X, Yang H J, et al. 2012b. Effect of protein and starch level in practical extruded diets on growth, feed utilization, body composition, and hepatic transaminases of juvenile grass carp, *Ctenopharyngodon idella*. Journal of the World Aquaculture Society, 43(2): 187-197.

Cheng Y, Meng Q S, Wang C X, et al. 2010. Leucine deprivation decreases fat mass by stimulation of lipolysis in white adipose tissue and upregulation of uncoupling protein 1(UCP1) in brown adipose tissue. Diabetes, 59(1): 17-25.

Cheng Z Y, Buentello A, Gatlin III D M. 2011. Effects of dietary arginine and glutamine on growth performance, immune responses and intestinal structure of red drum, *Sciaenops ocellatus*. Aquaculture, 319(1-2): 247-252.

Cheng Z Y, Gatlin III D M, Buentello A. 2012. Dietary supplementation of arginine and/or glutamine influences growth performance, immune responses and intestinal morphology of hybrid striped bass (*Morone chrysops* ×*Morone saxatilis*). Aquaculture, 362-363: 39-43.

Chesson P. 1994. Multispecies competition in variable environments. Theoretical Population Biology, 45(3): 227-276.

Chitra P, Edwin S C, Moorthy M. 2014. Effect of dietary vitamin E and selenium supplementation on Japanese quail broilers. Indian Journal of Veterinary and Animal Sciences Rresearch, 43(3): 195-205.

Choct M, Hughes R J, Wang J, et al. 1996. Increased small intestinal fermentation is partly responsible for the anti-nutritive activity of non-starch polysaccharides in chickens. British Poultry Science, 37(3): 609-621.

Choo P S, Smith T K, Cho C Y, et al. 1991. Dietary excesses of leucine influence growth and body composition of rainbow trout. The Journal of Nutrition, 121(12): 1932-1939.

Choo P S. 2008. Population status, fisheries and trade of sea cucumbers in Asia//Toral-Granda V, Lovatelli A, Vasconcellos M (eds). Sea cucumbers. A global review of fisheries and trade. Rome: FAO Fisheries and Aquaculture Technical Paper. No. 516: pp. 81-118.

Chu Z J, Gong Y, Lin Y C, et al. 2014. Optimal dietary methionine requirement of juvenile Chinese sucker, *Myxocyprinus asiaticus*. Aquaculture Nutrition, 20: 253-264.

Coloso R M, King K, Fletcher J W, et al. 2003. Phosphorus utilization in rainbow trout (*Oncorhynchus mykiss*) fed practical diets and its consequences on effluent phosphorus levels. Aquaculture, 220(1-4): 801-820.

Coloso R M, Murillo-Gurrea D P, Borlongan I G et al. 1999. Sulphur amino acid requirement of juvenile Asian sea bass *Lates calcarifer*. Journal of Applied Ichthyology, 15(2): 54-58.

Cooke J, Lanfear R, Downing A, et al. 2015. The unusual occurrence of green algal balls of *Chaetomorpha linum* on a beach in Sydney, Australia. Botanica Marina, 58(5): 401-407.

Courtney-Martin G, Pencharz P B. 2016. Sulfur amino acids metabolism from protein synthesis to glutathione//The Molecular Nutrition of Amino Acids and Proteins: A Volume in the Molecular Nutrition Series. Academic Press: 265-286.

Cuesta A, Ortuño J, Rodriguez A, et al. 2002. Changes in some innate defence parameters of seabream (*Sparus aurata* L.) induced by retinol acetate. Fish and Shellfish Immunology, 13(4): 279-291.

Cui H, Wang L, Yu Y. 2015. Production and characterization of alkaline proteases from a high yielding and moderately halophilic strain of SD11 marine bacteria. Journal of Chemistry, Article (1): 798304.

Das B K, Pradhan J, Sahu S. 2009. The effect of Euglena viridis on immune response of rohu, Labeo rohita (Ham.). Fish and Shellfish Immunology, 26(6): 871-876.

Davis D A, Lawrence A L, Gatlin III D M. 1993. Response of *Penaeus vannamei* to dietary calcium, phosphorus and calcium: phosphorus ratio. Journal of the World Aquaculture Society, 24(4): 504-515.

Deng J M, Mai K S, Ai Q H, et al. 2006. Effects of replacing fish meal with soy protein concentrate on feed intake and growth of juvenile Japanese flounder, *Paralichthys olivaceus*. Aquaculture, 258(1-4): 503-513.

Deshimaru O, Yone Y. 1978. Studies on a purified diet for prawn. XIII. Effect of dietary carbohydrate source on the growth and feed efficiency of prawn. Nippon Suisan Gakkaishi, 44(10): 1161-1163.

Dong Y, Yu S, Wang Q, et al. 2011. Physiological responses in a variable environment: relationships between metabolism, Hsp and thermotolerance in an intertidal-subtidal species. PLoS ONE, 6(10): e26446.

Draper H H, Hadley M. 1990. Malondialdehyde determination as index of lipid peroxidation. Methods in Enzymology, 186: 421-431.

Drukarch B, Jongenelen C A M, Schepens E, et al. 1996. Glutathione is involved in the granular storage of dopamine in rat PC12 pheochromocytoma cells: implications for the pathogenesis of Parkinson's disease. The Journal of Neuroscience, 16(19): 6038-6045.

Du R, Zang Y, Tian X. 2013. Growth, metabolism and physiological response of the sea cucumber, *Apostichopus japonicus* Selenka during periods of inactivity. Journal of Ocean University of China, 12(1): 146-154.

Elmada C Z, Huang W, Jin W, et al. 2016. The effect of dietary methionine on growth, antioxidant capacity, innate immune response and disease resistance of juvenile yellow catfish (*Pelteobagrus fulvidraco*). Aquaculture Nutrition, 22: 1163-1173.

Erickson K L, McNeill C J, Gershwin M E, et al. 1980. Influence of dietary fat concentration and saturation on immune ontogeny in mice. Journal of Nutrition, 110(8): 1555-1572.

Espe M, Veiseth-Kent E, Zerrahn J E, et al. 2016. Juvenile Atlantic salmon decrease white trunk muscle IGF-1 expression and reduce muscle and plasma free sulphur amino acids when methionine availability is low while liver sulphur metabolites mostly is unaffected by treatment. Aquaculture Nutrition, 22: 801-812.

FAO. 2022. The State of World Fisheries and Aquaculture 2022. In Towards Blue Transformation, Rome.

FBMA (Fisheries Bureau of Ministry of Agriculture). 2022. China Fishery Statistical Yearbook. China Agriculture Press, Beijing (in Chinese).

Felipe N F, Adhemar R O, Claudia F, et al. 2016. Effect of shrimp stocking density and graded levels of dietary methionine over the growth performance of *Litopenaeus vannamei* reared in a green-water system. Aquaculture, 463: 16-21.

Feng L, Xie N B, Liu Y, et al. 2013. Dietary phosphorus prevents oxidative damage and increases antioxidant enzyme activities in the intestine and hepatopancreas of juvenile Jian carp. Aquaculture Nutrition, 19(3): 250-257.

Fernández I, Hontoria F, Ortiz-Delgado J B, et al. 2008. Larval performance and skeletal deformities in farmed gilthead sea bream (*Sparus aurata*) fed with graded levels of Vitamin A enriched rotifers (*Brachionus plicatilis*). Aquaculture, 283(1-4): 102-115.

Ferraris Q, Hale J, Teigland E, et al. 2020. Phospholipid analysis in whey protein products using hydrophilic interaction high-performance liquid chromatography-evaporative light scattering detection in an industry setting. Journal of Dairy Science, 103(12): 11079-11085.

Finkelstein J D. 1998. The metabolism of homocysteine: Pathways and regulation. European Journal of Pediatrics, 157: S40-S44.

Flickinger E A, Schreijen E M W C, Patil A R, et al. 2003. Nutrient digestibilities, microbial populations, and protein catabolites as affected by fructan supplementation of dog diets. Journal of Animal Science, 81(8): 2008-2018.

Fournier V, Gouillou-Coustans M F, Métailler R, et al. 2003. Excess dietary arginine affects urea excretion but does not improve N utilisation in rainbow trout *Oncorhynchus mykiss* and turbot *Psetta maxima*. Aquaculture, 217(1-4): 559-576.

Fridovich I. 1989. Superoxide dismutases: an adaptation to a parmagnetic gas. The Journal of Biological Chemistry, 264(14): 7761-7764.

Fu X Y, Xue C H, Miao B C, et al. 2005a. Study of a highly alkaline protease extracted from digestive tract of sea cucumber (*Stichopus japonicus*). Food Research International, 38(3): 323-329.

Fu X Y, Xue C H, Miao B C, et al. 2005b. Characterization of proteases from the digestive tract of sea cucumber (*Stichopusj japonicus*): highalkaline protease activity. Aquaculture, 246(1-4): 321-329.

Fukuzawa K, Tokumura A, Ouchi S, et al. 1982. Antioxidant activities of tocopherols on Fe^{2+}-ascorbate-induced lipid peroxidation in lecithin liposomes. Lipids, 17: 511-513.

Fukuzawa K, Tokumura A. 1976. Glutathione peroxidase activity in tissues of vitamin E-deficient mice. Journal of Nutritional Science and Vitaminology, 22(5): 405-407.

Furuichi M, Yone Y. 1981. Change of blood sugar and plasma insulin levels of fishes in glucose tolerance test. Bulletin of the Japanese Society of Scientific Fisheries, 47(6): 761-764.

Furuita H, Yamamoto T, Shima T, et al. 2003. Effect of arachidonic acid levels in broodstock diet on larval and egg quality of Japanese flounder *Paralichthys olivaceus*. Aquaculture, 220(1-4): 725-735.

Gan L, Zhou L L, Li X X, et al. 2016. Dietary leucine requirement of juvenile Nile tilapia, *Oreochromis niloticus*. Aquaculture Nutrition, 22(5): 1040-1046.

Gao J. 2013. Interactive effects of vitamin C and E supplementation on growth, fatty acid composition, and lipid peroxidation of sea cucumber, *Apostichopus japonicus*, fed with dietary oxidized fish oil. Journal of World Wquaculture Society, 44 (4): 536-546.

Gao Q F, Wang Y S, Dong S L, et al. 2011. Absorption of different food sources by sea cucumber *Apostichopus japonicus* (Selenka) (Echinodermata: Holothuroidea): evidence from carbon stable isotope. Aquaculture, 319(1-2): 272-276.

Gao W, Liu Y J, Tian L X, et al. 2010. Effect of dietary carbohydrate-to-lipid ratios on growth performance, body composition, nutrient utilization and hepatic enzymes activities of herbivorous grass carp (*Ctenopharyngodon idella*). Aquaculture Nutrition, 16(3): 327-333.

Gavrilova G S, Sukhin I Y. 2011. Characteristics of the Japanese sea cucumber *Apostichopus japonicus*'s population in the sea of Japan (Kievka Bay). Oceanology, 51(3): 449-456.

German D P. 2011. Food acquisition and digestion|Digestive efficiency. Encyclopedia of Fish Physiology, 3: 1596-1607.

Ghosh K, Tagore D M, Anumula R, et al. 2013. Crystal structure of rat intestinal alkaline phosphatase-role of crown domain in mammalian alkaline phosphatases. Journal of Structural Biology, 184(2): 182-192.

Giannenas I, Doukas D, Karamoutsios A, et al. 2016. Effects of Enterococcus faecium, mannan oligosaccharide, benzoic acid and their mixture on growth performance, intestinal microbiota, intestinal morphology and blood lymphocyte subpopulations of fattening pigs. Animal Feed Science and Technology, 220: 159-167.

Giri I N A, Teshima S I, Kanazawa A, et al. 1997. Effects of dietary pyridoxine and protein levels on growth, vitamin B_6 content, and free amino acid profile of juvenile *Penaeus japonicus*. Aquaculture, 157(3-4): 263-275.

Goff J B, Gatlin III D M. 2004. Evaluation of different sulfuramino acid compounds in the diet of red drum, *Sciaenops ocellatus* and sparing value of cystine for methionine. Aquaculture, 241: 465-477.

Góth L. 1991. A simple method for determination of serum catalase activity and revision of reference range. Clinica Chimica Acta, 196: 143-151.

Graff I E, Hoie S, Totland G K, et al. 2015. Three different levels of dietary vitamin D3 fed to first feeding fry of Atlantic salmon (*Salmo salar* L.): effect on growth, mortality, calcium content and bone formation. Aquaculture Nutrition, 8(2): 103-111.

Griffin M E, Wilson K A, Brown P B. 1994. Dietary arginine requirement of juvenile hybrid striped bass. Journal of Nutrition, 124(6): 888-893.

Grillo M A, Colombatto S. 2008. S-adenosylmethionine and its products. Amino Acids, 34(2): 187-193.

Grimble G, Keohane P, Higgins B, et al. 1986. Effect of peptide chain length on amino acid and nitrogen absorption from two lactalbumin hydrolysates in the normal human jejunum. Clinical Science, 71(1): 65-69.

Guertin D A, Guntur K V P, Bell G W, et al. 2006. Functional genomics identifies TOR-regulated genes that control growth and division. Current Biology, 16(10): 958-970.

Gui D, Liu W B, Shao X P, et al. 2010. Effects of different dietary levels of cottonseed meal protein hydrolysate on growth, digestibility, body composition and serum biochemical indices in crucian carp (*Carassius auratus gibelio*). Animal Feed Science and Technology, 156(3-4): 112-120.

Guo F F, Cavener D R. 2007. The GCN2 eIF2α kinase regulates fatty-acid homeostasis in the liver during deprivation of an essential amino acid. Cell Metabolism, 5(2): 103-114.

Halver J E. 2002. The vitamins in fish nutrition//Halver J E, Hardy R W (eds.) Fish nutrition (3rd ed.). San Diego: Academic Press.

Hamre K. 2011. Metabolism, interactions, requirements and functions of vitamin E in fish. Aquaculture Nutrition, 17: 98-115.

Han K H, Shimada K, Hayakawa T, et al. 2014. Porcine Splenic Hydrolysate has antioxidant activity *in vivo* and *in vitro*. Korean Journal for Food Science of Animal Resources, 34(3): 325.

Hasegawa N, Sawaguchi S, Tokuda M, et al. 2014. Fatty acid composition in sea cucumber *Apostichopus japonicus* fed with microbially degraded dietary source. Aquaculture Research, 45(12): 2021-2031.

Hasek B, Boudreau A, Shin J, et al. 2013. Remodeling the integration of lipid metabolism between liver and adipose tissue by dietary methionine restriction in rats. Diabetes, 62: 3362-3372.

Haug T, Kjuul A K, Styrvold O B, et al. 2002. Antibacterial activity in *Strongylocentrotus droebachiensis* (Echinoidea), *Cucumaria frondosa* (Holothuroidea), and *Asterias rubens* (Asteroidea). Journal of Invertebrate Pathology, 81(2): 94-102

Haussler M R, Nagode L A, Rasmussen H. 1970. Induction of intestinal brush border alkaline phosphatase by vitamin D and identity with Ca-ATPase. Nature, 228(5277): 1199-1201.

He H Q, Lawrence A L, Liu R Y. 1992. Evaluation of dietary essentiality of fat-soluble vitamins, A, D, E and K for penaeid shrimp (*Penaeus vannamei*). Aquaculture, 103(2), 177-185.

Hemre G I, Deng D F, Wilson R P, et al. 2004. Vitamin A metabolism and early biological responses in juvenile sunshine bass (*Morone chrysops* × *M. saxatilis*) fed graded levels of vitamin A. Aquaculture, 235(1-4): 645-658.

Hemre G I, Mommsen T P, Krogdahl Å. 2002. Carbohydrates in fish nutrition: effects on growth, glucose metabolism and hepatic enzymes. Aquaculture Nutrition, 8(3): 175-194.

Hernandez L H, Hardy R W. 2020. Vitamin A functions and requirements in fish. Aquaculture Research, 51(8): 3061-3071.

Hernández-López J, Gollas-Galván T, Vargas-Albores F. 1996. Activation of the prophenoloxidase system of the brown shrimp (*Penaeus californiensis* Holmes).Comparative Biochemistry and

Physiology Part C: Pharmacology, Toxicology and Endocrinology, 113(1): 61-66.

Hibbs J B Jr, Taintor R R, Vavrin Z. 1987. Macrophage cytotoxicity: Role for L-arginine deiminase and imino nitrogen oxidation to nitrite. Science, 235(4787): 473-476.

Hodgkinson S M, Moughan P J, Reynolds G W, et al. 2000. The effect of dietary peptide concentration on endogenous ileal amino acid loss in the growing pig. British Journal of Nutrition, 83(4): 421-430.

Hong K J, Lee C H, Kim S W. 2004. *Aspergillus oryzae* GB-107 fermentation improves nutritional quality of soybeans and food soybean meals. Journal of Medicinal Food, 7(4): 430-435

Honorat C A, Almeida L C, da Silva Nunes C, et al. 2010. Effects of processing on physical characteristics of diets with distinct levels of carbohydrates and lipids: the outcomes on the growth of pacu (*Piaractus mesopotamicus*). Aquaculture Nutrition, 16(1): 91-99.

Hoshi T, Heinemann S H. 2001. Regulation of cell function by methionine oxidation and reduction. The Journal of Physiology, 531(1): 1-11.

Hou S Y, Jin Z W, Jiang W W, et al. 2019. Physiological and immunological responses of sea cucumber *Apostichopus japonicus* during desiccation and subsequent resubmersion. PeerJ, 7: e7427.

Hou Y R, Sun Y J, Gao Q F, et al. 2018. Bioturbation by sea cucumbers *Apostichopus japonicas* affects sediment phosphorus forms and sorption characteristics. Aquaculture Environment Interactions, 10: 201-211.

Hrčková M, Rusňáková M, Zemanovič J. 2002. Enzymatic hydrolysis of defatted soy flour by three different proteases and their effect on the functional properties of resulting protein hydrolysates. Czech Journal of Food Sciences, 20(1): 7-14.

Hu C J, Chen S M, Pan C H, et al. 2006. Effects of dietary vitamin A or β-carotene concentrations on growth of juvenile hybrid tilapia, *Oreochromis niloticus* × *O. aureus*. Aquaculture, 253(1-4): 602-607.

Hu H B, Zhang Y J, Mai K S, et al. 2015. Effects of dietary stachyose on growth performance, digestive enzyme activities and intestinal morphology of juvenile turbot (*Scophthalmus maximus* L.). Journal of Ocean University of China, 14(5): 905-912.

Hu J F, Zhang J M, Wu S J. 2021. The growth performance and non-specific immunity of juvenile grass carp (*Ctenopharyngodon idella*) affected by dietary alginate oligosaccharide. 3 Biotech, 11(2): 46.

Hu Z, Feng J, Song H, et al. 2022. Metabolic response of *Mercenaria mercenaria* under heat and hypoxia stress by widely targeted metabolomic approach. Science of the Total Environment, 809: 151172.

Huang C H, Chang R J, Huang S L, et al. 2003. Dietary vitamin E supplementation affects tissue lipid peroxidation of hybrid tilapia, *Oreochromis niloticus* × *O. aureus*. Comparative Biochemistry and Physiology Part B: Biochemistry and Molecular Biology, 134(2): 265-270.

Huang C H, Lin W Y. 2002. Estimation of optimal dietary methionine requirement for soft shell turtle, *Pelodiscus sinensis*. Aquaculture, 207(3-4): 281-287.

Huang C H, Lin W Y. 2015. Effects of dietary vitamin E level on growth and tissue lipid peroxidation of soft-shelled turtle, *Pelodiscus sinensis* (Wiegmann). Aquaculture Research, 35(10): 948-954.

Hughes S G, Rumsey G L, Nesheim M C. 1983. Dietary requirements for essential branched-chain amino acids by lake trout. Transactions of the American Fisheries Society, 112(6): 812-817.

Huo D, Sun L, Ru X, et al. 2018. Impact of hypoxia stress on the physiological responses of sea cucumber *Apostichopus japonicus*: respiration, digestion, immunity and oxidative damage. Peer J., 6: e4651.

Huo Y W, Jin M, Sun P, et al. 2017. Effect of dietary leucine on growth performance, hemolymph and hepatopancreas enzyme activities of swimming crab, *Portunus trituberculatus*. Aquaculture Nutrition, 23(6): 1341-1350.

Hurtado M A, Reza M, Ibarra A M, et al. 2009. Arachidonic acid (20: 4*n*-6) effect on reproduction, immunology, and prostaglandin E_2 levels in *Crassostrea corteziensis* (Hertlein, 1951). Aquaculture, 294(3-4): 300-305.

Iaccarino D, Uliano E, Agnisola C. 2009. Effects of arginine and/or lysine diet supplementation on nitrogen excretion in zebrafish. Comparative Biochemistry and Physiology part A: Molecular and Integrative Physiology, 154(1S): 36-37.

Ingledew W M. 1999. Yeast-could you base a business on this bug biotechnology in the feed industry. Nottingham: Nottingham University Press.

Ishizaki Y, Takeuchi T, Watanabe T, et al. 1998. A preliminary experiment on the effect of Artemia enriched with arachidonic acid on survival and growth of yellowtail. Fisheries Science, 64(2): 295-299.

Izquierdo M, Domínguez D, Jiménez J I, et al. 2019. Interaction between taurine, vitamin E and vitamin C in microdiets for gilthead seabream (*Sparus aurata*) larvae. Aquaculture, 498: 246-253.

Jiang J, Shi D, Zhou X Q, et al. 2016. Effects of lysine and methionine supplementation on growth, body composition and digestive function of grass carp (*Ctenopharyngodon idella*) fed plant protein diets using high-level canola meal. Aquaculture Nutrition, 22: 1126-1133.

Jiang S H, Dong S L, Gao Q F, et al. 2013. Comparative study on nutrient composition and growth of green and red sea cucumber, *Apostichopus japonicus* (Selenka, 1867), under the same culture conditions. Aquaculture Research, 44: 317-320.

Jiang Y B, Yin Q Q, Yang Y R. 2009. Effect of soybean peptides on growth performance, intestinal structure and mucosal immunity of broilers. Journal of Animal Physiology and Animal Nutrition. 93(6): 754-760.

Jie Y Z, Zhang J Y, Zhao L H, et al. 2014. The correlationship between the metabolizable energy content, chemical composition and color score in different sources of corn DDGS. Journal of Animal Science and Biotechnology, 4(3): 1-8.

Jin C, Rahman M M, Lee S M. 2013. Distillers dried grain from makgeolli by-product is useful as a dietary ingredient for growth of juvenile sea cucumber *Apostichopus japonicus*. Fisheries and Aquatic Sciences, 16(4): 279-283.

Jin J, Ma H, Zhou C, Luo M, et al. 2015. Effect of degree of hydrolysis on the bioavailability of corn gluten meal hydrolysates. Journal of the Science of Food and Agriculture, 95(12): 2501-2509.

Jobgen W S, Fried S K, Fu W J, et al. 2006. Regulatory role for the arginine-nitric oxide pathway in metabolism of energy substrates. The Journal of Nutritional Biochemistry, 17(9): 571-588.

Johnson M, Carey F, McMillan R M. 1983. Alternative pathways of arachidonate metabolism: prostaglandins, thromboxane and leukotrienes. Essays in Biochemistry, 19(5): 40-141.

Jones I D. 1975. Effect of processing by fermentation of nutrients//Harris R S, Karmas E (eds). Nutritional Evaluation of Food Processing. Westport: Avi Publishing Company: 324.

Josephkiii C, Wadeo W, Moti H, et al. 2011. Effects of dietary arachidonic acid on larval performance, fatty acid profiles, stress resistance, and expression of Na^+/K^+ ATPase mRNA in black sea bass *Centropristis striata*. Aquaculture, 319(1): 111-121.

Kader M A, Koshio S, Ishikawa M, et al. 2012. Can fermented soybean meal and squid by-product blend be used as fishmeal replacements for Japanese flounder (*Paralichthys olivaceus*). Aquaculture Research, 43(10): 1427-1438.

Kamal-Eldin A, Appelqvist L Å. 1996. The chemistry and antioxi-dant properties of tocopherols and tocotrienols. Lipids, 31(7): 671-701.

Kanazawa A, Teshima S, Sasaki M. 1984. Requirements of the juvenile prawn for calcium, phosphorus, magnesium, potassium, copper, manganese and iron. Memoirs of Faculty of Fisheries Kagoshima University, 33: 63-71.

Karasov W H. 1992. Tests of the adaptive modulation hypothesis for dietary control of intestinal nutrient transport.The American Journal of Physiology, 267: R496-R502.

Kaushik G, Debarati M T, Rushith A, et al. 2013. Crystal structure of rat intestinal alkaline phosphatase-role of crown domain in mammalian alkaline phosphatases. Journal of Structural Biology, 184: 182-192.

Kaushik S J, Fauconneau B, Terrier L, et al. 1988. Arginine requirement and status assessed by different biochemical indices in rainbow trout (*Salmo gairdneri* R.). Aquaculture, 70(1-2): 75-95.

Kaushik S J, Médale F. 1994. Energy requirements, utilization and dietary supply to salmonids. Aquaculture, 124(1-4): 81-97.

Kaushik S J. 1998. Whole body amino acid composition of European seabass (*Dicentrarchus labrax*), gilthead seabream (*Sparus aurata*) and turbot (*Psetta maxima*) with an estimation of their IAA requirement profiles. Aquatic Living Resources, 11(5): 355-358.

Kehrer J P. 1993. Free radicals as mediators of tissue injury and disease. Critical Reviews in Toxicology, 23(1): 21-48.

Kensei K. 2015. Encyclopedia of Astrobiology. 2nd ed. Heidelberg: Springer.

Khalil H S, Momoh T, Al-Kenawy D, et al. 2021. Nitrogen retention, nutrient digestibility and growth efficiency of Nile tilapia (*Oreochromis niloticus*) fed dietary lysine and reared in fertilized ponds. Aquaculture Nutrition, 27(6): 2320-2532.

Khosravi S, Bui H T D, Rahimnejad S. 2015. Effect of dietary hydrolysate supplementation on growth performance, non-specific immune response and disease resistance of olive flounder (*Paralichthys olivaceus*) challenged with *Edwardsiella tarda*. Aquaculture Nutrition, 21(3): 321-331.

Kihara M, Sakata T. 2002. Production of short-chain fatty acids and gas from various oligosaccharides by gut microbes of carp (*Cyprinus carpio* L.) in micro-scale batch culture.

Comparative Biochemistry and Physiology Part A: Molecular and Integrative Physiology, 132(2): 333-340.

Kim B N, Yang J L, Song Y S. 1999. Physiological functions of chongkukjang. Food Industry and Nutrition, 4(2): 40-46.

Kim J D, Lall S P P. 2000. Amino acid composition of whole body tissue of Atlantic halibut (Hippoglossus hippoglossus), yellowtail flounder (Pleuronectes ferruginea) and Japanese flounder (Paralichthys olivaceus). Aquaculture, 187(3-4): 367-373.

Kind P R H, King E J. 1954. Estimation of plasma phosphatase by determination of hydrolysed phenol with amino-antipyrine. Journal of Clinical Pathology, 7(4): 322-326.

Kitagawa S, Sugiyama M, Motoyama M T, et al. 2013. Soy peptides enhance yeast cell growth at low temperatures. Biotechnology Letters, 35: 375-382.

Kitajka K, Puskás L G, Zvara A, et al. 2002. The role of n-3 polyunsaturated fatty acids in brain: Modulation of rat brain gene expression by dietary n-3 fatty acids. Proceedings of the National Academy of Sciences of the United States of America, 99(5): 2619-2224.

Klatt S F, Danwitz A V, Hasler M, et al. 2016. Determination of the lower and upper critical concentration of Methionine + Cystine in diets of juvenile turbot (*Psetta maxima*). Aquaculture, 452: 12-23.

Ko S H, Go S, Okorie O E, et al. 2009. Preliminary study of the dietary a-tocopherol requirement in sea cucumber, *Apostichopus japonicus*. Journal of the World Aquaculture Society, 40(5): 659-666.

Kwak H, Austic R E, Dietert R R. 1999. Influence of dietary arginine concentration on lymphoid organ growth in chickens. Poultry Science, 78(11): 1536-1541.

Lall S P. 2003. The minerals//Halver J E, Hardy R W (eds). Fish Nutrition. San Diego: Academic Press: 259-308.

Lallès J D. 2019. Intestinal alkaline phosphatase in the gastrointestinal tract of fish: biology, ontogeny, and environmental and nutritional modulation. Reviews in Aquaculture, 12(2): 555-581.

Le T K, Fotedar R, Partridge G. 2014. Selenium and vitamin E interaction in the nutrition of yellowtail kingfish (*Seriola lalandi*): physiological and immune responses. Aquaculture Nutrition, 20: 303-313.

Lee J, Roh K B, Kim S C, et al. 2012. Soy peptide-induced stem cell proliferation: involvement of ERK and TGF-β1. The Journal of Nutritional Biochemistry, 23(10): 1341-1351.

Li B S, Han X J, Wang J Y, et al. 2021. Optimal dietary methionine requirement for juvenile sea cucumber *Apostichopus japonicus* Selenka. Aquaculture Research, 52(4): 1348-1358.

Li B S, Wang L L, Wang J Y, et al. 2020. Requirement of vitamin E of growing sea cucumber *Apostichopus japonicus* Selenka. Aquaculture Research, 51: 1284-1292.

Li J S, Li J L, Wu T T. 2009. Effects of non-starch polysaccharides enzyme, phytase and citric acid on activities of endogenous digestive enzymes of tilapia (*Oreochromis niloticus* × *Oreochromis aureus*). Aquaculture Nutrition, 15(4): 415-420.

Li J, Liang X F, Tan Q S, et al. 2014. Effects of vitamin E on growth performance and antioxidant status in juvenile grass carp *Ctenopharyngodon idellus*. Aquaculture, 430(1): 21-27.

Li L, Chen M, Storey K B. 2019. Metabolic response of longitudinal muscles to acute hypoxia in sea cucumber *Apostichopus japonicus* (Selenka): a metabolome integrated analysis. Comparative Biochemistry and Physiology Part D: Genomics and Proteomics, 29: 235-244.

Li M, Chen L Q, Qin J G, et al. 2013a. Growth performance, antioxidant status and immune response in darkbarbel catfish *Pelteobagrus vachelli* fed different PUFA/vitamin E dietary levels and exposed to high or low ammonia. Aquaculture, 406-407: 18-27.

Li P, Webb K A, Gatlin III D M. 2008. Evaluation of acid-insoluble ash as an indicator for digestibility determination with red drum, *Sciaenops ocellatus*, and hybrid striped bass, *Morone chrysops* × *M. saxatilis*. Journal of the World Aquaculture Society, 39(1): 120-125.

Li W X, Feng Z, Zhu w, et al. 2016. Cloning, expression and functional characterization of the polyunsaturated fatty acid elongase (ELOVL5) gene from sea cucumber (*Apostichopus japonicu*). Gene, 593(11): 217-224.

Li X F, Wang Y, Liu W B, et al. 2013b. Effects of dietary carbohydrate/lipid ratios on growth performance, body composition and glucose metabolism of fingerling blunt snout bream *Megalobrama amblycephala*. Aquaculture Nutrition, 19(5): 701-708.

Li Y, Ai Q H, Mai K S, et al. 2010. Dietary leucine requirement for juvenile large yellow croaker *Pseudosciaena crocea* (Richardson, 1846). Journal of Ocean University of China, 9(4): 371-375.

Liang H L, Mokrani A, Ji K, et al. 2018. Dietary leucine modulates growth performance, Nrf2 antioxidant signaling pathway and immune response of juvenile blunt snout bream (*Megalobrama amblycephala*). Fish and Shellfish Immunology, 73: 57-65.

Liao M L, Ren T J, Chen W, et al. 2017. Effects of dietary lipid level on growth performance, body composition and digestive enzymes activity of juvenile sea cucumber, *Apostichopus japonicus*. Aquaculture Research, 48(1): 92-101.

Liao M L, Ren T J, He L J, et al. 2014. Optimum dietary protein level for growth and coelomic fluid non-specific immune enzymes of sea cucumber *Apostichopus japonicus* juvenile. Aquaculture Nutrition, 20(4): 443-450.

Liao M L, Ren T J, He L J, et al. 2015. Optimum dietary proportion of soybean meal with fish meal, and its effects on growth, digestibility, and digestive enzyme activity of juvenile sea cucumber *Apostichopus japonicus*. Fisheries Science, 81: 915-922.

Liebler D C. 1993. The role of metabolism in the antioxidant function of vitamin E. Critical Reviews in Toxicology, 23(2): 147-169.

Liemburg-Apers D, Willems P, Koopman W, et al. 2015. Interactions between mitochondrial reactive oxygen species and cellular glucose metabolism. Archives of Toxicology, 89: 1209-1226.

Lin Y F, Shiau S Y. 2005. Dietary vitamin E requirement of grouper, *Epinephelus malabaricus*, at two lipid levels, and their effects on immune responses. Aquaculture, 248(1-4): 235-244.

Liu C, Han Y, Ren T, et al. 2017a. Effects of dietary lysine levels on growth, intestinal digestive enzymes, and coelomic fluid nonspecific immune enzymes of sea cucumber, *Apostichopus japonicus*, juveniles. Journal of the World Aquaculture Society, 48(2): 290-302.

Liu X Y, Wang L, Feng Z F, et al. 2017b. Molecular cloning and functional characterization of the fatty acid delta 6 desaturase (FAD6) gene in the sea cucumber *Apostichopus japonicu*. Aquaculture Research, 24(4): 1-13.

Liu Y, Dong S L, Tian X L, et al. 2009. Effects of dietary sea mud and yellow soil on growth and energy budget of the sea cucumber *Apostichopus japonicus* (Selenka). Aquaculture, 286(3-4): 266-270.

Liu Y, Dong S L, Tian X L, et al. 2010. The effect of different macroalgae on the growth of sea cucumbers (*Apostichopus japonicas* Selenka). Aquaculture Research, 41(11): e881-e885.

Liu Y, Wang W N, Wang A L, et al. 2007. Effects of dietary vitamin E supplementation on antioxidant enzyme activities in *Litopenaeus vannamei* (Boone, 1931) exposed to acute salinity changes. Aquaculture, 265(1-4): 351-358.

Louis P, Hold G L, Flint H J. 2014. The gut microbiota, bacterial metabolites and colorectal cancer. Nature Reviews Microbiology, 12(10): 661-672.

Lowry O H, Rosebrough N J, Farr A L, et al. 1951. Protein measurement with the folin-phenol reagent. The Journal of Biological Chemistry, 193(1): 265-275.

Loy D D, Lundy E L. 2019. Nutritional Properties and Feeding Value of Corn and Its Coproducts//Serna-Saldivar S O (eds). Corn: Chemistry and Technology, Third Edition. London: Woodhead Publishing and AACC International Press: 633-659.

Lu Y, Liang X P, Jin M, et al. 2016. Effects of dietary vitamin E on the growth performance, antioxidant status and innate immune response in juvenile yellow catfish (*Pelteobagrus fulvidraco*). Aquaculture, 464: 609-617.

Lund I, Steenfeldt S J, Hansen B W. 2007. Effect of dietary arachidonic acid, eicosapentaenoic acid and docosahexaenoic acid on survival, growth and pigmentation in larvae of common sole (*Solea solea* L.). Aquaculture, 273(4): 532-544.

Luo Z Y, Wang B J, Liu M, et al. 2014. Effect of dietary supplementation of vitamin C on growth, reactive oxygen species, and antioxidant enzyme activity of *Apostichopus japonicus* (Selenka) juveniles exposed to nitrite. Chinese Journal of Oceanology and Limnology, 32(4): 749-763.

Luo Z, Liu Y J, Mai K S, et al. 2004. Partial replacement of fish meal by soybean protein in diets for grouper *Epinephelus coioides* juveniles. Journal of Fisheries of China, 28(2): 175-181.

Luo Z, Liu Y J, Mai K S, et al. 2007. Effects of dietary arginine levels on growth performance and body composition of juvenile grouper *Epinephelus coioides*. Journal of Applied Ichthyology, 23(3): 252-257.

Luo Z, Tan X Y, Li X D, et al. 2012. Effect of dietary arachidonic acid levels on growth performance, hepatic fatty acid profile, intermediary metabolism and antioxidant responses for juvenile *Synechogobius hasta*. Aquaculture Nutrition, 18(3): 340-348.

Luo Z, Tan X Y, Liu X, et al. 2010. Dietary total phosphorus requirement of juvenile yellow catfish *Pelteobagrus fulvidraco*. Aquaculture International, 18(5): 897-908.

Luzzana U, Hardy R W, Halver J E. 1998. Dietary arginine requirement of fingerling coho salmon (*Oncorhynchus kisutch*). Aquaculture, 163(1-2): 137-150.

Lv Y, Liu H, Ren J, et al. 2013. The positive effect of soybean protein hydrolysates—calcium complexes on bone mass of rapidly growing rats. Food and Function, 4(8): 1245-1251.

Lynch C J, Adams S H. 2014. Branched-chain amino acids in metabolic signalling and insulin resistance. Nature Reviews Endocrinology, 10(12): 723-736.

Lynch S M, Frei B. 1993. Mechanisms of copper- and iron-dependent oxidative modification of human low density lipoprotein. Journal of Lipid Research, 34(10): 1745-1753.

Ma Y S, Wang L T, Sun X H, et al. 2013. Study on hydrolysis conditions of flavourzyme in soybean polypeptide alcalase hydrolysate. Advanced Materials Research, 652: 435-438.

Maalouf J, Nabulsi M, Vieth R, et al. 2008. Short- and long-term safety of weekly high-dose vitamin D3 supplementation in school children. The Journal of Chinical Endocrinology and Metabolism, 93(7): 2693-2701.

Mahdi N, Saeed K, Amir P S, et al. 2017. Effects of dietary vitamin E and selenium nanoparticles supplementation on acute stress responses in rainbow trout (Oncorhynchus mykiss) previously subjected to chronic stress. Aquaculture, 473: 215-222.

Mai K S, Mercer J P, Donlon J. 1995. Comparative studies on the nutrition of two species of abalone, Haliotis tuberculata L. and Haliotis discus hannai Ino. IV. Optimum dietary protein level for growth. Aquaculture, 136(1-2): 165-180.

Mai K S, Wan J L, Ai Q H, et al. 2006. Dietary methionine requirement of large yellow croaker, Pseudosciaena crocea R. Aquaculture, 253: 564-572.

Mamauaga R E P, Koshio S, Ishikawa M, et al. 2011. Soy peptide inclusion levels influence the growth performance, proteolytic enzyme activities, blood biochemical parameters and body composition of Japanese flounder, Paralichthys olivaceus. Aquaculture, 321: 252-258.

Maqsood S, Singh P, Samoon M H, et al. 2010. Effect of dietary chitosan on non-specific immune response and growth of Cyprinus carpio challenged with Aeromonas hydrophila. International Aquatic Research, 2: 77-85.

Maranesi M, Marchetti M, Bochicchio D, et al. 2005. Vitamin B_6 supplementation increases the docosahexaenoic acid concentration of muscle lipids of rainbow trout (Oncorhynchus mykiss). Aquaculture Research, 36(5): 431-438.

Martinez-Álvarez R M, Hidalgo M C, Domezain A, et al. 2002. Physiological changes of sturgeon Acipenser naccarii caused by increasing environmental salinity. Journal of Experimental Biology, 205(23): 3699-3706.

Martínez-Milián G, Olvera-Novoa M A, Toledo-Cuevas E M. 2021. Novel findings in sea cucumber's digestive capacities: enzymatic activities in the respiratory tree, implications for aquaculture. Journal of the world aquaculture society, 52(6): 1259-1272.

Massadeh M I, Sabra F M. 2011. Production and characterization of lipase from Bacillus stearothermophilus. African Journal of Biotechnology, 10(61): 13139-13146.

McBride S C, Lawrence J M, Lawrence A L, et al. 1998. The effects of protein concentration in prepared diets on growth, feeding rate, total organic absorption, and gross assimilation efficiency of the sea urchin Strongylocentrotus franciscanus. Journal of Shellfish Research, 17: 1562-1570.

Mcleod M J, Krismanich A P, Assoud A, et al. 2019. Characterization of 3-[(Carboxymethyl) thio]picolinic acid: a novel inhibitor of phosphoenolpyruvate carboxykinase. Biochemistry, 58(37): 3918-3926.

Mehdi A, Hasan G. 2012. Immune response of broiler chicks fed yeast derived mannanoligosaccharides and humate against new castle disease. World Applied Sciences Journal, 18(6): 779-785.

Meng X, Wang J T, Wan W J, et al. 2017. Influence of low molecular weight chitooligosaccharides on growth performance and non-specific immune response in Nile tilapia *Oreochromis niloticus*. Aquaculture International, 25(3): 1265-1277.

Merino G E, Jetzer T, Doizaki W M D, et al. 1975. Methionine-induced hepatic coma in dogs. The American Journal of Surgery, 130(1): 41-46.

Meseguer I, Aguilar V, González M J, et al. 1998. Extraction and colorimetric quantification of uronic acids of the pectic fraction in fruit and vegetables. Journal of Food Composition and Analysis, 11(4): 285-291.

Millamena O M, Bautista-Teruel M N, Kanazawa, A. 1996. Methionine requirement of juvenile tiger shrimp *Penaeus monodon* Fabricius. Aquaculture, 143(3-4): 403-410.

Millamena O M, Teruel M B, Kanazawa A, et al. 1999. Quantitative dietary requirements of postlarval tiger shrimp, Penaeus monodon, for histidine, isoleucine, leucine, phenylalanine and tryptophan. Aquaculture, 179(1-4): 169-179.

Min Y N, Yang H L, Xu Y X, et al. 2016. Effects of dietary supplementation of synbiotics on growth performance, intestinal morphology, sIgA content and antioxidant capacities of broilers. Journal of Animal Physiology and Animal Nutrition, 100(6): 1073-1080.

Mohamed J S. 2001. Dietary pyridoxine requirement of the Indian catfish, Heteropneustes fossilis. Aquaculture, 194(3-4): 327-335.

Mohanta K N, Mohanty S N, Jena J K. 2007. Protein-sparing effect of carbohydrate in silver barb, *Puntius gonionotus* fry. Aquaculture Nutrition, 13(4): 311-317.

Mohanta K N, Mohanty S N, Jena J, et al. 2009. Carbohydrate level in the diet of silver barb Puntius gonionotus (Bleeker) fingerlings: effect on growth, nutrient utilization and whole body composition. Aquaculture Research, 40(8): 927-937.

Mohapatra M, Sahu N P, Chaudhari A. 2003. Utilization of gelatinized carbohydrate in diets of *Labeo rohita* fry. Aquaculture Nutrition, 9(3): 189-196.

Molina-Poveda C, Morales M E. 2004. Use of a mixture of barley-based fermented grains and wheat gluten as an alternative protein source in practical diets for *Litopenaeus vannamei* (Boone). Aquaculture Research, 35(12): 1158-1165

Moreau R, Dabrowski K. 2003. Alpha-tocopherol downregulates gulonolactone oxidase activity in sturgeon. Free Radical Biology and Medicine, 34(10): 1326-1332.

Moren M, Opstad I, Berntssen M H G, et al. 2004. An optimum level of vitamin A supplements for Atlantic halibut(*Hippoglossus hippoglossus* L.) juveniles. Aquaculture, 235(1-4): 587-599.

Moriarty D J W. 1982. Feeding of Holothuria atra and Stichopus chloronotus on bacteria, organic carbon and organic nitrogen in sediments of the Great Barrier Reef. Marine and Freshwater Research, 33(2): 255-263.

Mostafizur R M, Choi J, Lee S M. 2015. Influences of dietary distillers dried grain level on growth performance, body composition and biochemical parameters of juvenile olive flounder (*Paralichthys olivaceus*). Aquaculture Research, 46(1): 39-48.

Moughan P J, Butts C A, Rowan A M, et al. 2005. Dietary peptides increase endogenous amino acid losses from the gut in adults. The American Journal of Clinical Nutrition, 81(6): 1359-1365.

Mourente G, Tocher D R, Diaz-Salvago E, et al. 1999, Study of the *n*-3 highly unsaturated fatty acids requirement and antioxidant status of *Dentex dentex* larvae at the *Artemia* feeding stage. Aquaculture, 179(1-4): 291-307.

Muñoz M, Cedeño R, Rodríguez J, et al. 2000. Measurement of reactive oxygen intermediate production in haemocytes of the penaeid shrimp, *Penaeus vannamei*. Aquaculture, 191(1-3): 89-107.

Murillo-Gurrea D P, Coloso R M, Borlongan I G, et al. 2001. Lysine and arginine requirements of juvenile Asian sea bass (*Lates calcarifer*). Journal of Applied Ichthyology, 17(2): 49-53.

Nazeer R A, Kumar N S, Jai Ganesh R. 2012. *In vitro* and in vivo studies on the antioxidant activity of fish peptide isolated from the croaker (*Otolithes ruber*) muscle protein hydrolysate. Peptides, 35(2): 261-268.

Nazemroaya S, Yazdanparast R, Nematollahi M A, et al. 2015. Ontogenetic development of digestive enzymes in sobaity sea bream *Sparidentex hasta* larvae under culture condition. Aquaculture, 448: 545-551.

Nishimura J, Masaki T, Arakawa M et al. 2010. Isoleucine prevents the accumulation of tissue triglycerides and upregulates the expression of PPARα and uncoupling protein in diet-induced obese mice. The Journal of Nutrition, 140(3): 496-500.

Niu J, Lemme A, He J Y, et al. 2018. Assessing the bioavailability of the Novel Met-Met product (AQUAVI® MetMet) compared to DL-methionine (DL-Met) in white shrimp (*Litopenaeus vannamei*). Aquaculture, 484: 322-332.

Nordrum S, Krogdahl Å, Røsjø C, et al. 2000. Effects of methionine, cysteine and medium chain triglycerides on nutrient digestibility, absorption of amino acids along the intestinal tract and nutrient retention in Atlantic salmon *Salmo salar* L. under pair-feeding regime. Aquaculture, 186: 341-360.

Oehme M, Grammes F, Takle H, *et al.* 2010. Dietary supplementation of glutamate and arginine to Atlantic salmon (*Salmo salar* L.) increases growth during the first autumn in sea. Aquaculture, 310(1-2): 156-163.

Oh K J, Han H S, Kim M J, et al. 2013. CREB and FoxO1: two transcription factors for the regulation of hepatic gluconeogenesis. BMB Reports, 46(12): 567-574.

Okorie O E, Ko S H, Go S, et al. 2008. Preliminary study of the optimum dietary ascorbic acid level in sea cucumber, *Apostichopus japonicus* (Selenka). Journal of the World Aquaculture Society, 39(6): 758-765.

Okorie O E, Ko S H, Go S, et al. 2011. Preliminary study of the optimum dietary riboflavin level in sea cucumber, *Apostichopus japonicus* (Selenka). Journal of the World Aquaculture Society, 42(5): 657-666.

Oliva-Teles A. 2012. Nutrition and health of aquaculture fish. Journal of Fish Diseases, 35(2): 83-108.

Oliveira G T, Rossi I C, Kucharski L C, et al. 2004. Hepatopancreas gluconeogenesis and glycogen content during fasting in crabs previously maintained on a high-protein or carbohydrate-rich diet. Comparative Biochemistry and Physiology Part A-Molecular and Integrative Physiology, 137(2): 383-390.

Onderci M, Sahin N, Sahin K, et al. 2003. Antioxidant properties of chromium and zinc. Biological

Trace Element Research, 92(2): 139-149.

Otero-Villanueva M M, Kelly M S, Burnell G. 2004. How diet influence energy partitioning in the regular echinoid *Psammechinus miliaris*; constructing an energy budget. Journal of Experimental Marine Biology and Ecology, 304(2): 159-181.

Øverland M, Krogdahl Å, Shurson G, et al. 2013. Evaluation of distiller's dried grains with solubles (DDGS) and high protein distiller's dried grains (HPDDG) in diets for rainbow trout (*Oncorhynchus mykiss*). Aquaculture, 416: 201-208.

Pandey A, Satoh S. 2008. Effects of organic acids on growth and phosphorus utilization in rainbow trout *Oncorhynchus mykiss*. Fisheries Science, 74(4): 867-874.

Pei S R, Dong S L, Wang F, et al. 2012. Effects of density on variation in individual growth and differentiation in endocrine response of Japanese sea cucumber (*Apostichopus japonicus* Selenka). Aquaculture, 356: 398-403.

Pekmezci D. 2011. Vitamin E and immunity. Vitamins and Hormones, 86: 179-215.

Peng M, Xu W, Ai Q, et al. 2013. Effects of nucleotide supplementation on growth, immune responses and intestinal morphology in juvenile turbot fed diets with graded levels of soybean meal (*Scophthalmus maximus* L.). Aquaculture, 392: 51-58.

Peng S, Chen L, Qin J G, et al. 2009. Effects of dietary vitamin E supplementation on growth performance, lipid peroxidation and tissue fatty acid composition of black sea bream (*Acanthopagrus schlegeli*) fed oxidized fish oil. Aquaculture Nutrition, 15(3): 329-337.

Peragón J, Barroso J B, García-Salguero L, et al. 1999. Carbohydrates affect protein-turnover rates, growth, and nucleic acid content in the white muscle of rainbow trout (*Oncorhynchus mykiss*). Aquaculture, 179(1-4): 425-437.

Percudani R, Peracchi A. 2003. A genomic overview of pyridoxal-phosphate-dependent enzymes. EMBO Reports, 4(9): 850-854.

Percudani R, Peracchi A. 2009. The B_6 database: A tool for the description and classification of vitamin B_6-dependent enzymatic activities and of the corresponding protein families. BMC Bioinformatics, 10: 273.

Peskin A V, Winterbourn C C. 2000. A microtiter plate assay for superoxide dismutase using a water-soluble tetrazolium salt (WST-1). Clinica Chimica Acta, 293(1-2): 157-166.

Pilkis S J, Claus T H. 1991. Hepatic gluconeogenesis/glycolysis: regulation and structure/function relationships of substrate cycle enzymes. Annual Review of Nutrition, 11: 465-515.

Polakof S, Panserat S, Soengas J L, et al. 2012. Glucose metabolism in fish: a review. Journal of Comparative Physiology B, 182: 1015-1045.

Pöortner H O. 2002. Climate variations and the physiological basis of temperature dependent biogeography: systemic to molecular hierarchy of thermal tolerance in animals, Comp. Fish Physiology and Biochemistry. 132(2): 739-761.

Powell C D, Chowdhury M A K, Bureau D P. 2017. Assessing the bioavailability of L-methionine and a methionine hydroxyl analogue (MHA-Ca) compared to DL-methionine in rainbow trout (*Oncorhynchus mykiss*). Aquaculture Research, 48: 332-346.

Prescott S M, Topham M K. 2013. Diacylglycerol kinases and phosphatidic acid phosphatases.

Encyclopedia of Biological Chemistry, 1: 593-597.

Preston C M, McKracken K J, McAllister A. 2000. Effect of diet form and enzyme supplementation on growth, efficiency and energy utilisation of wheat-based diets for broilers. British Poultry Science, 41(3): 324-331.

Purcell S W, Hair C A, Mills D J. 2012. Sea cucumber culture, farming and sea ranching in the tropics: progress, problems and opportunities. Aquaculture, 368: 68-81.

Radford C A, Marsden I D, Davison W, et al. 2005. Haemolymph glucose concentrations of juvenile rock lobsters, Jasus edwardsii, feeding on different carbohydrate diets. Comparative Biochemistry and Physiology Part A: Molecular and Integrative Physiology, 140(2): 241-249.

Ramaswamy M, Thangavel P, Panneer Selvam N. 1999. Glutamic oxaloacetic transaminase (GOT) and glutamic pyruvic transaminase (GPT) enzyme activities in different tissues of *Sarotherodon mossambicus* (Peters) exposed to a carbamate pesticide, carbaryl. Pesticide Science, 55(12): 1217-1221.

Ravi J, Devaraj K V. 1991. Quantitative essential amino acid requirements for growth of catla, *Catla catla* (Hamilton). Aquaculture, 96: 281-291.

Rawles S D, Gatlin D M. 1998. Carbohydrate utilization in striped bass (*Morone saxatilis*) and sunshine bass (*M. chrysops* ♀ × *M. saxatilis* ♂). Aquaculture, 161(1-4): 201-212.

Refstie S, Baeverfjord G, Seim R R, et al. 2010. Effects of dietary yeast cell wall β-glucans and MOS on performance, gut health, and salmon lice resistance in Atlantic salmon (*Salmo salar*) fed sunflower and soybean meal. Aquaculture, 305(1-4): 109-116.

Refstie S, Sahlström S, Bråthen E, et al. 2005. Lactic acid fermentation eliminates indigestible carbohydrates and antinutritional factors in soybean meal for Atlantic salmon (*Salmo salar*). Aquaculture, 246(1-4): 331-345.

Refstie S, Storebakken T, Roem A J. 1998. Feed consumption and conversion in Atlantic salmon (*Salmo salar*) fed diets with fish meal, extracted soybean meal or soybean meal with reduced content of oligosaccharides, trypsin inhibitors, lectins and soya antigens. Aquaculture, 162(3-4): 301-312.

Ren M C, Ai Q H, Mai K S. 2011. Effect of dietary carbohydrate level on growth performance, body composition, apparent digestibility coefficient and digestive enzyme activities of juvenile cobia, *Rachycentron canadum* L. Aquaculture Research, 42(10): 1467-1475.

Ren M C, Liao Y J, Xie J, et al. 2013. Dietary arginine requirement of juvenile blunt snout bream, *Megalobrama amblycephala*. Aquaculture, 414-415: 229-234.

Ren T J, Liao M L, Han Y Z, et al. 2016. Effectiveness of L-ascorbyl-2-polyphosphate as an ascorbic acid source for sea cucumber, *Apostichopus japonicus*. Aquaculture Research, 47(8): 2594-2606.

Rengpipat S, Rukpratanporn S, Piyatiratitivorakul S, et al. 2000. Immunity enhancement in black tiger shrimp (*Penaeus monodon*) by a probiont bacterium (Bacillus S11). Aquaculture, 191(4): 271-288.

Robbins K R, Norton H W, Baker D H. 1979. Estimation of nutrient requirements from growth data. The Journal of Nutrition, 109(10): 1710-1714.

Rocha C B, Portelinha M K, Fernandes J M, et al. 2014. Dietary phosphorus requirement of pejerrey fingerlings (*Odontesthes bonariensis*). Revista Brasileira de Zootecnia, 43: 55-59.

Rodríguez L A, Borges P, Caballero M J, et al. 2017. Effect of different dietary vitamin E levels on growth, fish composition, fillet quality and liver histology of meagre (*Argyrosomus regius*). Aquaculture, 468: 175-183.

Rodriguez-Matas M C, Lisbona F, Gómez-Ayala A E I, et al. 1998. Influence of nutritional iron deficiency development on some aspects of iron, copper and zinc metabolism. Laboratory Animals, 32(3): 298-306.

Roem A J, Kohler C C, Stickney R R. 1990. Vitamin E requirements of the blue tilapia, *Oreochromis aureus* (Steindachner), in relation to dietary lipid level. Aquaculture, 87(2): 155-164.

Rosalki S B, Rau D, Lehmann D, et al. 1970. Gamma-glutamyl transpeptidase in chronic alcoholism. Lancet, 2(7683): 1139.

Roy P. K, Lall S P. 2003. Dietary phosphorus requirement of juvenile haddock *(Melanogram mus aeglefinus* L.). Aquaculture, 221(1-4): 451-468.

Ruchimat T, Masumoto T, Hosokawa H, et al. 1997. Quantitative methionine requirement of yellowtail (*Seriola quinqueradiata*). Aquaculture, 150(1-2): 113-122.

Rudnicki M, Silveira M M, Pereira T V, et al. 2007. Protective effects of *Passiflora alata* extract pretreatment on carbon tetrachloride induced oxidative damage in rats. Food and Chemical Toxicology, 45(4): 656-661.

Salah El-Deen, Rogers W A. 1993. Changes in total protein and transaminase activities of grass carp exposed to diquat. Journal of Aquatic Animal Health, 5(4): 280-286.

Salini M J, Wade N M, Araújo B C, et al. 2016. Eicosapentaenoic acid, arachidonic acid and eicosanoid metabolism in juvenile barramundi *Lates calcarifer*. Lipids, 51(8): 973-988.

Samadi P, Grégoire L, Rouillard C, et al. 2006. Docosahexaenoic acid reduces levodopa induced dyski- nesias in1-methyl 4-phenyl-1, 2, 3, 6 tetrahydropyridine monkeys. Annals of Neurology, 59(2): 282-288.

Sanchez W, Aït-Aïssa S, Palluel O, et al. 2007. Preliminary investigation of multi-biomarker responses in three-spined stickleback (*Gasterosteus aculeatus* L.) sampled in contaminated streams. Ecotoxicology, 16(2): 279-287.

Sánchez-Machado D I, López-Cervantes J, López-Hernández J, et al. 2004. Fatty acids, total lipid, protein and ash contents of processed edible seaweeds. Food Chemistry, 85(3): 439-444.

Santiago C B, Lovell R T. 1988. Amino acid requirements for growth of Nile Tilapia. The Journal of Nutrition, 118(12): 1540-1546.

Sargent J R, Bell J G, Bell M V, et al. 1995, Requirement criteria for essential fatty acids. Journal of Applied Ichthyology, 11(3-4): 183-198.

Sau S K, Paul B N, Mohanta K N, et al. 2004. Dietary vitamin E requirement, fish performance and carcass composition of rohu (*Labeo rohita*) fry. Aquaculture, 240(1-4): 359-368.

Seo J Y, Lee S M. 2011. Optimum dietary protein and lipid levels for growth of juvenile sea cucumber *Apostichopus japonicus*. Aquaculture Nutrition, 17(2): e56-e61.

Seo J Y, Shin I S, Lee S M. 2011. Effect of dietary inclusion of various plant ingredients as an alternative for *Sargassum thunbergii* on growth and body composition of juvenile sea cucumber *Apostichopus japonicus*. Aquaculture Nutrition, 17(5): 549-556.

Shahkar E, Hamidoghli A, Yun H, et al. 2018. Effects of dietary vitamin E on hematology, tissue α-tocopherol concentration and non-specific immune responses of Japanese eel, *Anguilla japonica*. Aquaculture, 484: 51-57.

Shan L L, Li X Q, Zheng X M, et al. 2017. Effects of feed processing and forms of dietary methionine on growth and IGF-1 expression in Jian carp. Aquaculture Research, 48: 56-67.

Shen H M, Chen X R, Chen W Y, et al. 2017. Influence of dietary phosphorus levels on growth, body composition, metabolic response and antioxidant capacity of juvenile snakehead (*Channa argus* × *Channa maculata*). Aquaculture nutrition, 23(4): 662-670.

Shi C, Dong S L, Wang F, et al. 2013. Effects of four fresh microalgae in diet on growth and energy budget of juvenile sea cucumber *Apostichopus japonicas* (Selenka). Aquaculture, (416-417): 296-301.

Shi C, Dong S, Wang F, et al. 2015. Effects of the sizes of mud or sand particles in feed on growth and energy budgets of young sea cucumber (*Apostichopus japonicus*). Aquaculture, 440: 6-11.

Shiau S Y, Hsu C Y. 2002. Vitamin E sparing effect by dietary vitamin C in juvenile hybrid tilapia, *Oreochromis niloticus* × *O. aureus*. Aquaculture, 210(1-4): 335-342.

Shiau S Y, Hwang J Y. 1994. The dietary requirement of juvenile grass shrimp (*Penaeus monodon*) for vitamin D. The Journal of Nutrition, 124(12): 2445-2450.

Shiau S Y, Wu M H. 2003. Dietary vitamin B6 requirement of grass shrimp, *Penaeus monodon*. Aquaculture, 225(1-4): 397-404.

Shimeno S, Ming D C, Takeda M. 1993. Metabolic response to dietary carbohydrate to lipid ratios in *Oreochromis niloticus*. Bulletin of the Chemical Society of Japan, 59(5): 827-833.

Skalli A, Hidalgo M C, Abellán E, et al. 2004. Effects of the dietary protein/lipid ratio on growth and nutrient utilization in common dentex (*Dentex dentex* L.) at different growth stages. Aquaculture, 235(1-4): 1-11.

Slater M J, Carton A G. 2010. Sea cucumber habitat differentiation and site retention as determined by intraspecific stable isotope variation. Aquaculture Research, 41(10): e695-e702.

Slater M J, Jeffs A G, Carton A G. 2009. The use of the waste from green-lipped mussels as a food source for juvenile sea cucumber, *Australostichopus mollis*. Aquaculture, 292(3-4): 219-224.

Slater M J, Lassudrie M, Jeffs A G. 2011. Method for determining apparent digestibility of carbohydrate and protein sources for artificial diets for juvenile sea cucumber, *Australostichopus mollis*. Journal of the World Aquaculture Society, 42(5): 714-725.

Sloan N A. 1984. Echinoderm fisheries of the world: A review//Keegan B E, O'Conner B D S (eds). Echinodermata (Proceedings of the Fifth International Echinoderm Conference). Rotterdam: A A Balkema Publishers: 109-124.

Söderhäll K, Cerenius L. 1998. Role of the prophenoloxidase-activating system in invertebrate immunity. Current Opinion in Immunology, 10(1): 23-28.

Song X Y, Xu Q, Zhou Y, et al. 2017. Growth, feed utilization and energy budgets of the sea cucumber *Apostichopus japonicus*, with different diets containing the green tide macroalgae *Chaetomorpha linum*, and the seagrass *Zostera marina*. Aquaculture, 470: 157-163.

Song Z D, Li P Y, Wang J Y, et al. 2016. Effects of seaweed replacement by hydrolyzed soybean meal on growth, metabolism, oxidation resistance and body composition of sea cucumber *Apostichopus japonicus*. Aquaculture, 463: 135-144.

Song Z, Li H, Wang J, et al. 2014. Effects of fishmeal replacement with soy protein hydrolysates on growth performance, blood biochemistry, gastrointestinal digestion and muscle composition of juvenile starry flounder (*Platichthys stellatus*). Aquaculture, 426-427: 96-104.

Song Z, Li P, Hu S, et al. 2023. Influence of dietary phosphorus on the growth, feed utilization, proximate composition, intestinal enzymes, and oxidation resistance of sea cucumber *Apostichopus japonicas*. Aquaculture nutrition, (1): 2266191.

Sørensen M, Penn M, El-Mowafi A, et al. 2011. Effect of stachyose, raffinose and soya-saponins supplementation on nutrient digestibility, digestive enzymes, gut morphology and growth performance in Atlantic salmon (*Salmo salar*, L). Aquaculture, 314(1-4): 145-152.

SPSS Inc. 2013. SPSS 16. 0 Student Version for Windows. Prentice Hall, Upper Saddle River, New Jersey.

Stadtman E R, Berlett B S. 1999. Reactive oxygen-mediated pro-tein oxidation in aging and disease. Chemical Research in Toxicology, 10(5): 485-494.

Staykov Y, Spring P, Denev S, et al. 2007. Effect of a mannan oligosaccharide on the growth performance and immune status of rainbow trout (*Oncorhynchus mykiss*). Aquaculture International, 15(2): 153-161.

Stebbing A R D. 1982. Hormesis-the stimulation of growth by low levels of inhibitors. Science of the Total Environment, 22(3): 213-234.

Stone D A J, Allan G L, Anderson A J. 2003. Carbohydrate utilization by juvenile silver perch, *Bidyanus bidyanus* (Mitchell). III. The protein-sparing effect of wheat starch-based carbohydrates. Aquaculture Research, 34(2): 123-134.

Sugiura S H, Dong F M, Hardy R W. 2000. Primary responses of rainbow trout to dietary phosphorus concentrations. Aquaculture Nutrition, 6(4): 235-245.

Sugiura S H, Hardy R W, Roberts R J. 2004. The pathology of phosphorus deficiency in fish-a review. Journal of Fish Diseases, 27(5): 255-265.

Sugiura S, Higashitani A, Sasaki T. 2011. Effects of dietary phosphorus restriction on fillet fat deposition and hepatic lipid metabolism in rainbow trout (*Oncorhynchus mykiss*) and crucian carp (*Carassius auratus grandoculis*). Aquaculture Science, 59(1): 109-122.

Sun H L, Liang M Q, Yan J P, et al. 2004. Nutrient requirements and growth of the sea cucumber, *Apostichopus japonicus*, advances in sea cucumber aquaculture and management. FAO Fisheries Technical Paper No. 463: 327-331.

Sun J M, Zhang L B, Pan Y, et al. 2015. Feeding behavior and digestive physiology in sea cucumber *Apostichopus japonicus*. Physiology and Behavior, 139: 336-343.

Sun L M, Chen L Q, Li E C, et al. 2013. Effects of dietary methionine supplementation on feeding, growth and antioxidant ability of juvenile Chinese mitten crab, *Eriocheir sinensis*. Acra Hydrobiologica Sinica, 37(2): 336-343.

Sun W H, Leng K L, Lin H, et al. 2010. Analysis and evaluation of chief nutrient composition in different parts of *Stichopus japonicus*. Chinese Journal of Animal Nutrition, 22(1): 212-220.

Sveinsdóttir H, Thorarensen H, Gudmundsdóttir Á. 2006. Involvement of trypsin and chymotrypsin activities in Atlantic cod (*Gadus morhua*) embryogenesis. Aquaculture, 260(1-4): 307-314.

Takeuchi T, Dedi J, Haga Y, et al. 1998. Effect of vitamin A compounds on bone deformity in larval Japanese flounder (*Paralichthys olivaceus*). Aquaculture, 169(3-4): 155-165.

Tan B P, Mai K S, Liufu Z G. 2002. Dietary phosphorus requirement of young abalone *Haliotis discus Hannai* ino. Chinese Journal of Oceanology and Limnology, 20(1): 22-31.

Tan Q S, Wang F, Xie S Q, et al. 2009. Effect of high dietary starch levels on the growth performance, blood chemistry and body composition of gibel carp (*Carassius auratus* var. gibelio). Aquaculture Research, 40(9): 1011-1018.

Tan X H, Lin H Z, Huang Z, et al. 2016. Effects of dietary leucine on growth performance, feed utilization, non-specific immune responses and gut morphology of juvenile golden pompano *Trachinotus ovatus*. Aquaculture, 465: 100-107.

Tang D G, Chen Y Q, Honn K V. 1996. Arachidonate lipoxygenases as essential regulators of cell survival and apoptosis. Proceedings of the National Academy of Sciences of the United States of America, 93(11): 5241-5246.

Tang Q, Wang C, Xie C, et al. 2012. Dietary available phosphorus affected growth performance, body composition, and hepatic antioxidant property of juvenile yellow catfish *Pelteobagrus fulvidraco*. The Scientific World Journal, (1): 987570.

Tang Z R, Yin Y L, Nyachoti C M, et al. 2005. Effect of dietary supplementation of chitosan and galacto-mannan-oligosaecharide on serum parameters and the insulin-like growth factor-I mRNA expression in early weaned piglets. Domestic Animal Endocrinology, 28(4): 430-441.

Tanner D K, Leonard E N, Brazner J C. 1999. Microwave digestion method for phosphorus determination of fish tissue. Limnology and Oceanography, 44(3): 708-709.

Teixeira C P, Barros M M, Pezzato L E, et al. 2012. Growth performance of Nile tilapia, *Oreochromis niloticus*, fed diets containing levels of pyridoxine and haematological response under heat stress. Aquaculture Research, 43(8): 1081-1088.

Thomsen C, Rasmussen O, Lousen T, et al. 1999. Differential effects of saturated and monounsaturated fatty acids on postprandial lipemia and incretin responses in healthy subjects. The American Journal of Clinical Nutrition, 69(6): 1135-1143.

Tian J, Ji H, Oku H, et al. 2014. Effects of dietary arachidonic acid (ARA) on lipid metabolism and health status of juvenile grass carp, *Ctenopharyngodon idellus*. Aquaculture, 430(15): 57-65.

Topham M K. 2013. Diacylglycerol kinases and phosphatidic acid phosphatases. Encyclopedia of Biological Chemistry: 659-663.

Torrecillas S, Makol A, Caballero M J, et al. 2007. Immune stimulation and improved infection resistance in European sea bass (*Dicentrarchus labrax*) fed mannan oligosaccharides. Fish and Shellfish Immunology, 23(5): 969-981.

Tortuero F, Fernández E, Rupérez P, et al. 1997. Raffinose and lactic acid bacteria influence caecal fermentation and serum cholesterol in rats. Nutrition Research, 17(1): 41-49.

Tran-Duy A, Smit B, Van Dam A A, et al. 2008. Effects of dietary starch and energy levels on maximum feed intake, growth and metabolism of Nile tilapia, *Oreochromis niloticus*.

Aquaculture, 277(3-4): 213-219.

Trenzado C E., Morales A E., Palma, J M., Higuer M. 2009. Blood antioxidant defenses and hematological adjustments in crowded/uncrowded rainbow trout (Oncorhynchus mykiss) fed on diets with different levels of antioxidant vitamins and HUFA. Comparative Biochemistry and Physiology Part C: Toxicology and Pharmacology, 149(3): 440-447.

Trichet V V. 2010. Nutrition and immunity: an update. Aquaculture Research, 41(3): 356-372.

Trushenski J, Schwarz M, Lewis H, et al. 2011. Effect of replacing fish oil with soybean oil on production performance and fillet lipid and acid composition of juvenile cobia *Rachycentron canadum*. Aquaculture, 17(9): e437-e447.

Tseng C K. 1983. Common seaweeds of China. Beijing: Science Press.

Uauy R, Stringel G, Thomas R, et al. 1990. Effect of dietary nucleosides on growth and maturation of the developing gut in the rat. Journal of Pediatric Gastroenterology and Nutrition, 10(4): 497-503.

Uyan O, Koshio S, Ishikawa M, et al. 2007. Effects of dietary phosphorus and phospholipid level on growth, and phosphorus deficiency signs in juvenile Japanese flounder, *Paralichthys olivaceus*. Aquaculture, 267(1-4): 44-54.

Uyan O, Koshio S, Teshima S I, et al. 2006. Growth and phosphorus loading by partially replacing fishmeal with tuna muscle by-product powder in the diet of juvenile Japanese flounder, *Paralichthys olivaceus*. Aquaculture, 257(1-4): 437-445.

Valko M, Leibfritz D, Moncol J, et al. 2007. Free radicals and antioxidants in normal physiological functions and human disease. International Journal of Biochemistry and Cell Biology, 39(1): 44-84.

van Vranken J G, Bricker D K, Dephoure N, et al. 2014. SDHAF4 promotes mitochondrial succinate dehydrogenase activity and prevents neurodegeneration. Cell Metabolism, 20(2): 241-252.

Vargas-Albores F, Yepiz-Plascencia G. 2000. Beta glucan binding protein and its role in shrimp immune response. Aquaculture, 191(1-3): 13-21.

Vásquez-Torres W, Arias-Castellanos J A. 2013. Effect of dietary carbohydrates and lipids on growth in cachama (*Paractus brachypomus*). Aquaculture Research, 44(11): 1768-1776.

Vielma J, Koskela J, Ruohonen K, et al. 2003. Optimal diet composition for European whitefish (*Coregonus lavaretus*): carbohydrate stress and immune parameter responses. Aquaculture, 225(1-4): 3-16.

Vrolijk M F, Opperhuizen A, Jansen E H J M, et al. 2017. The vitamin B6 paradox: Supplementation with high concentrations of pyridoxine leads to decreased vitamin B6 function. Toxicology in Vitro, 44: 206-212.

Wakita C, Honda K, Shibata T, et al. 2011. A method for detection of 4-hydroxy-2-nonenal adducts in proteins. Free Radical Biology and Medicine, 51(1): 1-4.

Walton M J., Cowey C B., Adron J W. 1982. Methionine metabolism in rainbow trout fed diets of differing methionine and cystine content. Journal of Nutrition, 112: 1525-1535.

Wamberg, S, Engel, K, Kildeberg, P. 1987. Methionine-induced acidosis in the weanling rat. Acta Physiologica Scandinavica, 129: 575-583.

Wan M, Mai K S, Ma H M, et al. 2004. Effects of dietary selenium and vitamin E on antioxidant enzyme activities in abalone, *Haliotis discus hannai* ino. Acta Hydrobiologica Sinica, 28(5): 496-503.

Wang F, Dong S L, Huang G Q, et al. 2003. The effect of light color on the growth of Chinese shrimp *Fenneropenaeus chinensis*. Aquaculture, 228(1-4): 351-360.

Wang J Y, Zhu S G, Xu C F. 2007. Biochemistry Third Edition (Volume 2). Beijing: Higher Education Press, 1 (In Chinese with English abstract).

Wang J, Ren T, Han Y, et al. 2015a. Effects of dietary vitamin C supplementation on lead-treated sea cucumbers, *Apostichopus japonicus*. Ecotoxicology and Environmental Safety, 118: 21-26.

Wang J, Xu Y P, Li X Y, et al. 2015b. Vitamin E requirement of sea cucumber (*Apostichopus japonicus*) and its' effects on nonspecific immune responses. Aquaculture Research, 46(7): 1628-1637.

Wang J, Yang R W, Liu J B. 2014. Effects of soybean antioxidant peptides (SAP) on SOD, GSH-Px, CAT activity and MDA level in vivo. Advanced Materials Research, 1025-1026: 476-481.

Wang L, Ye L, Hua Y, et al. 2019. Effects of dietary dl-methionyl-dl-methionine (Met-Met) on growth performance, body composition and haematological parameters of white shrimp (*Litopenaeus vannamei*) fed with plant protein-based diets. Aquaculture Research, 50: 1718-1730.

Wang P, Li X, Xu Z, et al. 2021. The digestible phosphorus requirement in practical diet for largemouth bass (*Micropterus salmoides*) based on growth and feed utilization. Aquaculture and Fisheries, 7(6): 632-638.

Wang Y B, He Z L. 2009. Effect of probiotics on alkaline phosphatase activity and nutrient level in sediment of shrimp, *Penaeus vannamei*, ponds. Aquaculture, 287(1-2): 94-97.

Wang Y, Geng Y, Shi X, et al. 2022. Effects of dietary phosphorus levels on growth performance, phosphorus utilization, and intestinal calcium and phosphorus transport-related gene expression of juvenile Chinese soft-shelled turtle (*Pelodiscus sinensis*). Animals, 12(22): 3101.

Wang Y, Liu Y J, Tian L X, et al. 2005. Effects of dietary carbohydrate level on growth and body composition of juvenile tilapia, *Oreochromis niloticus* × *O. aureus*. Aquaculture Research, 36(14): 1408-1413.

Wang Z, Mai K S, Xu W, et al. 2016. Dietary methionine level influences growth and lipid metabolism via GCN2 pathway in cobia (*Rachycentron canadum*). Aquaculture, 454: 148-156.

Warden R A, Strazzari M J, Dunkley P R, et al. 1996. Vitamin a-deficient rats have only mild changes in Jejunal structure and function. The Journal of Nutrition, 126(7): 1817-1826.

Wassef E A, Masry E M H, Mikhail F R. 2015. Growth enhancement and muscle structure of striped mullet, *Mugil cephalus* L. fingerlings by feeding algal meal-based diets. Aquaculture Research, 32: 315-322.

Watanabe T, Kiron V, Satoh S. 1997. Trace minerals in fish nutrition. Aquaculture, 151(1-4): 185-207.

Waterland R A. 2006. Assessing the effects of high methionine intake on DNA methylation. The Journal of Nutrition, 136(6): 1706S-1710S.

Webster C D, Tidwell J H, Goodgame L S, et al. 1992. Use of soybean meal and distillers grains with solubles as partial or total replacement of fish meal in diets for channel catfish, *Ictalurus punctatus*. Aquaculture, 106(3-4): 301-309.

Webster C D, Tidwell J H, Goodgame L S, et al. 1993. Growth, body composition, and organoleptic evaluation of channel catfish fed diets containing different percentages of distillers' grains with solubles. Progressive Fish-Culturist, 55(2): 95-100.

Wee K L. 1991. Use of nonconventional feedstuffs of plant origin as fish feeds–is it practical and economically feasible? In: Fish Nutrition Research in Asia. Proceedings of the Fourth Asian Fish Nutrition Workshop (DeSilva SS ed), Asian Fish Soc, Manila, Philippines, 13-31.

Welker T L, Lim C, Klesius P, et al. 2014. Evaluation of distiller's dried grains with solubles from different grain sources as dietary protein for hybrid tilapia, *Oreochromis niloticus* (♀) ×*Oreochromis aureus* (♂). Journal of the World Aquaculture Society, 45(6): 625-637

Welker T L, Lim C, Yildirim-Aksoy M, et al. 2007. Immune response and resistance to stress and Edwardsiella ictaluri challenge in channel catfish, Ictalurus punctatus, fed diets containing commercial whole-cell yeast or yeast subcomponents. Journal of the World Aquaculture Society, 38: 24-35.

Wen B, Gao Q F, Dong S, et al. 2016a. Utilization of different macroalgae by sea cucumber *Apostichopus japonicas* revealed by carbon stable isotope analysis. Aquaculture Environment Interactions, 8: 171-178.

Wen B, Gao Q F, Dong S L, et al. 2016b. Effects of different feed ingredients on growth, fatty acid profiles, lipid peroxidation and aminotransferases activities of sea cucumber *Apostichopus japonicus* (Selenka). Aquaculture, 454(6): 176-183.

Wen B, Gao Q F, Dong S L, et al. 2016c. Absorption of different macroalgae by sea cucumber *Apostichopus japonicus*, (Selenka): evidence from analyses of fatty acid profiles. Aquaculture, (451): 421-428.

Wen J, Jiang W, Feng L, et al. 2015. The influence of graded levels of available phosphorus on growth performance, muscle antioxidant and flesh quality of young grass carp (*Ctenopharyngodon idella*). Animal Nutrition, 1(2): 77-84.

Wiggins H S. 1984. Nutritional value of sugars and related compounds undigested in the small gut. Proceedings of the Nutrition Society, 43(1): 69-75.

Wilson R P, Poe W E, Robinson E H. 1980. Leucine, isoleucine, valine and histidine requirements of fingerling channel catfish. The Journal of Nutrition, 110(4): 627-633.

Wilson R P. 1994. Utilization of dietary carbohydrate by fish. Aquaculture, 124(1-4): 67-80.

Winston G W, Di Giulio R T. 1991. Prooxidant and antioxidant mechanisms in aquatic organisms. Aquatic Toxicology, 19(2): 137-161.

Witten P E, Owen M A G, Fontanillas R, et al. 2016. A primary phosphorus-deficient skeletal phenotype in juvenile Atlantic salmon *Salmo salar*: the uncoupling of bone formation and mineralization. Journal of Fish Biology, 88(2): 690-708.

Wu B, Xia S, Rahman M M, et al. 2015. Substituting seaweed with corn leaf in diet of sea cucumber (*Apostichopus japonicus*): effects on growth, feed conversion ratio and feed digestibility. Aquaculture, 444: 88-92.

Wu C L, Chen L, Lu Z B, et al. 2017. The effects of dietary leucine on the growth performances, body composition, metabolic abilities and innate immune responses in black carp *Mylopharyngodon piceus*. Fish and Shellfish Immunology, 67: 419-428.

Wu Z X, Yu Y M, Chen X, et al. 2014. Effect of prebiotic konjac mannanoligosaccharide on growth performances, intestinal microflora, and digestive enzyme activities in yellow catfish, *Pelteobagrus fulvidraco*. Fish Physiology and Biochemistry, 40(3): 763-771.

Xia B, Gao Q F, Wang J Y, et al. 2015a. Effects of dietary carbohydrate level on growth, biochemical composition and glucose metabolism of juvenile sea cucumber *Apostichopus japonicus* (Selenka). Aquaculture, 448: 63-70.

Xia B, Ren Y, Wang F, et al. 2017. A comparative study on growth, protein turnover and energy budget of green and white color morphs of sea cucumber *Apostichopus japonicus* (Selenka). Aquaculture Environment Interactions, 9: 405-414.

Xia B, Wang J Y, Gao Q F, et al. 2015b. The nutritional contributions of dietary protein sources to tissue growth and metabolism of sea cucumber *Apostichopus japonicus* (Selenka): evidence from nitrogen stable isotope analysis. Aquaculture, 435: 237-244.

Xia S D, Yang H S, Li Y, et al. 2012a. Effects of different seaweed diets on growth, digestibility, and ammonia-nitrogen production of the sea cucumber *Apostichopus japonicus* (Selenka). Aquaculture, 338-341: 304-308.

Xia S D, Zhao P, Chen K, et al. 2012b. Feeding preferences of the sea cucumber *Apostichopus japonicus* (Selenka) on various seaweed diets. Aquaculture, 344-349: 205-209.

Xiao W W, Feng L, Kuang S Y, et al. 2012. Lipid peroxidation, protein oxidant and antioxidant status of muscle and serum for juvenile Jian carp (*Cyprinus carpio* var. Jian) fed grade levels of methionine hydroxyl analogue. Aquaculture Nutrition, 18: 90-97.

Xie D, Han D, Zhu X, et al. 2017. Dietary available phosphorus requirement for on- growing gibel carp (*Carassius auratus gibelio* var. CAS III). Aquaculture Nutrition, 23(5): 1104-1112.

Xie J J, Lemme A, He J Y, et al. 2018. Fishmeal levels can be successfully reduced in white shrimp (*Litopenaeus vannamei*) if supplemented with DL-Methionine (DL-Met) or DL-Methionyl-DL-Methionine (Met-Met). Aquaculture Research, 28: 1144-1152.

Xu C, Yu H, Zhang Q, et al. 2021. Dietary phosphorus requirement of coho salmon (*Oncorhynchus kisutch*) alevins cultured in freshwater. Aquaculture Nutrition, 27(6): 2427-2435.

Xu G L, Xing W, Li T L, et al. 2018. Effects of dietary raffinose on growth, non-specific immunity, intestinal morphology and microbiome of juvenile hybrid sturgeon (*Acipenser baeri* Brandt ♀ × *A. schrenckii* Brandt ♂). Fish and Shellfish Immunology, 72: 237-246.

Xu H G, Ai Q H, Mai K S, et al. 2010. Effects of dietary arachidonic acid on growth performance, survival, immune response and tissue fatty acid composition of juvenile Japanese seabass, *Lateolabrax japonicas*. Aquaculture, 307(1): 75-82.

Xu L, Chen X, Wen H, et al. 2022. Dietary phosphorus requirement of red swamp crayfish (*Procambarus clarkia*). Aquaculture Research, 53(4): 1293-1303.

Xu Q Y, Xu H, Wang C, et al. 2011. Studies on dietary phosphorus requirement of juvenile Siberian sturgeon *Acipenser baerii*. Journal of Applied Ichthyology, 27(2): 709-714.

Yamamoto T, Unuma T, Alkiyama T. 2000. The influence of dietary protein sources on tissue free amino acid levels of fingerling rainbow trout. Japanese Society of Fisheries Science, 66(2):

310-320

Yan Q, Xie S, Zhu X, et al. 2007. Dietary methionine requirement for juvenile rockfish, *Sebastes schlegeli*. Aquaculture Nutrition, 13(3): 163-169.

Yang H S, Yuan X T, Zhou Y, et al. 2005. Effects of body size and water temperature on food consumption and growth in the sea cucumber *Apostichopus japonicus* (Selenka) with special reference to aestivation. Aquaculture Research, 36(11): 1085-1092.

Yang P, Hu H B, Liu Y, et al. 2018. Dietary stachyose altered the intestinal microbiota profile and improved the intestinal mucosal barrier function of juvenile turbot, *Scophthalmus maximus* L. Aquaculture, 486: 98-106.

Yang Q, Liang H, Maulu S, et al. 2021. Dietary phosphorus affects growth, glucolipid metabolism, antioxidant activity and immune status of juvenile blunt snout bream (*Megalobrama amblycephala*). Animal Feed Science and Technology, 274: 114896.

Yang S D, Lin T S, Liu F G, et al. 2006. Influence of dietary phosphorus levels on growth, metabolic response and body composition of juvenile silver perch (*Bidyanus bidyanus*). Aquaculture, 253(1-4): 592-601.

Yanik S, Aras M H, Erkilic S, et al. 2016. Histopathological features of bisphosphonates related osteonecrosis of the jaw in rats with and without vitamin D supplementation. Archives of Oral Biology, 65: 59-65.

Ye J, Wang Z, Wang K. 2011. Effect of partial fish meal replacement by soybean meal on the growth performance and biochemical indices of juvenile Japanese flounder (*Paralichthys olivaceus*). Aquaculture International, 19(1): 143-153.

Ye W J, Tan X Y, Chen Y D, et al. 2009. Effects of dietary protein to carbohydrate ratios on growth and body composition of juvenile yellow catfish, *Pelteobagrus fulvidraco* (Siluriformes, Bagridae, Pelteobagrus). Aquaculture Research, 40(12): 1410-1418.

Yildirim-Aksoy M, Lim C, Li M H, et al. 2008. Interaction between dietary levels of vitamins C and E on growth and immune responses in channel catfish, *Ictalurus punctatus* (Rafinesque). Aquaculture Research, 39(11): 1198-1209.

Yingst J Y. 1976. The utilization of organic matter in shallow marine sediments by an epibenthic deposit-feeding holothurian. Journal of Experimental Marine Biology and Ecology, 23(1): 55-69.

Yoon T H, Lee D H, Won S G, et al. 2015. Optimal incorporation level of dietary alternative phosphate ($MgHPO_4$) and requirement for phosphorus in juvenile Far Eastern catfish (*Silurusasotus*). Asian-Australasian Journal of Animal Sciences, 28(1): 111.

Yoshida T, Kruger R, Inglis V. 1995. Augmentation of non-specific protection in African catfish, *Clarias gariepinus*(Burchell), by the long term oral administration of immunostimulants. Journal of Fish Diseases, 18(2): 195-198.

Yoshizawa F. 2004. Regulation of protein synthesis by branched-chain amino acids in vivo. Biochemical and Biophysical Research Communications, 313(2): 417-422.

Yu H B, Gao Q F, Dong S L, et al. 2015a. Changes in fatty acid profiles of sea cucumber *Apostichopus japonicus* (*Selenka*) induced by terrestrial plants in diets. Aquaculture, 442(7): 119-124.

Yu H B, Gao Q F, Dong S L, et al. 2015b. Utilization of corn meal and extruded soybean meal by sea cucumber *Apostichopus japonicus* (Selenka): insights from carbon stable isotope analysis. Aquaculture, 435: 106-110.

Yu H B, Gao Q F, Dong S L, et al. 2016. Regulation of dietary glutamine on the growth, intestinal function, immunity and antioxidant capacity of sea cucumber *Apostichopus japonicus* (Selenka). Fish and Shellfish Immunology, 50: 56-65.

Yu H R, Ai Q H, Mai K S, et al. 2013. L-methionine requirement of large yellow croaker (*Pseudosciaena crocea* R.) larvae. Acta Hydrobiologica sinica, 37(6): 1094-1102.

Yuan X T, Yang H S, Zhou Y, et al. 2006. The influence of diets containing dried bivalve feces and/or powdered algae on growth and energy distribution in sea cucumber *Apostichopus japonicus* (Selenka) (Echinodermata: Holothuroidea). Aquaculture, 256: 457-467.

Yuan Y C, Lin Y C, Yang H J, et al. 2013. Evaluation of fermented soybean meal in the practical diets for juvenile Chinese sucker, *Myxocyprinus asiaticus*. Aquaculture Nutrition, 19(1): 74-83.

Zafar N, Khan M A. 2018. Determination of dietary phosphorus requirement of stinging catfish *Heteropneustes fossilis* based on feed conversion, growth, vertebrate phosphorus, whole body phosphorus, haematology and antioxidant status. Aquaculture Nutrition, 24(5): 1577-1586.

Zamora L N, Jeffs A G. 2011. Feeding, selection, digestion, and absorption of the organic matter from mussel waste by juveniles of the deposit-feeding sea cucumber, *Australostichopus mollis*. Aquaculture, 317(1-4): 223-228.

Zang Y, Tian X, Dong S, et al. 2012. Growth, metabolism and immune responses to evisceration and the regeneration of viscera in sea cucumber, *Apostichopus japonicus*. Aquaculture, 358-359: 50-60.

Zhang C X, Mai K S, Ai Q H, et al. 2006. Dietary phosphorus requirement of juvenile Japanese seabass, *Lateolabrax japonicus*. Aquaculture, 255(1-4): 201-209.

Zhang Q G, Liang M Q, Xu H G. 2019. Dietary methionine requirement of juvenile tiger puffer (*Takifugu rubripes*). Progress in Fishery Sciences, 40(4): 1-10.

Zhang Q, Mai K S, Zhang W B, et al. 2011. Effects of dietary selenoyeast and vitamin E on growth, immunity and disease resistance of sea cucumbers (*Apostichopus japonicus* Selenka). Chinese Journal of Animal Nutrition, 23(10): 1745-1755.

Zhang R Q, Chen Q X, Zheng W Z, et al. 2000. Inhibition kinetics of green crab (*Scylla serrata*) alkaline phosphatase activity by dithiothreitol or 2-mercaptoethanol. International Journal of Biochemistry and Cell Biology, 32(8): 865-872.

Zhang X M, Zhou Y, Liu P, et al. 2015. Temporal pattern in biometrics and nutrient stoichiometry of the intertidal seagrass *Zostera japonica* and its adaptation to air exposure in a temperate marine lagoon (China): implications for restoration and management. Marine Pollution Bulletin, 94(1-2): 103-113.

Zhang Y Y, Guo K Y, LeBlanc R E, et al. 2007. Increasing dietary leucine intake reduces diet-induced obesity and improves glucose and cholesterol metabolism in mice via multimechanisms. Diabetes, 56(6): 1647-1654.

Zhao H X, Cao J M, Wang A L, et al. 2012. Effect of long-term administration of dietary β-1, 3-glucan on growth, physiological, and immune responses in *Litopenaeus vannamei* (Boone, 1931). Aquaculture International, 20: 145-158.

Zhao M, Luo J, Zhou Q, et al. 2021. Influence of dietary phosphorus on growth performance, phosphorus accumulation in tissue and energy metabolism of juvenile swimming crab (*Portunus trituberculatus*). Aquaculture Reports, 20: 100654.

Zhao Y C, Zhang Q, Yuan L, et al. 2017. Effects of dietary taurine on the growth, digestive enzymes, and antioxidant capacity in juvenile sea cucumber, *Apostichopus japonicus*. Journal of the World Aquaculture Society, 48(3): 478-487.

Zhao Y, Wang H, Wang H, et al. 2022. Metabolic response of the sea cucumber *Apostichopus japonicas* during the estivation-arousal cycles. Frontiers in Marine Science, 9: 980221.

Zhou F, Shao Q J, Xiao J X, et al. 2011a. Effects of dietary arginine and lysine levels on growth performance, nutrient utilization and tissue biochemical profile of black sea bream, *Acanthopagrus schlegelii*, fingerlings. Aquaculture, 319(1-2): 72-80.

Zhou F, Song W X, Shao Q J, et al. 2011b. Partial replacement of fish meal by fermented soybean meal in diets for black sea bream, *Acanthopagrus schlegelii*, Juveniles. Journal of World Aquaculture Society, 42(2): 184-197.

Zhou F, Xiao J X, Hua Y, et al. 2011c. Dietary L-methionine requirement of juvenile black sea bream (*Sparus macrocephalus*) at a constant dietary cystine level. Aquaculture Nutrition, 17: 469-481.

Zhou Q C, Zeng W P, Wang H L, et al. 2012a. Dietary arginine requirement of juvenile yellow grouper *Epinephelus awoara*. Aquaculture, 350-353: 175-182.

Zhou Q C, Zeng W P, Wang H L, et al. 2012b. Dietary arginine requirement of juvenile Pacific white shrimp, *Litopenaeus vannamei*. Aquaculture, 364-365: 252-258.

Zhou S F, Sun Z W, Ma L Z, et al. 2010. Effect of feeding enzymolytic soybean meal on performance, digestion and immunity of weaned pigs. Asian-Australasian Journal of Animal Sciences, 24(1): 103-109.

Zhu J, Xu W, Zhang W, et al. 2014. Optimal dietary methionine requirement of red swamp crayfish (*Procambarus clarkii*). Journal of Fishery Sciences of China, 21: 300-309.

Zimmermann N, Rothenberg M E. 2006. The arginine-arginase balance in asthma and lung inflammation. European Journal of Pharmacology, 533(1-3): 253-262.

Zorov D B, Juhaszova M, Sollott S J. 2014. Mitochondrial reactive oxygen species (ROS) and ROS-Induced ROS release. Physiological Reviews, 94(3): 909-950.

Zuo R T, Ai Q H, Mai K S, et al. 2012. Effects of dietary n-3 highly unsaturated fatty acids on growth, nonspecific immunity, expression of some immune related genes and disease resistance of large yellow croaker (*Larmichthys crocea*) following natural infestation of parasites (*Cryptocaryon irritans*). Fish and Shellfish Immunology, 32(2): 249-258.

附　　录

计算方法：

增重量（WG，g）=$W_t - W_0$；

增重率（WGR，%）=$(W_t - W_0)/W_0 \times 100\%$；

特定生长率（SGR，%/d）=$(\ln W_t - \ln W_0)/t \times 100\%$；

存活率（SR，%）=$(N_t/N_0) \times 100\%$；

饲料转化率（FCR）=（终末干重–初始干重）/饲料摄食量；

蛋白质效率（PER）=（终末干重–初始干重）/蛋白质摄食量；

摄食率 [FI，g/(g·d)] =饲料摄食量/[（初始干重+末干重）/2×时间（d）]；

排粪率 [FPR，g/(g·d)] =排粪量/[（初始干重+末干重）/2×时间（d）]；

饲料干物质表观消化率（ADC，%）=[1–（饲料中 Y_2O_3 含量/粪便中 Y_2O_3 含量）×（粪便量/饲料摄食量）]；

饲料营养成分表观消化率（%）=100×[1–（饲料中 Y_2O_3 含量/粪便中 Y_2O_3 含量）×（粪便中营养成分含量/饲料中营养成分含量）]；

脏壁比（VBR，%）=$W_v/W_b \times 100\%$；

肠壁比（IBR，%）=$W_i/W_b \times 100\%$；

肠体比（IWR，%）=$(W_i/W_t) \times 100\%$；

肠长比（IBL）=L_g/L_b；

体壁指数（BI，%）=体壁重/体质量×100%；

肠道指数（II，%）=肠重/体壁重×100%；

式中，N_t 为刺参终末头数；N_0 为刺参初始头数；W_t 为刺参终末体质量（g）；W_0 为刺参初始体质量（g）；d 为养殖实验天数（d）；W_i 为取样刺参肠道质量（g）；W_b 为取样刺参体壁质量（g）；L_g 为取样刺参肠道长度（cm）；L_b 为取样刺参体长度（cm）；W_v 为刺参内脏质量（g）。

测定方法：

实验饲料及刺参体壁水分、干物质含量采用 105℃恒重法测定 [《饲料中水分的测定》（GB/T 6435—2014）]，粗蛋白含量采用凯氏定氮法测定 [《饲料中粗蛋白的测定 凯氏定氮法》（GB/T 6432—2018）]，粗脂肪含量采用索氏抽提法测定 [《饲料中粗脂肪的测定》（GB/T 6433—2006）]，粗灰分含量采用 550℃马弗炉灼烧法测定 [《饲料中粗灰分的测定》（GB/T 6438—2007）]，饲料能量测定采用燃烧法，使用量热仪（IKA，C6000，德国）测定，饲料中氨基酸测定参照国标［《实验动物 配合饲料 氨基酸的测定》（GB/T 14924.10—2008）］，采用全自动氨基酸测定仪（Hitachi L-8900，日本）测定，脂肪酸的测定采用气相色谱法［《食品安全国家标准 食品中脂肪酸的测定》（GB 5009.168—2016）］，采用气相色谱仪（GC-2010，岛津，日本）测定。